COLLEGE TRIGONOMETRY

MUSTAFA A. MUNEM

WILLIAM TSCHIRHART

JAMES P. YIZZE

MACOMB COUNTY COMMUNITY COLLEGE

WORTH PUBLISHERS, INC.

COLLEGE TRIGONOMETRY

444 PARK AVENUE SOUTH, NEW YORK, NEW YORK 10016

PRINTED IN THE UNITED STATES OF AMERICA

LIBRARY OF CONGRESS CATALOG CARD NO. 73-85131

ISBN: 0-87901-028-2

SECOND PRINTING JUNE 1975

DESIGN BY MALCOLM GREAR DESIGNERS, INC.

PREFACE

PURPOSE: This book is designed to provide the preparation necessary for students who intend to continue their study in mathematics, technical mathematics, and various engineering fields. The book can be used in a three-hour one-semester course or a four-hour one-quarter course.

PREREQUISITES: It is assumed that the students who use this textbook have had the equivalent of two years of high school algebra and one year of high school geometry, or a comparable background.

OBJECTIVES: The primary objective of the book is to present trigonometry in a sound pedagogical manner and to provide all the material that the average student will require to understand the subject and to apply what he or she has learned to specific situations. This is accomplished by interspersing the text with many worked-out examples. Geometric interpretations supplement explanations whenever possible. Definitions, properties, and theorems are carefully stated. Problems are coordinated with examples that are worked out in the book. The problems have been arranged according to their degree of difficulty. Theory has been deemphasized throughout, and technique, drill, and applications are emphasized. Applied problems and examples relating to surveying and other engineering disciplines are included.

FEATURES: Color has been used to highlight definitions, properties, and theorems and to clarify key features of the graphs. Each chapter ends with a review problems set to help students understand the material of the chapter.

CONTENTS: Chapter 1 introduces some important preliminary material dealing with sets and functions.

Chapter 2 presents trigonometric functions of angles.

Chapter 3 introduces circular functions by means of the unit circle and relates trigonometric functions to circular functions.

Chapter 4 deals with graphs and inverse trigonometric functions.

Chapter 5 treats trigonometric identities and equations.

Chapter 6 is devoted to solving triangles using right triangle trigonometry, the law of cosines, and the law of sines. Here students can relate the solving of triangles to application of trigonometric principles in surveying and other engineering fields.

Chapter 7 covers complex numbers, and DeMoivre's theorem is used to find the powers and roots of complex numbers.

In Chapter 8, vectors in the plane and elementary properties of rotations are presented as applications of geometry and trigonometry.

ADDITIONAL AIDS: A semi-programmed work book (study guide), containing fill-in statements and problems, chapter reviews, and tests for each chapter is available to accompany the text. The study guide conforms with the book, section by section. Also available are a syllabus and teachers manual to help teachers set the pace of the course.

ACKNOWLEDGEMENTS: A general expression of appreciation is due many people who have helped in the preparation and review of this book. In particular, we are indebted to Professors William Brett, Louis Contey, and Michael Sentlowitz at Rockland Community College, Richard G. Denning at Southern Technical Institute of Technology, Joseph L. Mazanec at Delta College, Henry Nace at Lawrence Institute of Technology, Robert L. Wilson Jr. at the University of Wisconsin, Madison, Winston B. Richter at Miami Dade Junior College, North Campus, Carl W. Richards, formerly at U.C.L.A., Charles R. Stone and Warren H. Mason at DeKalb Community College, and George W. Schultz at St. Petersburg Junior College, Clearwater Campus. Robert C. Andrews of Worth Publishers, who coordinated this project, merits special thanks.

Mustafa A. Munem
William Tschirhart
James P. Yizze
Warren, Michigan
January 1974

CONTENTS

CHAPTER 1

Real Number Sets and Functions

CHAPTER 1

Real Number Sets and Functions

1 Introduction

In this chapter we shall review some important topics—sets, real numbers, and functions—that will be important to understand later when we take up the study of trigonometry. We begin with a survey of the symbols and terms used in set language.

2 Set Language

A *set* is a collection of objects. The objects are called the *elements* or *members* of the set. The symbol $\in$ is used to indicate that an element "belongs to" a set and the symbol $\notin$ is used to indicate that a particular element "does not belong" to a set. For example, if set B is the set of all counting numbers from 1 to 7, we can write B as $B = \{1, 2, 3, 4, 5, 6, 7\}$. Here $1 \in B$, $2 \in B$, and $7 \in B$, whereas $0 \notin B$ and $8 \notin B$.

The set that does not have any elements is called the *null set* or *empty set.* It is denoted by $\emptyset$ or by $\{ \}$.

We can describe sets by using the identifying properties of the elements of the set. This set description, called *set builder notation,* takes the form $A = \{x \mid x \text{ has property } p\}$, which is read "$A$ is the set of all elements x such that x has the property p."

For example, if B is the set of all counting numbers from 1 to 100, we can write $B = \{x \mid x \text{ is a counting number and } 1 \leq x \leq 100\}$, which reads "set B is the set of all x such that x represents a counting number between 1 and 100, including 1 and 100." Also, set B can be described as $B = \{1, 2, 3, \ldots, 100\}$, where the three dots are used in the same way as the term "etc." is used—to mean "and so on."

2.1 Set Relations

If set $A = \{1, 2, 3, 4, 5, 6\}$ and $B = \{2, 5, 3\}$, then each member of set B is also a member of set A, and we say that B is a *subset* of A. In general, a set S is a *subset* of a set T, written $S \subseteq T$, if every element of S is an element of T. The empty set is considered to be a subset of every set.

Note that if $A \subseteq B$, A is not necessarily a "smaller" set than B. For example, if $A = \{1, 2, 3\}$ and $B = \{x \mid x$ is a counting number less than $4\}$, then $A \subseteq B$ and $B \subseteq A$. In fact, it can be seen that A and B are really the same set. This example illustrates the definition of *equality of sets*. If $S \subseteq T$ and $T \subseteq S$, we consider S and T to be equal sets and we write $S = T$.

If $A = \{1, 2, 3\}$ and $B = \{2, 3, 1, 7, 9\}$, $A \subseteq B$. Note, however, that $A \neq B$. This relation, in which A is said to be a *proper subset* of set B, is generally written $A \subset B$ (the horizontal bar is usually omitted). In general, S is a *proper subset* of T, written $S \subset T$, if all members of S are in T and T has at least one member not in set S.

If $A = \{2, 3, 4\}$ and $C = \{7, 8\}$, then A and C have no elements in common. Pairs of sets that have no elements in common are called *disjoint sets*.

When the selection of elements of subsets is limited to a fixed set, the limiting set is called a *universal set* or a *universe*. A universal set customarily represents the complete set or the largest set from which all other sets in that same discussion are formed. The choice of the universal set depends upon the situation being considered.

EXAMPLE

Describe set A, where $A = \{x \mid x$ is a number greater than 2 and x is a member of universal set $U\}$.

a) $U = \{1, 2, 3, \frac{4}{3}, \frac{11}{8}\}$
b) U is the set of all counting numbers.
c) $U = \{0, 1, 2\}$

SOLUTION

a) $A = \{3\}$
b) $A = \{x \mid x$ is a counting number greater than $2\}$ or, equivalently, $A = \{3, 4, 5, \ldots\}$
c) $A = \emptyset$

2.2 Set Operations

Consider a universal set $U = \{1, 2, 3, 4, 5, 6, 7, 8\}$. From U we can form $A = \{1, 2, 3, 4\}$ and $B = \{1, 3, 7\}$. How can sets A and B be used to form other sets? One way is simply to combine all the elements of A and B to form $\{1, 2, 3, 4, 7\}$. This example suggests the operation that we call *set union*.

In general, the *union* of S and T, written $S \cup T$ and read "S union T," is defined as

$$S \cup T = \{x \mid x \in S \text{ or } x \in T \text{ (or both)}\}$$

Hence, in the example,

$$\{1, 2, 3, 4\} \cup \{1, 3, 7\} = \{1, 2, 3, 4, 7\}$$

Another way to use A and B to form another set is to form the set $\{1, 3\}$, which is the set of all elements common to A and B. This is an example of *set intersection.*

In general, S *intersect* T, written $S \cap T$, is defined as

$$S \cap T = \{x \mid x \in S \text{ and (simultaneously) } x \in T\}$$

For example,

$$\{1, 2, 3, 4\} \cap \{1, 3, 7\} = \{1, 3\}$$

If $A = \{1, 2, 3, 4\}$ and $B = \{5, 6, 7\}$, then $A \cap B = \emptyset$ and A and B are disjoint sets. In general, S and T are disjoint sets whenever $S \cap T = \emptyset$.

EXAMPLES

1 Determine $A \cup B$ and $A \cap B$ if $A = \{1, 2, 3, 4, 5\}$ and $B = \{2, 5, 6, 7\}$.

SOLUTION

$$A \cup B = \{1, 2, 3, 4, 5, 6, 7\} \qquad \text{and} \qquad A \cap B = \{2, 5\}$$

2 Let $A = \{x \mid x \text{ is a counting number}\}$ and let $B = \{x \mid x \text{ is an even counting number}\}$, that is, $B = \{2, 4, 6, 8, \ldots\}$. Form $A \cup B$ and $A \cap B$.

SOLUTION Note that $B \subset A$.

$$A \cup B = \{x \mid x \text{ is a counting number or } x \text{ is an even counting number (or both)}\}$$

Therefore,

$$A \cup B = \{x \mid x \text{ is a counting number}\} = A$$
$$A \cap B = \{x \mid x \text{ is a counting number and (simultaneously) } x \text{ is an even counting number}\}$$

Therefore,

$$A \cap B = \{x \mid x \text{ is an even counting number}\} = B$$

2.3 Real Numbers

The language of sets can be used to review the important real number sets.

1 *Positive Integers* (N)

The *natural numbers* or *counting numbers* 1, 2, 3, 4, 5, 6, 7, . . . make up the fundamental number set. This set, which we will designate as set N, is most often referred to as the set of *positive integers.* $N = \{1, 2, 3, \ldots\}$ is the set of positive integers.

2 *Negative Integers* (I_n)

The set of *negative integers* consists of the negatives of all the positive integers. If we denote the set of negative integers by I_n, then $I_n = \{-1, -2, -3, \ldots\}$.

3 *Integers* (I)

The set of *integers* is the set formed by the union of the positive integers, the negative integers, and the number zero. If we use I to denote the set of integers, we have

$$\begin{aligned} I &= \{1, 2, 3, 4, \ldots\} \cup \{-1, -2, -3, -4, \ldots\} \cup \{0\} \\ &= \{\ldots, -4, -3, -2, -1, 0, 1, 2, 3, 4, \ldots\}. \end{aligned}$$

4 *Rational Numbers* (Q)

A *rational number* is any number that can be written as the ratio of two integers (the denominator cannot be zero). For example, 3, $2\frac{1}{2}$, and 43 percent are rational numbers because each can be written as the ratio of two integers:

$$\begin{aligned} 3 &= \tfrac{3}{1} \\ 2\tfrac{1}{2} &= \tfrac{5}{2} \\ 43\% &= \tfrac{43}{100} \end{aligned}$$

If we use Q to denote the set of rational numbers, we have

$$Q = \left\{q \mid q = \frac{a}{b}, a \in I, b \in I, b \neq 0\right\}$$

The rational numbers can also be described by investigating their decimal representations. For example, let us consider the specific rational number $\frac{3}{7}$. The number can be represented, using $3 \div 7$, as

$$\tfrac{3}{7} = 0.428571428571\overline{428571}$$

where the bar identifies the block of digits that repeats an "infinite" number of times.

In general, every rational number can be represented by an eventually repeating decimal. The converse of this statement also holds; that is, every eventually repeating decimal represents a rational number.

Rational numbers such as 2 and $\frac{14}{2}$, in which the repeating block is the digit 0, are sometimes called *terminating decimals.*

In summary, a rational number is a number that can be considered from two different viewpoints: either as a ratio of two integers or as a repeating decimal.

5 *Irrational Numbers* (L)

So far, we have seen that the set of rational numbers is the set of numbers represented by repeating decimals and the set of repeating decimals represents rational numbers. But there are decimals which do not repeat, for example, the decimal 1.01001000100001 . . . , where there is one more "0" after each "1" than there is before the "1." Also, $\sqrt{2}$ is not rational, for it can be shown that $\sqrt{2}$ cannot be represented as a ratio of two integers. $1 + \sqrt{2}$, $3 - \sqrt{2}$, $3 + \sqrt{3}$, $\sqrt[3]{3}$, π, and $\sqrt{5}$ are other examples of numbers that cannot be represented as the ratio of two integers. Such numbers are called *irrational numbers.*

6 *Real Numbers* (R)

The set of *real numbers* is, in a sense, the set of all numbers that can be written as decimal numbers. Consequently, the set of real numbers consists of two sets of numbers—the rational numbers, which are represented by repeating decimals, and the irrational numbers, which are represented by nonrepeating decimals. Using R to denote the set of real numbers, we can express R by using set notation as follows:

$R = Q \cup L$ where L represents the set of irrational numbers

Note that $Q \cap L = \emptyset$.

PROBLEM SET 1

1 Use set builder notation, $\{x \mid x \text{ has property } p\}$, to describe each of the following sets.

a) The set of even counting numbers
b) The set of counting numbers divisible by 5, that is, all counting numbers that have a zero remainder when divided by 5
c) The set of counting numbers greater than 2 and less than 13

2 Which of the following statements are true and which are false?

a) $3 \in \{3, 4\}$
b) $\{a\} \subseteq \{a, \{a\}, \{\{a\}\}\}$
c) $\{3\} \subset \{3, 4\}$
d) $\{2\} \in \{2, \{2\}\}$
e) $\{3\} \subseteq \{3, 4\}$
f) $\{1, 3\} \subseteq \{1, 5, 15\}$
g) $\{0\}$ is empty
h) $\{a, b\} = \{b, a\}$

3 Indicate which of the set relations (proper subset, equal, disjoint) hold between each pair of the following sets.

a) The empty set; $\{2, 3, 4\}$
b) The set of all counting numbers greater than 3; the set of all counting numbers less than 3
c) The set of all counting numbers greater than 3; $\{x \mid x > 3, x \in R\}$
d) The set of all positive even counting numbers less than 11; $\{2, 4, 6, 8, 10\}$

4 List all the subsets of each of the following sets. Indicate which of the subsets are proper subsets.

a) $\{2\}$
b) $\{2, 3\}$
c) $\{a, b, c\}$
d) $\{\emptyset, 0\}$
e) $\{5, 6, 7, 8\}$

5 Use $A = \{1, 2, 4\}$, $B = \{2, 3, 5, 7\}$, and $C = \{6, 3, 5, 8\}$ to form each of the following sets.

a) $A \cap B$
b) $B \cup A$
c) $B \cap C$
d) $B \cup C$
e) $A \cap C$
f) $C \cup B$
g) $C \cap A$
h) $A \cap \emptyset$
i) $C \cup A$
j) $B \cup \emptyset$
k) $A \cap (B \cup C)$
l) $A \cup (B \cap C)$

6 Explain why each of the following numbers is rational numbers.

a) -5
b) $2\frac{4}{13}$
c) 0.2
d) $7.3621\overline{621}$
e) $3.46\overline{46}$
f) 17 percent
g) $\frac{4}{13}$
h) 1.999

7 Using I_n, I, Q, L, and R as defined in Section 2.3, describe each of the following sets.

a) $L \cap Q$
b) $I_n \cup N$
c) $I \cap N$
d) $L \cup R$
e) $Q \cup R$

3 Order

The *number line* (Figure 1), which represents the set of real numbers as the set of all points on a straight line, provides us with a vehicle for introducing the notion of order.

Figure 1

−6 −5 −4 −3 −2 −1 0 1 2 3 4 5 6

3.1 Trichotomy

The real numbers may be classified on the number line as follows: The numbers located to the right of zero are the *positive real numbers,* and they are arranged in order of increasing magnitude, starting at zero and moving to the right. The numbers located to the left of zero are the *negative real numbers,* and they are arranged in decreasing order starting at zero and moving to the left. Accordingly, we have the following principle:

TRICHOTOMY

If a is a real number, then one and only one of the following situations can hold

1 a is positive.

2 a is negative.

3 a is zero.

Geometrically, any number whose corresponding point on the number line lies to the right of the corresponding point of a second number is said to be of *greater magnitude,* or, simply, *greater than* (denoted by "$>$") the second number. The second number is also said to be *less than* (denoted by "$<$") the first number. For example,

$$7 > 3 \quad \text{and} \quad 3 < 7$$
$$1 > -5 \quad \text{and} \quad -5 < 1$$

More formally, we have the following definition.

Definition

For any two real numbers a and b, a is less than b ($a < b$), or, equivalently, b is greater than a ($b > a$) if $b - a$ is a positive number.

For example, $3 < 7$ (or $7 > 3$), since the difference $7 - 3$ is 4, a positive number. Also, $-5 < -3$ (or $-3 > -5$) since $-3 - (-5) = 2$, which is positive. We can also indicate that a number is positive or negative by the use of an inequality relation. If the number a is positive, then $a > 0$ (Figure 2); that is, a lies to the right of 0. Similarly, if the number b is negative (Figure 2), then $b < 0$; that is, b lies to the left of zero.

Figure 2

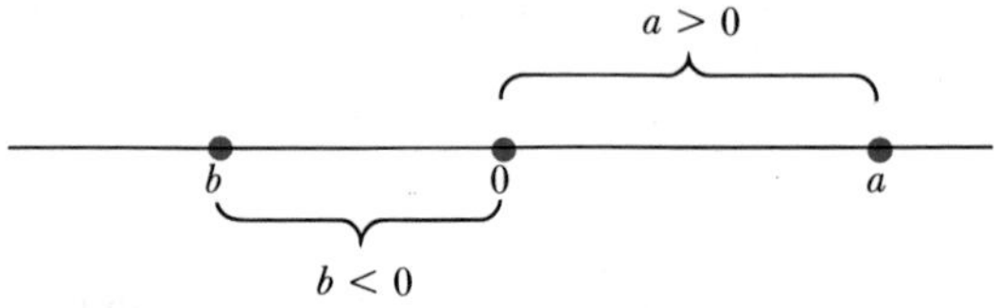

3.2 Properties of Order Relations

Here we list a number of important properties for order relations.

1 If $a < b$ and c is any number, then $a + c < b + c$.

2 If $a < b$ and $c > 0$, then $ac < bc$.

3 If $a < b$ and $c < 0$, then $ac > bc$.

4 If $a < b$ and $b < c$, then $a < c$.

Similar properties of the equality relation also hold:

1 If $a = b$ and c is any number, then $a + c = b + c$

2 If $a = b$ and c is any number, then $ac = bc$

3 If $a = b$ and $b = c$, then $a = c$

We can combine the two sets of properties as follows. We interpret the symbol "$\leqslant$" to mean *less than or equal* and "$\geqslant$" to mean *greater than or equal.* That is, if $a \leqslant b$, then either $a < b$ or $a = b$. This gives us the following results:

PROPERTY 1 (ADDITION PROPERTY)

If $a \leqslant b$ and c is any number, then $a + c \leqslant b + c$.

PROPERTY 2 (MULTIPLICATION PROPERTY)

i) If $a \leqslant b$ and $c > 0$, then $ac \leqslant bc$.

ii) If $a \leq b$ and $c < 0$, then $ac \geq bc$.

PROPERTY 3 (TRANSITIVE PROPERTY)

If $a \leq b$ and $b \leq c$, then $a \leq c$.

EXAMPLES

1 Solve $3x - 2 < 10$.

SOLUTION

$3x - 2 < 10$
$3x - 2 + 2 < 10 + 2$ (Property 1)
$3x < 12$
$\frac{1}{3}(3x) < \frac{1}{3}(12)$ (Property 2, i)
$x < 4$

2 Solve $x + 2 \geq 7x - 1$.

SOLUTION

$x + 2 \geq 7x - 1$
$x + 2 + (-2) \geq 7x - 1 + (-2)$ (Property 1)
$x \geq 7x - 3$
$x + (-7x) \geq 7x + (-7x) - 3$ (Property 1)
$-6x \geq -3$
$-\frac{1}{6}(-6x) \leq -\frac{1}{6}(-3)$ (Property 2, ii)
$x \leq \frac{1}{2}$

3 Show that if $x \neq 0$, then $x^2 > 0$.

PROOF. If $x \neq 0$, then $x > 0$ or $x < 0$.
If $x > 0$, then $x \cdot x > x \cdot 0$ or $x^2 > 0$ (Property 2, i).
On the other hand, if $x < 0$, then $x \cdot x > x \cdot 0$ or $x^2 > 0$ (Property 2, ii).
Hence, for $x \neq 0$, $x^2 > 0$.

3.3 Interval Notation

Some standard notations are used in connection with the order relation to describe certain subsets of the number line which we call *intervals.*

1 *Bounded intervals*

We shall assume that a and b are real numbers such that $a < b$. The *open interval from a to b*, denoted (a, b), is defined as

$(a, b) = \{x \mid a < x < b\}$ (Figure 3)

Figure 3

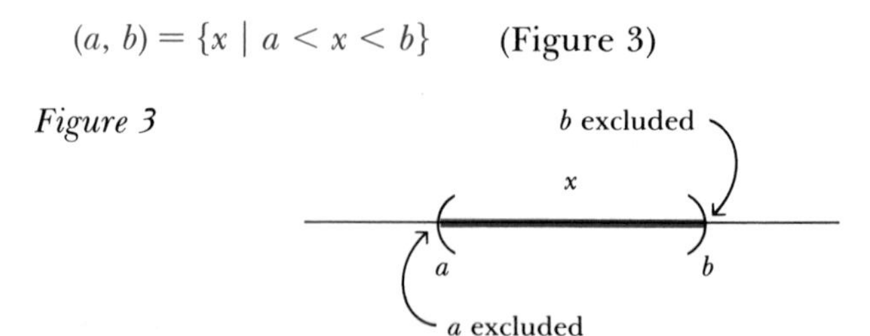

Notice that $a \notin (a, b)$ and $b \notin (a, b)$.

The *closed interval from a to b*, denoted $[a, b]$, is defined as

$[a, b] = \{x \mid a \leqslant x \leqslant b\}$ (Figure 4)

Figure 4

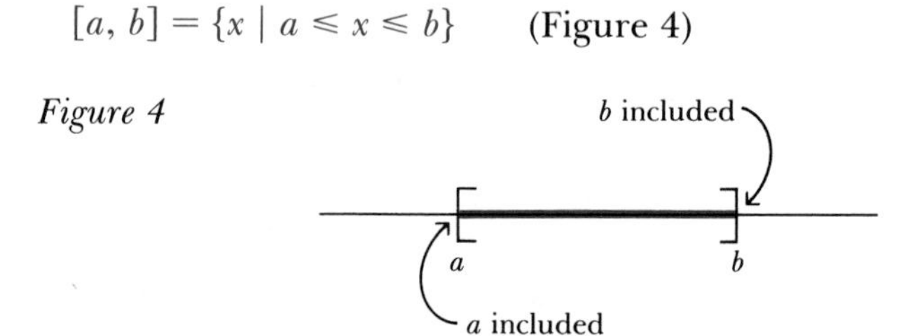

The closed interval includes the end points, whereas the open interval does not include the end points. For example, $[0, 1]$ is the set of all real numbers between 0 and 1, including 0 and 1, whereas $(0, 1)$ is the set of all real numbers between 0 and 1, excluding 0 and 1 (Figures 5*a* and 5*b*).

Figure 5

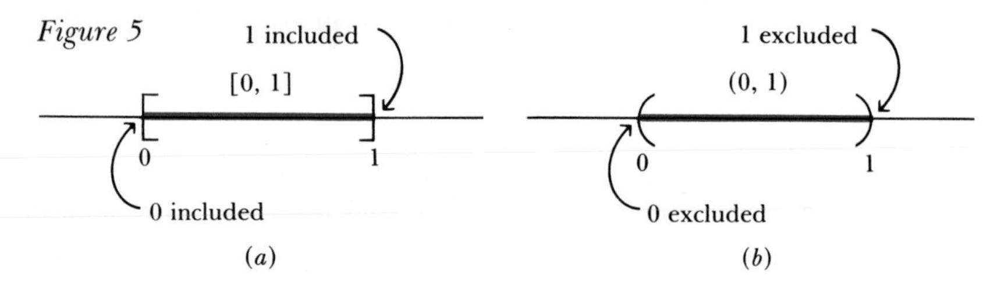

An interval from a to b including one end point but excluding the other end point is said to be *half-open* (or *half-closed*). This can happen in one of two ways.

i $[a, b) = \{x \mid a \leqslant x < b\}$ (Figure 6)

Figure 6

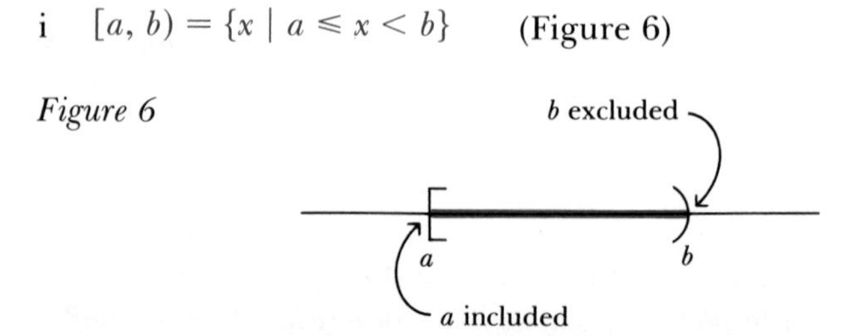

ii $(a, b] = \{x \mid a < x \leqslant b\}$ (Figure 7)

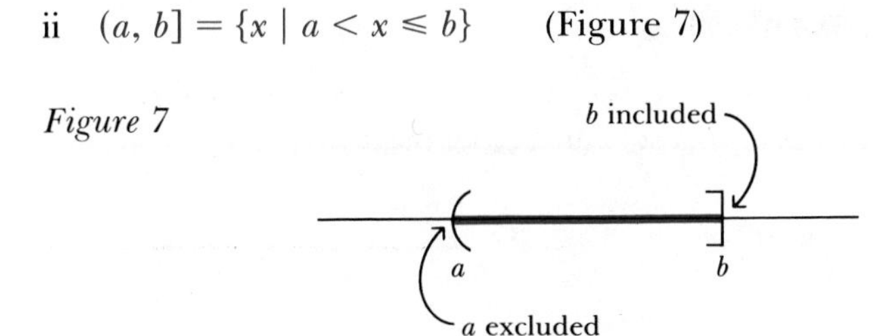

Figure 7

2 *Unbounded intervals*

We use the symbols ∞ and $-\infty$ (*infinity* and *minus infinity*) to describe unbounded intervals as follows. If a is a real number (∞ and $-\infty$ are *not* real numbers), then

i $[a, \infty) = \{x \mid x \geqslant a\}$ (Figure 8)

Figure 8

ii $(-\infty, a] = \{x \mid x \leqslant a\}$ (Figure 9)

Figure 9

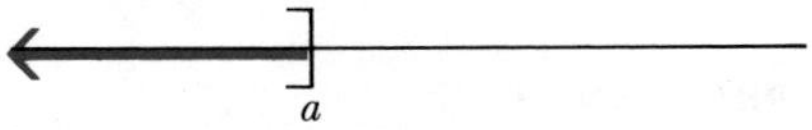

iii $(a, \infty) = \{x \mid x > a\}$ (Figure 10)

Figure 10

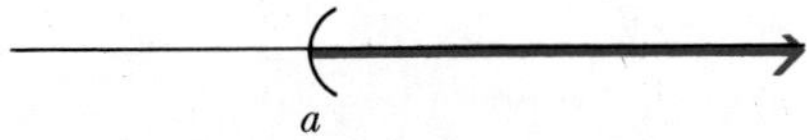

iv $(-\infty, a) = \{x \mid x < a\}$ (Figure 11)

Figure 11

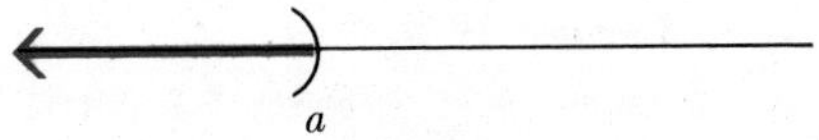

For example,

$$(-\infty, 3) \cap (0, \infty) = (0, 3) \qquad \text{(Figure 12)}$$

Figure 12

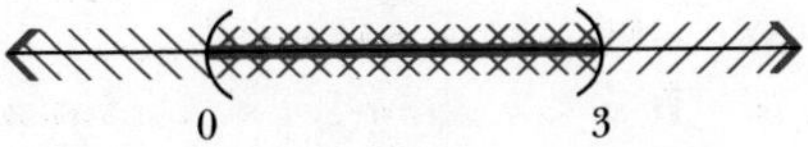

Finally, we can use interval notation $(-\infty, \infty)$ to denote the set of all real numbers R.

EXAMPLES

1 If $2 \leq x \leq 4$, the set of values of x determines the interval in Figure 13 and this set can be denoted by $[2, 4]$.

Figure 13

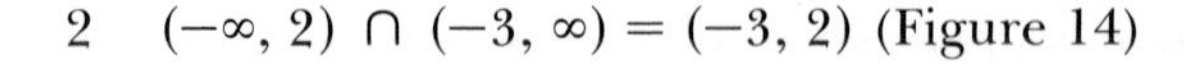

2 $(-\infty, 2) \cap (-3, \infty) = (-3, 2)$ (Figure 14)

Figure 14

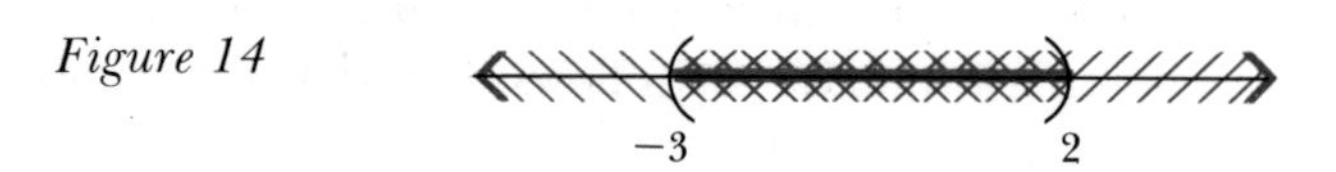

3 $\{x \mid x > 1 \text{ or } x < -2\} = \{x \mid x > 1\} \cup \{x \mid x < -2\} = (1, \infty) \cup (-\infty, -2)$ (Figure 15)

Figure 15

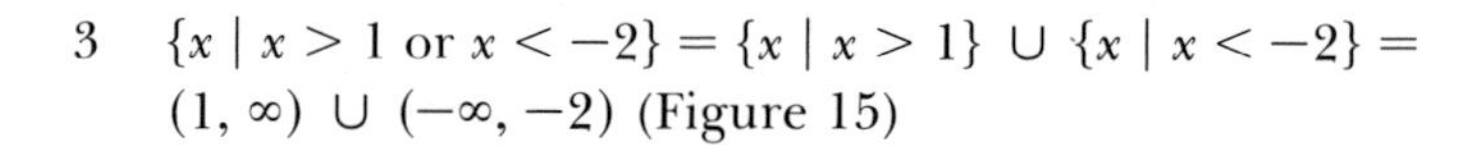

PROBLEM SET 2

1 Locate each of the following real numbers on the number line. Which numbers are rational numbers?

a) $-2\frac{1}{7}$ b) 1.8 percent
c) $\sqrt{7}$ d) -3.22
e) $\frac{17}{2}$ f) $-\frac{1}{2}$
g) $2\sqrt{2}$ h) 3.77
i) 6.99 j) 6.999

2 Indicate the order property which validates each of the following statements.

a) Since $-3 < 2$, then $-3 - x < 2 - x$.
b) If $x < y$, then $-2x > -2y$.
c) If $-\frac{1}{2} < x$ and $x < y$, then $y > -\frac{1}{2}$.
d) If $x > 0$, then $-2/x < -1/x$.
e) If a is a real number, then either $a = 3$ or $a < 3$ or $a > 3$.

3 Determine whether or not the following statements are true.

a) $-[2 - (4 - 5)] < 0$
b) If $a < b$, then $1/a < 1/b$ for $a, b > 0$.
c) If $0 < a < 1$, then $a^3 < a$.
d) If $a < b < c$, then $(a + b + c)/3 < c$.
e) If $a < b$, then $-1/a < -1/b$ for $a > 0$ and $b > 0$.

4 Use interval notation and set operations to represent each of the following sets.

a) 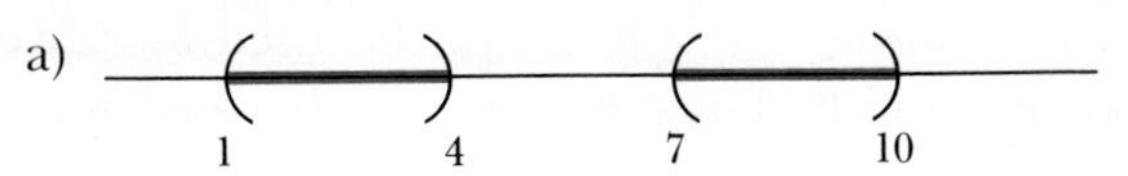

b) 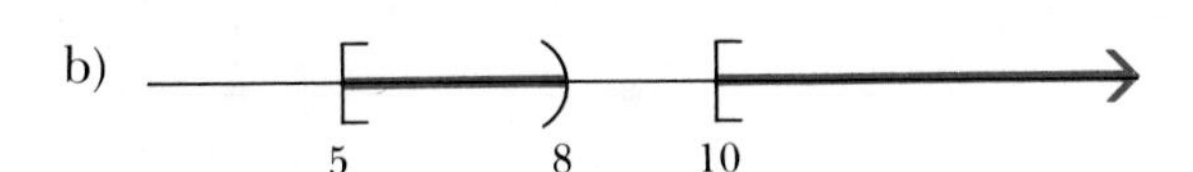

c) 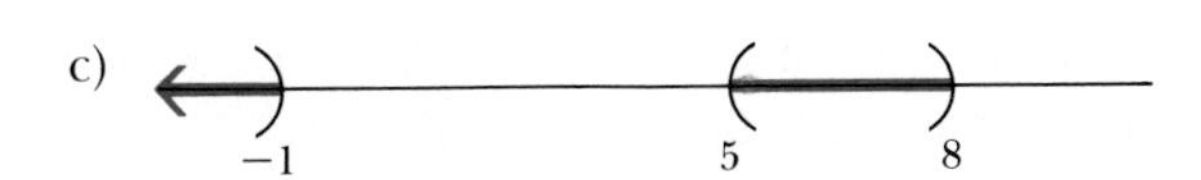

d) 

e) 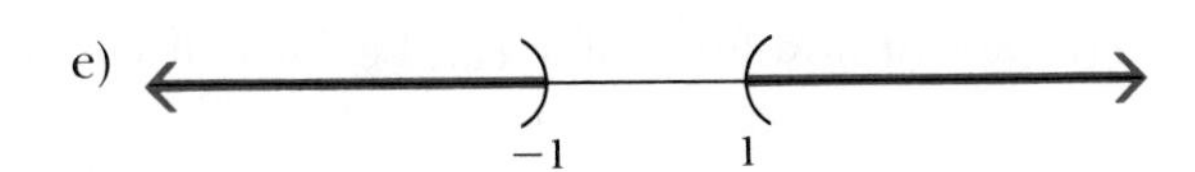

5 Solve the following inequalities and represent the solution in interval form. Also show the solution set on the real line.

a) $5x < 5$ b) $5x + 6 < 13$
c) $2x - 7 \geqslant 8$ d) $3x - 5 \leqslant 7$
e) $(3x - 5)/9 > 0$ f) $x/3 + 2 \leqslant x/4 - 2x$
g) $3x + 1 \leqslant 0$ h) $3(x + 2) \leqslant \frac{3}{5}(x + 1)$
i) $5 + x < -x + 3$ j) $x + 6 < 4 - 3x$

4 Absolute Value

If a and b are real numbers such that $a \leqslant b$, then the "distance" between a and b is considered to be the nonnegative number $b - a$ (Figure 1).

Figure 1

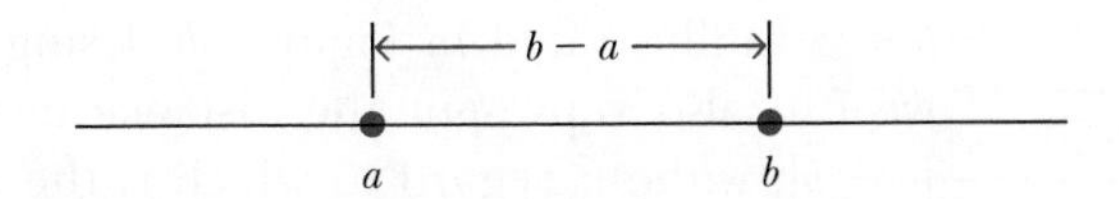

For example, the distance between -1 and 3 is given by $3 - (-1) = 4$, and the distance between -5 and -2 is equal to $(-2) - (-5) = 3$ (Figure 2).

Figure 2

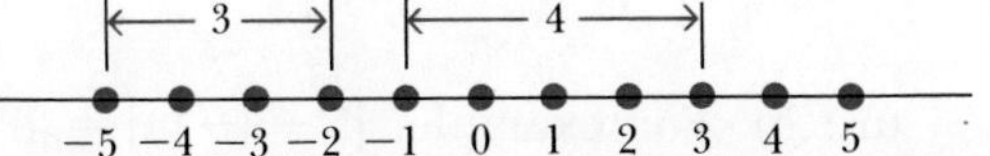

Suppose that we are interested in finding the distance between 0 and any real number x. For convenience, the notation $|x|$, which is read the *absolute value of* x, will be used to represent the distance between x and 0. Thus, if $x > 0$, we have $|x| = x - 0 = x$ (Figure 3).

Figure 3

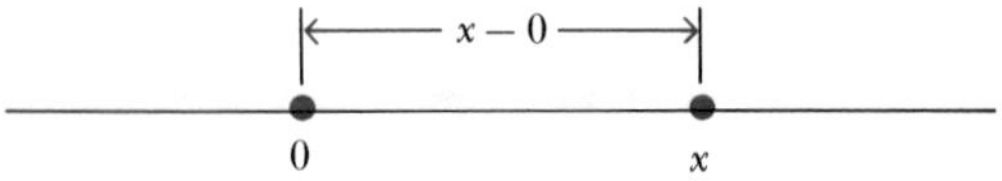

If $x = 0$, $|x| = 0 - 0 = 0$. If $x < 0$, $|x| = 0 - x = -x$ (Figure 4). Note that for $x < 0$, $-x > 0$.

Figure 4

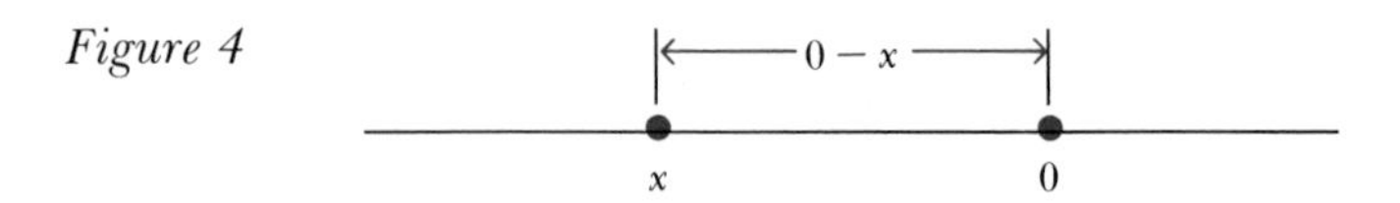

This concept of absolute value can be formalized as follows:

Definition

For x a real number, the *absolute value of* x, denoted by $|x|$, is defined by

$$|x| = \begin{cases} x & \text{if } x \geq 0 \\ -x & \text{if } x < 0 \end{cases}$$

Hence, from the definition, $|3| = 3$, because $3 > 0$, $|0| = 0$, and $|-4| = -(-4) = 4$ because $-4 < 0$.

4.1 Properties of Absolute Value Equalities

PROPERTY 1

The distance between two real numbers a and b is given by $|a - b|$.

If a and b are real numbers, then the distance between a and b is $b - a$ if $a \leq b$ as illustrated in Figure 5a, or the distance is $a - b$ if $b \leq a$ as illustrated in Figure 5b. Using the absolute value notation, we can also represent the distance between the points a and b by $|a - b|$, without regard to which is the smaller number, a or b (Figure 5). For example, $|5 - (-1)| = |6| = 6$ and $|-1 - 5| = |-6| = 6$,

Figure 5

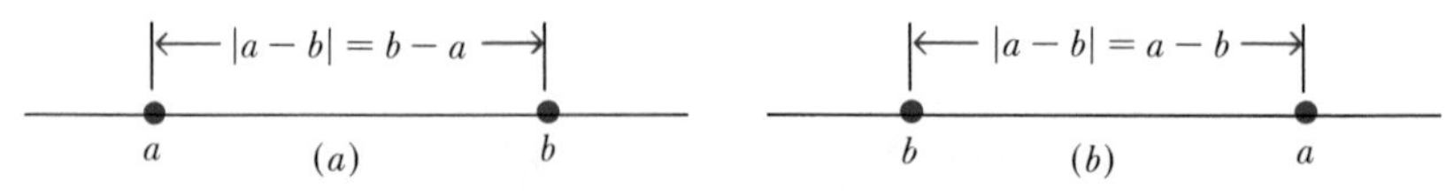

so the distance between 5 and -1 is 6 and between -1 and 5 is 6 (Figure 6).

Figure 6

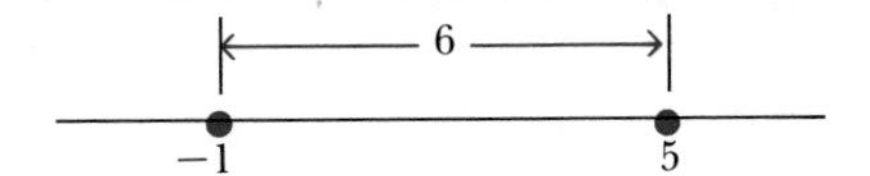

PROPERTY 2

It can also be illustrated geometrically that $|-a| = |a|$, since the distance between 0 and a is the same as the distance between 0 and $-a$ (Figure 7).

Figure 7

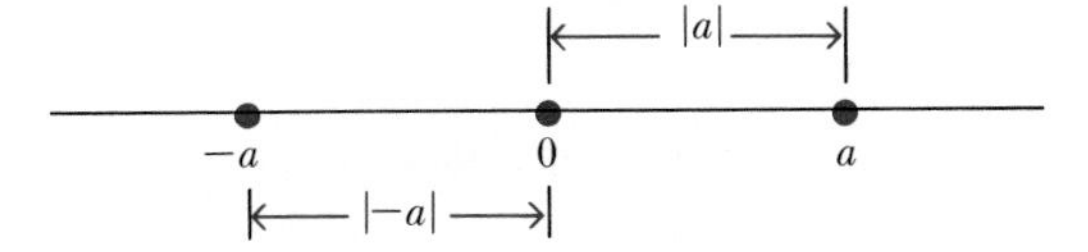

For example, the distance from 0 to 5 is 5 units, and the distance from 0 to -5 is also 5 units. Using absolute value notation, we have $|-5| = |5|$ (Figure 8).

Figure 8

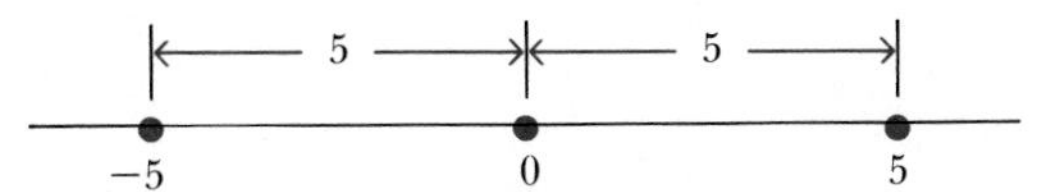

EXAMPLES

1 Find the value of $|a|/a$, for $a \neq 0$.

SOLUTION. For $a > 0$, $|a| = a$, so that

$$\frac{|a|}{a} = \frac{a}{a} = 1$$

For $a < 0$, $|a| = -a$, so that

$$\frac{|a|}{a} = \frac{-a}{a} = -1$$

Hence,

$$\frac{|a|}{a} = \begin{cases} 1 & \text{if } a > 0 \\ -1 & \text{if } a < 0 \end{cases}$$

2 Prove that $|a|^2 = a^2$, for all real numbers a.

PROOF. For $a \geqslant 0$, $|a| = a$, so that

$$|a|^2 = (a)^2 = a^2$$

For $a < 0$, $|a| = -a$, so that

$$|a|^2 = (-a)^2 = a^2$$

Hence, for all possible values of a, $|a|^2 = a^2$.

We shall now solve equations involving absolute values. Remember that to deal with a mathematical concept, you must know its definition. The procedure for solving such equations is illustrated in the following examples.

EXAMPLES

Solve each of the following absolute value equations.

1 $|x| = 3$

SOLUTION. By definition, we know that $|3| = 3$ and $|-3| = 3$. Therefore, $x = 3$ or $x = -3$. No other values would satisfy the equation $|x| = 3$. The solution set is $\{-3, 3\}$.

2 $|3x - 4| = 5$

SOLUTION. Since $|-5| = |5| = 5$, $3x - 4 = 5$ or $3x - 4 = -5$, so $x = 3$ or $x = -\frac{1}{3}$. The solution set is $\{-\frac{1}{3}, 3\}$.

4.2 Properties of Absolute Value Inequalities

We know that $|x|$ represents the distance between 0 and x. Now we can use this geometric interpretation, together with the results above, to get a clear understanding of absolute value inequalities of the forms $|x| < a$ or $|x| > a$, where a is a positive number.

GEOMETRIC INTERPRETATION 1. Quite simply, $|x| < a$ means that the distance between 0 and x is less than a units; or, equivalently, x is within a units of 0 (Figure 9). Using interval notation, $x \in (-a, a)$.

Figure 9

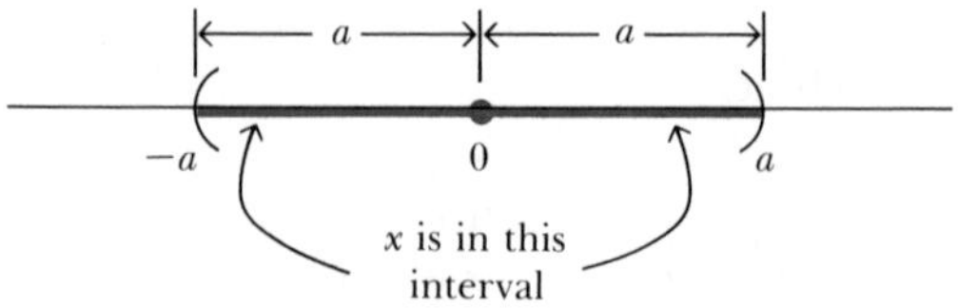

Using inequalities, this means that

$$-a < x < a.$$

Hence, we have the following property.

PROPERTY 1

If $|x| < a$, then $-a < x < a$, where $a > 0$; that is,

$$\{x \mid |x| < a\} = \{x \mid -a < x < a\} = (-a, a)$$

EXAMPLES

Solve the following absolute value inequalities, and illustrate the solution on the number line.

1 $|x| < 3$.

SOLUTION. By Property 1,

$$\{x \mid |x| < 3\} = \{x \mid -3 < x < 3\} = (-3, 3) \qquad \text{(Figure 10)}$$

Figure 10

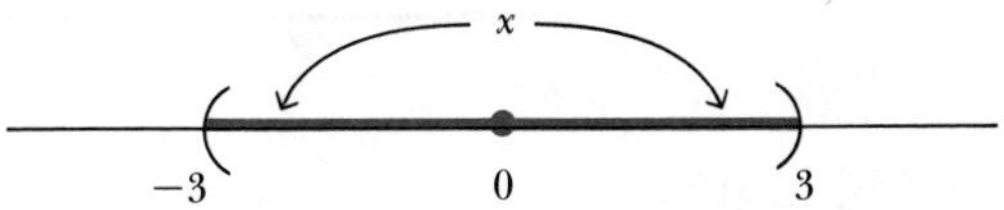

2 $|3x - 2| < 8$.

SOLUTION. By Property 1,

$$\begin{aligned}\{x \mid |3x - 2| < 8\} &= \{x \mid -8 < 3x - 2 < 8\} \\ &= \{x \mid -6 < 3x < 10\} \\ &= \{x \mid -2 < x < \tfrac{10}{3}\} \\ &= (-2, \tfrac{10}{3}) \qquad \text{(Figure 11)}\end{aligned}$$

Figure 11

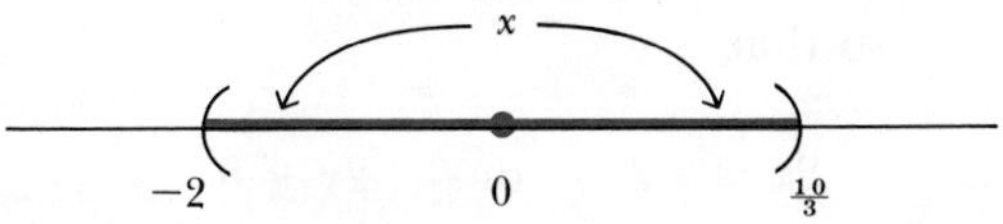

GEOMETRIC INTERPRETATION 2. $|x| > a$ means that the distance between 0 and x is more than a units, or, equivalently, x is more than a units from 0 (Figure 12).

Figure 12

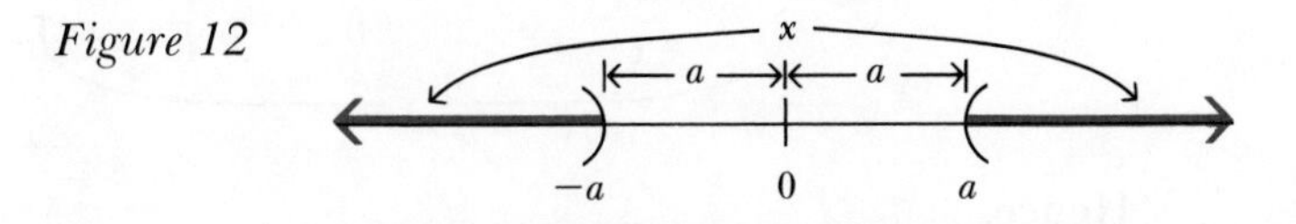

This suggests the next property.

PROPERTY 2

If $|x| > a$, then $x < -a$ or $x > a$, where $a > 0$; that is,

$$\begin{aligned}\{x \mid |x| > a\} &= \{x \mid x < -a\} \cup \{x \mid x > a\} \\ &= (-\infty, -a) \cup (a, \infty)\end{aligned}$$

EXAMPLES

Solve the absolute value inequalities, and illustrate the solution on the number line.

1 $|x| > 7$.

SOLUTION. By Property 2,

$\{x \mid |x| > 7\} = \{x \mid x < -7\} \cup \{x \mid x > 7\} = (-\infty, -7) \cup (7, \infty)$ (Figure 13)

Figure 13

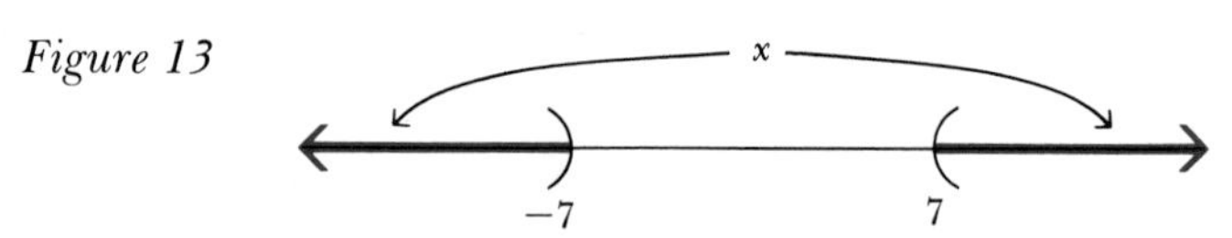

2 $|2x - 3| \geq 5$.

SOLUTION. $|2x - 3| \geq 5$ means that

$$2x - 3 \leq -5 \quad \text{or} \quad 2x - 3 \geq 5 \quad \text{(Figure 14)}$$

Figure 14

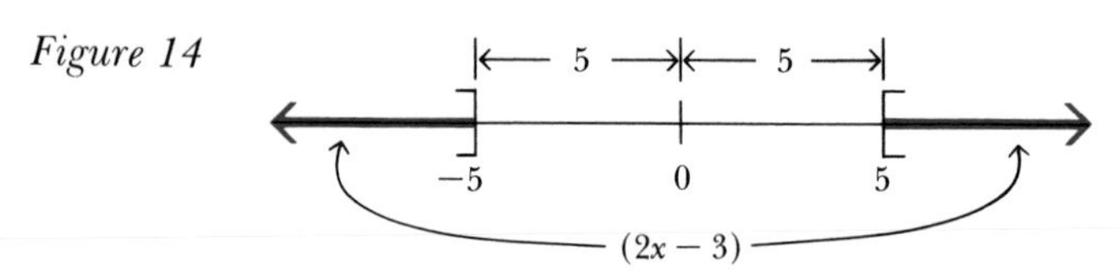

so that

$$2x \leq -2 \quad \text{or} \quad 2x \geq 8$$

that is,

$$x \leq -1 \quad \text{or} \quad x \geq 4 \quad \text{(Figure 15)}$$

Figure 15

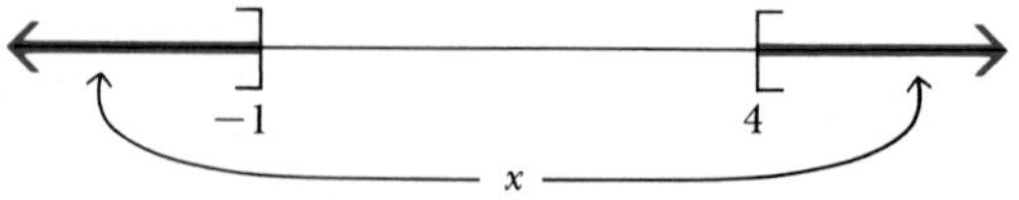

Hence,

$\{x \mid |2x - 3| \geq 5\} = \{x \mid x \leq -1\} \cup \{x \mid x \geq 4\} = (-\infty, -1] \cup [4, \infty)$

3 $|2x + 7| \geqslant 11$.

SOLUTION. $|2x + 7| \geqslant 11$ means that

$$2x + 7 \leqslant -11 \qquad \text{or} \qquad 2x + 7 \geqslant 11 \qquad \text{(Figure 16)}$$

Figure 16

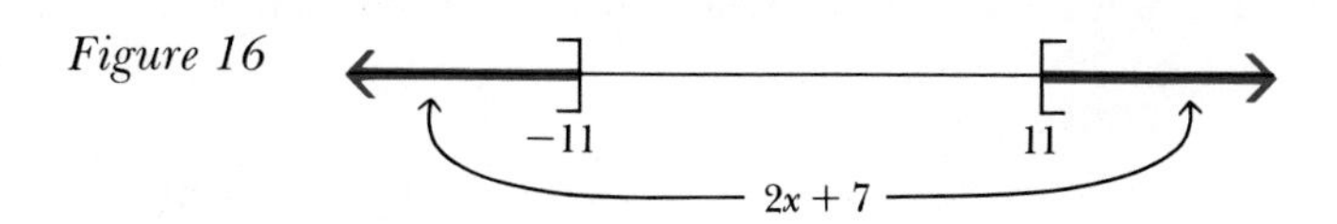

so that

$$2x \leqslant -18 \qquad \text{or} \qquad 2x \geqslant 4$$

that is,

$$x \leqslant -9 \qquad \text{or} \qquad x \geqslant 2$$

Hence,

$$\{x \mid |2x + 7| \geqslant 11\} = \{x \mid x \leqslant -9\} \cup \{x \mid x \geqslant 2\}$$
$$= (-\infty, -9] \cup [2, \infty) \qquad \text{(Figure 17)}$$

Figure 17

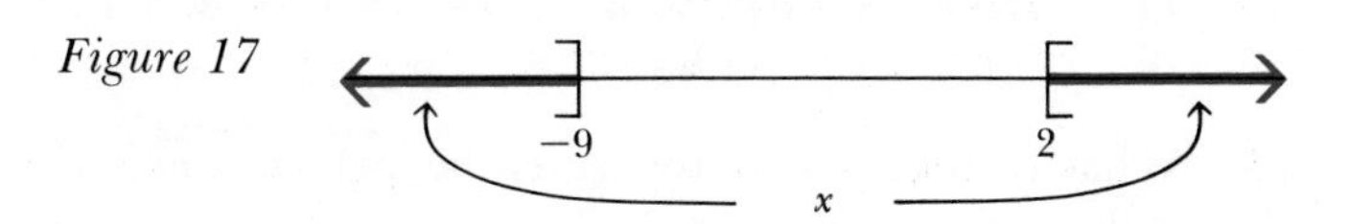

4 Express $\{x \mid -2 < x < 4\}$ using an absolute value inequality.

SOLUTION. Interval $(-2, 4)$ is of length 6 and has midpoint 1 (Figure 18). Now we saw in Section 4.1 that $|x - 1|$ represents the distance between x and 1; hence, $|x - 1| < 3$ for those x in the set, that is,

$$\{x \mid -2 < x < 4\} = \{x \mid |x - 1| < 3\}$$

Figure 18

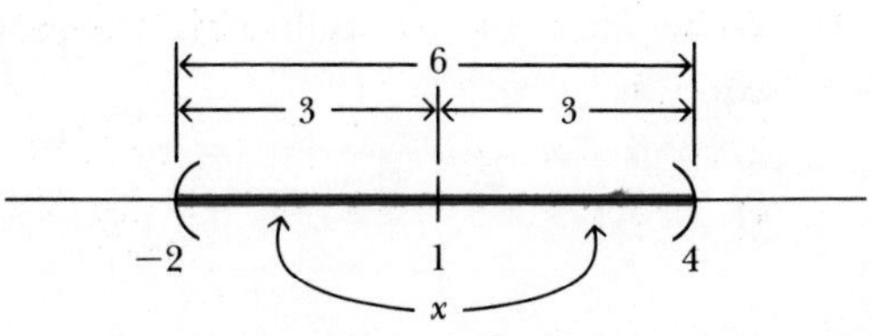

PROBLEM SET 3

1 If $x = 5$ and $y = -7$, compute each of the following numbers.
a) $|x| + |y|$ b) $|x + y|$

c) $|x - y|$
d) $|x| - |y|$
e) $|xy|$
f) $|x||y|$
g) $|y|^2$
h) $|x/y|$
i) $|3x| + |-4y|$
j) $3|x| - 4|y|$

2 Solve the following absolute value equalities.

a) $|x| = 5$
b) $|x| = 2$
c) $|x - 2| = 3$
d) $|x| = -2$
e) $|x + 5| = 6$
f) $|2x + 1| = 3$
g) $|x| = |3 - 2x|$
h) $|-3x| = 15$
i) $|x - 5| = |-3x + 7|$
j) $2|x + 1| = 3|1 - x|$

3 Solve the following absolute value inequalities and represent the solutions on the real line.

a) $|x| < 3$
b) $|x| > 3$
c) $|x - 1| \leqslant 3$
d) $|2x - 1| < 5$
e) $|3x - 1| > 1$
f) $|4 - 7x| \leqslant 3$
g) $|2x + 1| \geqslant 5$
h) $|5x - 2| > 0$
i) $|x + 5| < |x + 1|$
j) $|4x - 2| \leqslant 1$
k) $|5 - 1/x| < 1$

4 a) Compute $x/|x|$ if x is a nonzero real number.
b) For what values of x, if any, is $|x^3| = x^3$?
c) Under what conditions does $|x + y| = |x| + |y|$?
d) Under what conditions does $|x| = |y|$?

5 What restrictions, if any, must be put on x in order for each of the following statements to be true?

a) $|5x| = 5x$
b) $|x - 2| = 2 - x$
c) $|x + 2| = |x| + |2|$
d) $|1/x| = |x|$
e) $|3x| > 0$

6 For what values of x, if any, is each of the following statements true?

a) $|x - 2| \leqslant x + 2$
b) $|x - 5| \leqslant |x - 3|$
c) $|x| > |x - 1|$
d) $|x| < -3$
e) $|x| > -3$

7 Write each of the following inequalities as an absolute value inequality.

a) $-6 < x < 3$
b) $1.99 < x < 2.0$
c) $-3.1 < x < -3$

5 Cartesian Coordinate System and The Distance Formula

The real line provides us with a geometric representation of real numbers as points on a line and we used this geometric representation to investigate the order of the real numbers and the notion of the distance between points on a line. The Cartesian coordinate system enables us to represent "ordered pairs" of real numbers as points in a plane.

5.1 Ordered Pairs

The elements of a set do not have to be listed in any particular order. The set consisting of the two objects a and b could be written as $\{a, b\}$ or as $\{b, a\}$; in other words, $\{a, b\} = \{b, a\}$. If we wish to identify the order, we need some new notation.

We define (a, b) to be the *ordered pair* consisting of the *first element* a and the *second element* b. Two ordered pairs are considered to be equal when they have equal first members and equal second members. For example, $(1, 2) \neq (2, 1)$, even though each pair contains the same entries. Likewise, $(4, 3) \neq (4, 4)$ and $(7, 8) \neq (-7, 8)$, whereas $(9, x) = (y, 8)$ if and only if $x = 8$ and $y = 9$.

5.2 Cartesian Coordinate System

The set of all ordered pairs of real numbers can be represented as the set of points in a plane, using the *Cartesian coordinate system.* The system is based on two perpendicular lines each of which is a number line. These two lines are called the *coordinate axes* and are often referred to as the *horizontal axis* or *x axis* and the *vertical axis* or *y axis* (Figure 1).

Figure 1

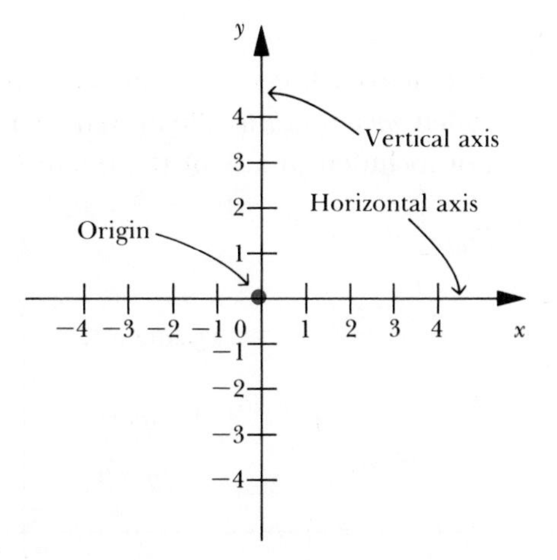

Given an ordered pair of real numbers (x, y), we can use the coordinate system to represent (x, y) as a point in the plane as follows.

The *abscissa* x is located on the x axis. Then a line is drawn perpendicular to this axis at point x; the *ordinate* y is located on the y axis at point y and a line is drawn perpendicular to the y axis at point y.

The intersection of the two lines we have just constructed is the point in the plane used to represent the ordered pair (x, y) (Figure 2).

Figure 2

For example, the ordered pairs $(1, 1)$, $(7, 5)$, and $(-\frac{1}{2}, 0)$ are represented in Figure 3.

Figure 3

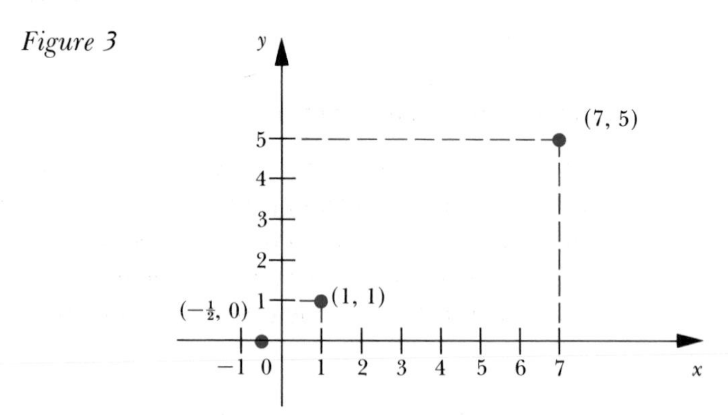

The coordinate axes divide the plane into four disjoint regions called *quadrants,* as illustrated in Figure 4. The coordinate axes are not included in any of the quadrants.

Figure 4

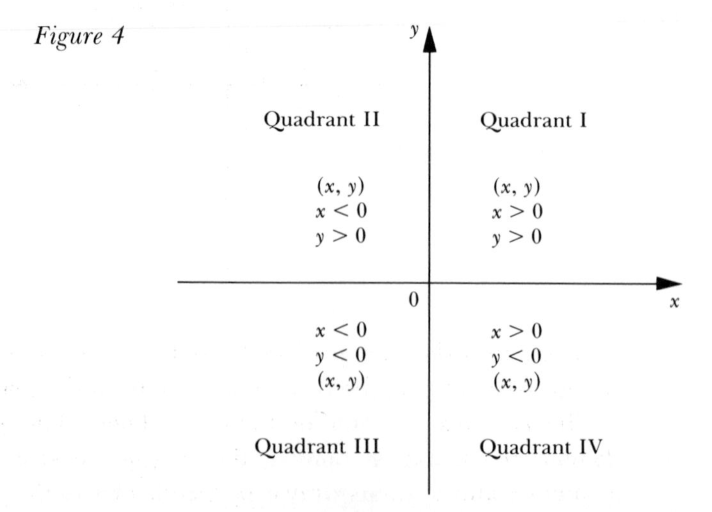

The set of all points in the plane whose coordinates correspond to the ordered pairs of a given set is called the *graph* of the set of ordered pairs. Locating these points in the plane by using the Cartesian coordinate system is called *graphing* the set of ordered pairs.

EXAMPLES

1 Graph each of the points $(1, -2)$, $(3, 4)$, $(-2, -3)$, and $(2, 0)$ and indicate which quadrant, if any, contains the points.

SOLUTION. These points are graphed in Figure 5. $(1, -2)$ lies in quadrant IV. $(3, 4)$ lies in quadrant I. $(-2, -3)$ lies in quadrant III. $(2, 0)$ does not lie in any quadrant.

Figure 5

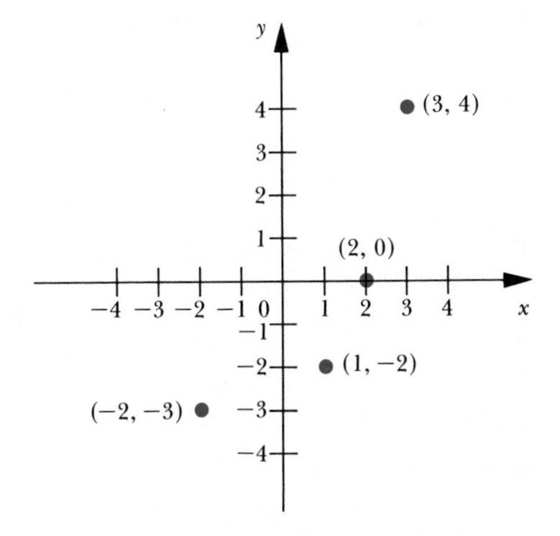

2 Graph $\{(x, y) \mid 0 < x < 2\} \cap \{(x, y) \mid 2 < y < 3\}$.

SOLUTION. The shaded region bounded by the dotted lines (Figure 6) is the graph of $\{(x, y) \mid 0 < x < 2\} \cap \{(x, y) \mid 2 < y < 3\}$. Notice the use of dotted lines, which indicates that points on the boundary of the rectangle are excluded.

Figure 6

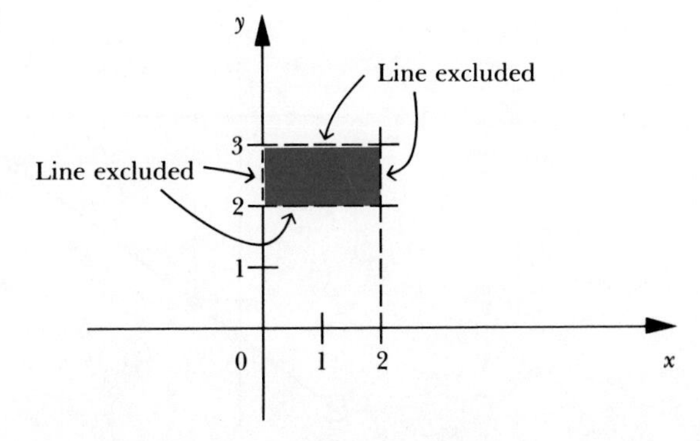

5.3 Distance Formula

It is possible to use the coordinates of any two given points to get an exact value for the distance between them by using the *distance formula,* which is derived as follows.

Given any two points P_1 and P_2 with coordinates (x_1, y_1) and (x_2, y_2), respectively, a formula for the distance d between P_1 and P_2 in terms of coordinates x_1, y_1, and x_2, y_2 can be derived by considering the following three cases.

1 If the two points lie on the same vertical line, that is, $x_1 = x_2$, then $d = |y_1 - y_2|$ (Figure 7).

2 If the two points lie on the same horizontal line, that is, $y_1 = y_2$, then $d = |x_1 - x_2|$ (Figure 8).

Figure 7

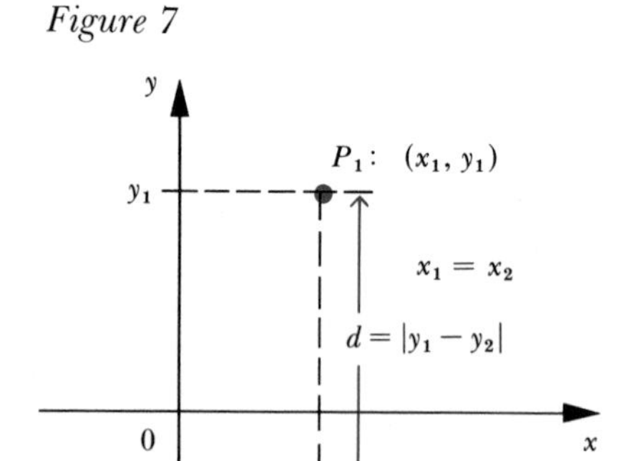

Figure 8

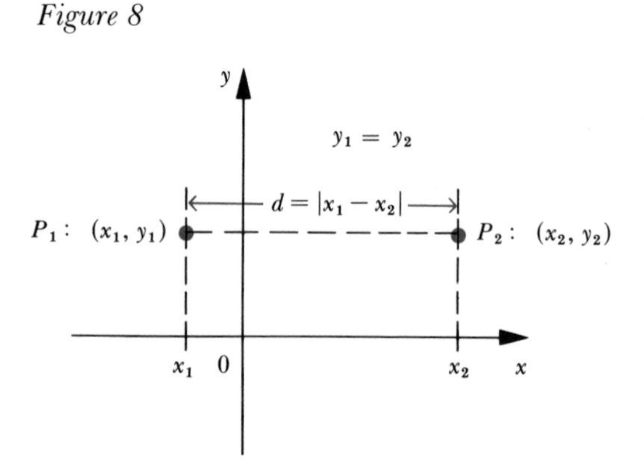

If the two points lie on a line that is neither horizontal nor vertical, then a right triangle (one with a 90-degree angle) is determined (Figure 9).

Figure 9

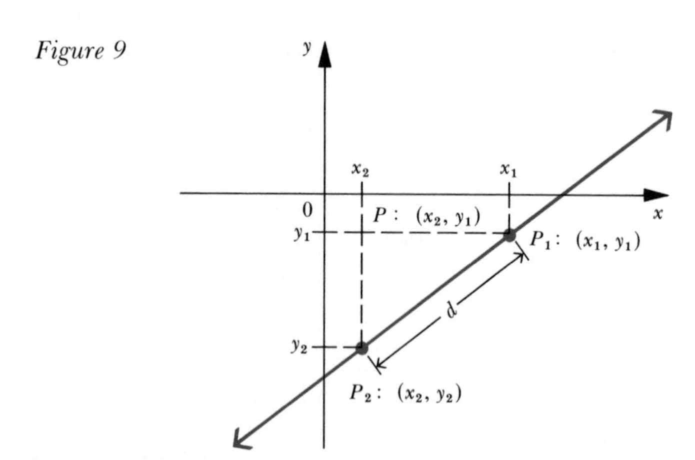

We find the length of the short sides of this right triangle by using 1 and 2 above. Thus, the horizontal side measures exactly $|x_1 - x_2|$ and the vertical side measures exactly $|y_1 - y_2|$.

Now we use the Pythagorean theorem, which is

$$(\text{length of segment } PP_1)^2 + (\text{length of segment } PP_2)^2 = d^2$$

That is,

$$|x_1 - x_2|^2 + |y_1 - y_2|^2 = d^2$$

so that

$$(x_1 - x_2)^2 + (y_1 - y_2)^2 = d^2 \qquad \text{(Why?)}$$

Hence,

$$d = \sqrt{(x_1 - x_2)^2 + (y_1 - y_2)^2}$$

We call this the *distance formula* for the points (x_1, y_1) and (x_2, y_2). Notice that the latter formula is also applicable in the special cases where P_1 and P_2 are on the same vertical line or the same horizontal line. (Why?) Since $(a - b)^2 = (b - a)^2$, the "order" of subtracting the abscissas or the ordinates is irrelevant.

EXAMPLES

1 Graph the points $(2, -4)$ and $(-2, -1)$ and then find the distance d between them.

SOLUTION. The points $P_1: (2, -4)$ and $P_2: (-2, -1)$ are graphed in Figure 10 and the distance d is given by

Figure 10

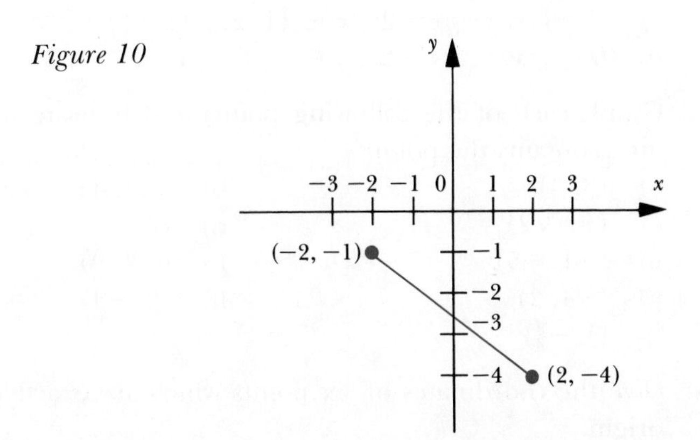

$$\begin{aligned} d &= \sqrt{[2-(-2)]^2 + [-4-(-1)]^2} \\ &= \sqrt{4^2 + (-3)^2} \\ &= \sqrt{16+9} = \sqrt{25} = 5 \end{aligned}$$

2 The *unit circle* is the circle with center at the origin and radius 1. Use the distance formula to derive the equation of the unit circle.

SOLUTION. Suppose that (x, y) is any point on the unit circle. For illustrative purposes (x, y) will be placed in the first quadrant (Figure 11). Then, by the distance formula,

$$1 = \sqrt{(x-0)^2 + (y-0)^2}$$

so that $1 = x^2 + y^2$. In other words, $\{(x, y) \mid x^2 + y^2 = 1\}$ is the set of all points on the unit circle, so $x^2 + y^2 = 1$ is the equation of the unit circle.

Figure 11

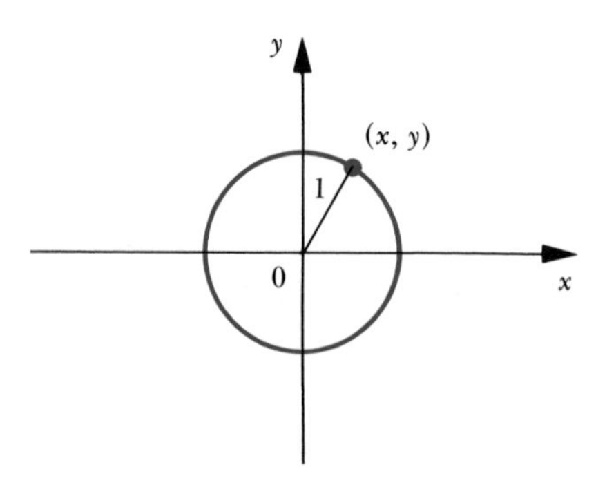

PROBLEM SET 4

1 List the members of each of the following sets of ordered pairs.
a) $A = \{(x, y) \mid y = 3x + 1, x \in \{-1, 0, 1, 2\}\}$
b) $B = \{(x, y) \mid y = -2x + 3, x \in \{-2, -1, 0, 1\}\}$
c) $C = \{(x, y) \mid y = 2x, x \in \{1, 2, 3, 4, 5\}\}$
d) $D = \{(x, y) \mid y = 2x, x \in \{-1, 0, 1, 2, 3, 4, 5, 6\}\}$

2 Graph each of the following points and indicate what quadrant, if any, contains the points.
a) $(3, 3)$ b) $(-2, 4)$
c) $(\pi, \sqrt{2})$ d) $(0, 7)$
e) $(-1, -5)$ f) $(-3, 0)$
g) $(-4, 2)$ h) $(2, -4)$
i) $(\frac{1}{2}, -\frac{3}{2})$

3 Give the coordinates of six points which are exactly 1 unit from the origin.

4 Graph each of the following sets.

a) $\{(x, y) \mid x > 1 \text{ and } y < 2\}$
b) $\{(x, y) \mid x > 2\} \cap \{(x, y) \mid x < 5\}$
c) $\{(x, y) \mid x < 0\} \cap \{(x, y) \mid y > 0\}$
d) $\{(x, y) \mid y < 0\}$
e) $\{(x, y) \mid x < 0\} \cap \{(x, y) \mid y < 0\}$
f) $\{(x, y) \mid -1 \leqslant x < 0\} \cap \{(x, y) \mid y \geqslant 0\}$
g) $\{(x, y) \mid -2 \leqslant x \leqslant 2 \text{ and } 0 \leqslant y \leqslant 2\}$
h) $\{(x, y) \mid x \geqslant 2\} \cap \{(x, y) \mid y \geqslant 1\}$
i) $\{(x, y) \mid y < 3\}$

5 Plot the following pairs of points in a Cartesian coordinate system and find the distance between each.

a) $(-\frac{1}{2}, 1), (2, 3)$
b) $(-3, -4), (-5, -7)$
c) $(5, 0), (-7, 3)$
d) $(t, 8), (t, 7)$
e) $(1, 1), (-3, 2)$
f) $(-3, -4), (8, 0)$

6 First plot the following pairs of points in a Cartesian coordinate system; then use the distance formula to determine if the triangle whose vertices are $(3, 1)$, $(4, 3)$, and $(6, 2)$ is an isosceles triangle (a triangle with two equal sides).

7 Given points $P_1: (a, b)$, $P_2: (c, d)$, and $P_3: ((a + c)/2, (b + d)/2)$.

a) Find the lengths of segments P_1P_3 and P_2P_3 in terms of a, b, c, and d by using the distance formula.
b) How do these two lengths compare?
c) What can you conclude about the geometric position of P_3 with respect to P_1 and P_2?
d) By using the formula above, find the coordinates of the midpoints of the following pairs of points. Also, use the distance formula to check the midpoints.
   i) $(5, 6), (-7, 8)$
   ii) $(-4, 7), (-3, 0)$
   iii) $(3, 8), (8, 3)$

6 Relations and Functions

The use of the word "relation" here will not always conform to the everyday usage of the term.

Definition

A *relation* is a set of ordered pairs. The *domain* of a relation is the set of all first members of the ordered pairs, and the *range* of a relation is the set of all second members of the ordered pairs.

Since we will usually consider only those relations that are formed from real numbers, we can use the Cartesian coordinate system to represent relations as sets of points in a plane. This kind of representation provides us with a "geometric picture" of the relation.

EXAMPLES

1 For the relations R_1 and R_2 below, let $U = \{1, 2, 3\}$. Enumerate the members of the relation; indicate the domain and range and graph the relation.

a) $R_1 = \{(x, y) \mid y = x, x \in U \text{ and } y \in U\}$

b) $R_2 = \{(x, y) \mid y > x, x \in U \text{ and } y \in U\}$

SOLUTION

a) $R_1 = \{(1, 1), (2, 2), (3, 3)\}$ so that the domain is the set $\{1, 2, 3\}$ and the range is the set $\{1, 2, 3\}$. The graph consists of three points (Figure 1).

Figure 1

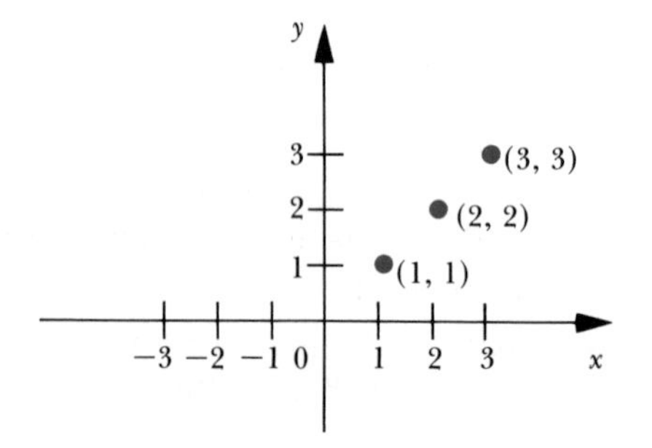

b) $R_2 = \{(x, y) \mid y > x\} = \{(1, 2), (1, 3), (2, 3)\}$, so that the domain of R_2 is $\{1, 2\}$ and the range of R_2 is $\{2, 3\}$. The graph consists of three points (Figure 2).

Figure 2

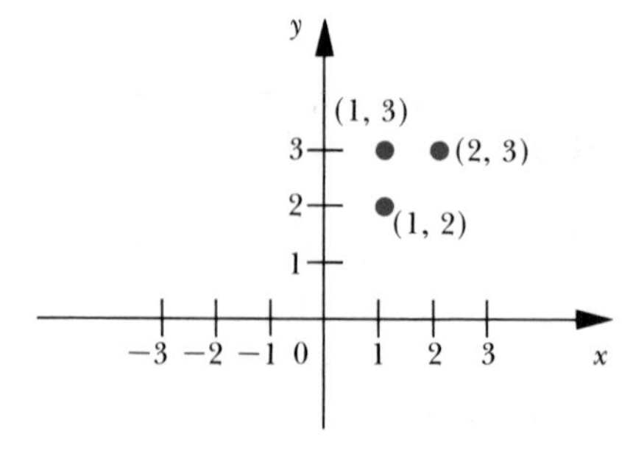

2 Graph the relation $\{(x, y) \mid y = 3x, x \in R\}$.

SOLUTION. Since we cannot list all members of this relation, we plot enough points to determine the pattern of the graph (Figure 3*a*). Here we use a table to list those members of the relation that were graphed.

Figure 3

x	$y = 3x$
-1	-3
0	0
1	3
2	6

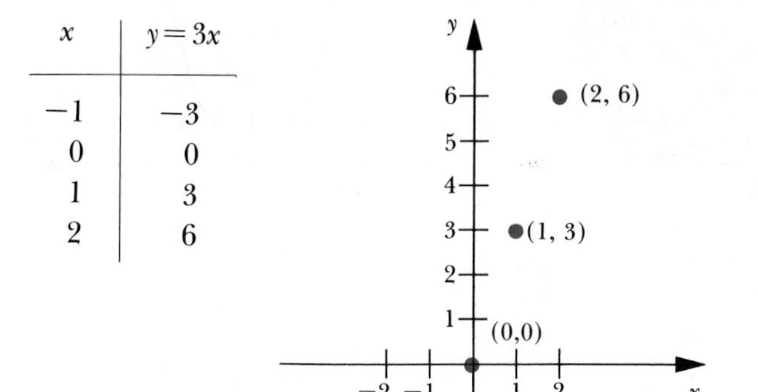

(*a*)

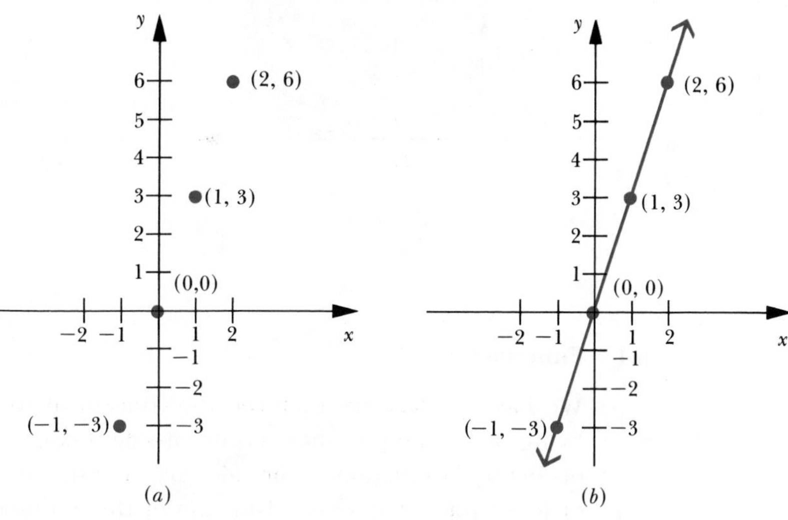

(*b*)

If we were to continue to graph members of this relation the pattern would continue in that the points would form a linear pattern. In fact, the graph would actually turn out to be a line (Figure 3*b*).

Quite often the description of a relation is abbreviated by omitting the set notation. When this is the case, the set of ordered pairs is implied. For example, $x < y$ is an abbreviated way of writing the relation $\{(x, y) \mid x < y, x \in R\}$.

Moreover, the domain and/or range of a relation is not always given explicitly. When this is the case, it may be necessary to derive the domain and/or range. For example, if $y = 1/x$, the domain consists of all real numbers except 0 since division by zero is not defined. Finally, if the universal set is not given, we shall assume the universal set to be R, the set of real numbers.

EXAMPLE

Graph $y = x^2$ and indicate the domain and range.

SOLUTION. $y = x^2$ represents the relation $\{(x, y) \mid y = x^2\}$. Since any real number can be squared and since the square of a real number is always nonnegative, the domain is R and the range is the set $\{y \mid y \geq 0\}$. This means that the graph "lies" in the region above and on the x axis (Figure 4*a*). Finally, we can use a table to determine the pattern of the graph, which we complete "by inspection" (Figure 4*b*).

Figure 4

x	$y = x^2$
0	0
1	1
−1	1
2	4
−2	4

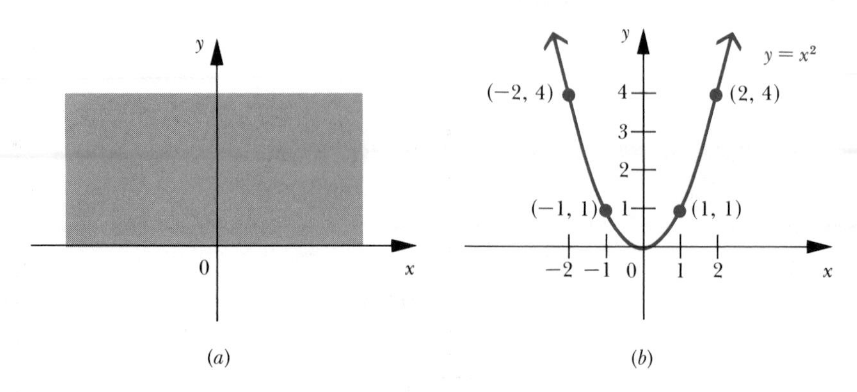

6.1 Functions

We have already encountered the concept of functions in everyday living. For example, the amount of sales tax charged on a purchase of \$5.00 is a function of the sales tax rate; the number of books to be ordered for a course is a function of the number of students in the course; the number of congressional representatives for a particular state is a function of the population of the state.

A function suggests some kind of correspondence. In each of the examples above, there is an established correspondence between numbers—the amount of sales tax corresponds to the cost, the number of books to the number of students, and the number of representatives to the number of people. In general, we have the following definition.

Definition 1

A *function* is a correspondence that assigns to each member in a certain set, called the *domain* of the function, exactly one member in a second set, called the *range* of the function.

The concept of a function can also be defined by using the notion of a relation.

Definition 2

A *function* is a relation in which no two different ordered pairs have the same first member. The set of all first members (of the ordered pairs) is called the *domain* of the function. The set of all second members (of the ordered pairs) is called the *range* of the function. For each ordered pair (x, y) of the function we say that x in the domain has y as the *corresponding* member in the range.

From Definition 2 we conclude that all functions are relations; however, not all relations are functions. For example, $\{(1, 1), (2, 2), (3, 7), (3, 5)\}$ is a relation, with domain $\{1, 2, 3\}$ and range $\{1, 2, 7, 5\}$; it is not a function because $(3, 7)$ and $(3, 5)$ have the same first members (see Definition 2). By contrast, $\{(1, 2), (3, 4), (4, 4)\}$ is a relation that is a function with domain $\{1, 3, 4\}$ and range $\{2, 4\}$. Note that in the latter example, two pairs have the same second member; this does *not* violate the definition of a function.

Hence, if any two different ordered pairs of a relation have the same first member, the relation is not a function. In other words, if a domain member appears with more than one range member, the relation is not a function. Geometrically, this means that if the graph of a relation has more than one point with the same abscissa, the relation is not a function. Thus, if it is possible to draw a vertical line that intersects the graph of a relation at more than one point, then the relation is not a function.

Consider the graphs of the relations in Figure 5. The relation $\{(x, y) \mid y = \sqrt{25 - x^2}\}$ in Figure 5*a* represents a function because the graph does not have two different points with the same x coordinates, whereas the relation $\{(x, y) \mid x = \sqrt{25 - y^2}\}$ in Figure 5*b* is not a function because there are two different points with the same x coordinates—for example, the two points $(3, 4)$ and $(3, -4)$.

Figure 5

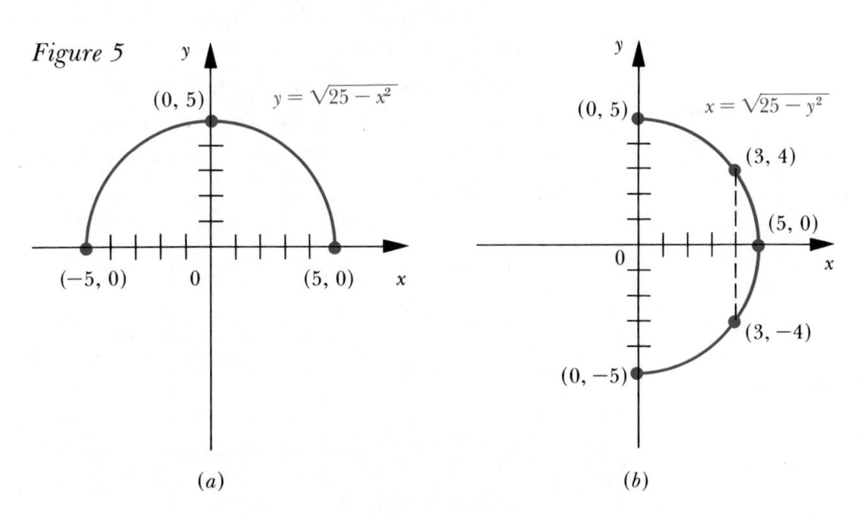

EXAMPLES

1 In each of the following parts, indicate whether or not the relation is a function. What is the domain and range? Graph the relation.

a) $\{(1, 2), (2, 3), (3, 4), (4, 4), (5, 6)\}$

b) $y = 3$ c) $x = 1$

d) $y = 2x$

SOLUTION

a) $\{(1, 2), (2, 3), (3, 4), (4, 4), (5, 6)\}$ is a function with domain $\{1, 2, 3, 4, 5\}$ and range $\{2, 3, 4, 6\}$ (Figure 6). Note that we do not draw lines connecting these points because these are the only points that belong to the relation.

Figure 6

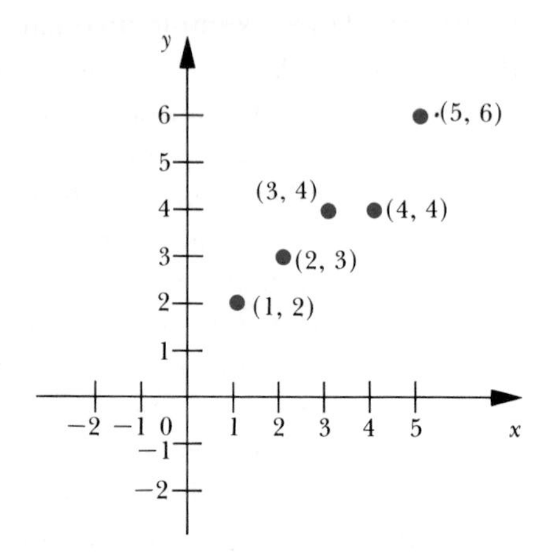

b) $y = 3$ is an abbreviated way of writing the relation $\{(x, y) \mid y = 3\}$. The domain is the set of all real numbers because there is no restriction on x, whereas the range is $\{3\}$. The relation is a function because no two different ordered pairs have the same first members. The graph is the line parallel to the x axis (Figure 7). This function is called a *constant* function because regardless of which domain element we choose, the range element is always the number 3.

Figure 7

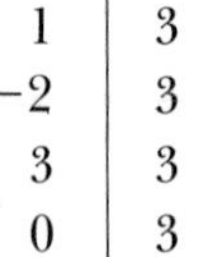

x	y
1	3
−2	3
3	3
0	3

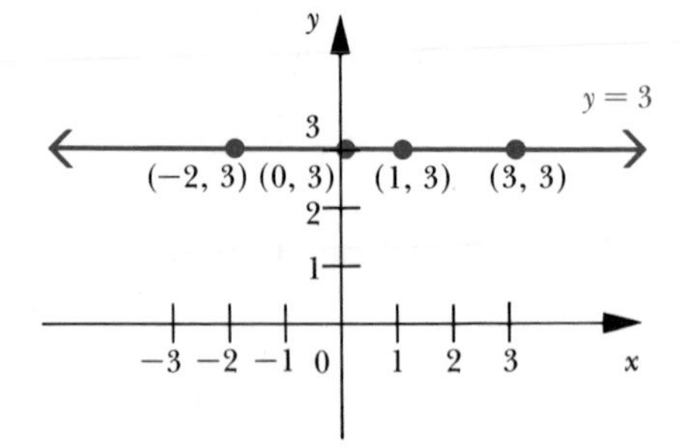

c) $x = 1$ is the relation $\{(x, y) \mid x = 1\}$. This relation is not a function because, for example, $(1, 3)$ and $(1, 4)$ are members of the set. The domain of the relation is $\{1\}$, and the range is the set of all real numbers. The graph is the line parallel to the y axis (Figure 8).

Figure 8

x	y
1	0
1	1
1	2
1	−1
1	−2

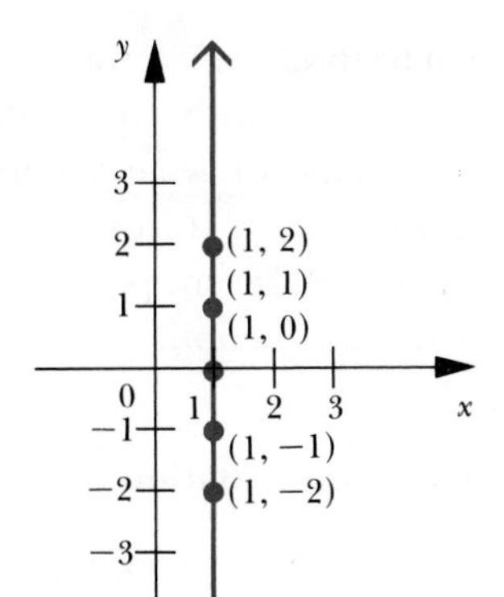

d) $y = 2x$ represents the set of ordered pairs $\{(x, y) \mid y = 2x\}$. It is a function and has the set of all real numbers as both its domain and its range. The graph is a line. Hence, we call $y = 2x$ a linear equation (Figure 9). It can be seen geometrically that this set of ordered pairs is a function, because no two points have the same abscissa.

Figure 9

x	y
0	0
1	2
−1	−2
2	4
3	6

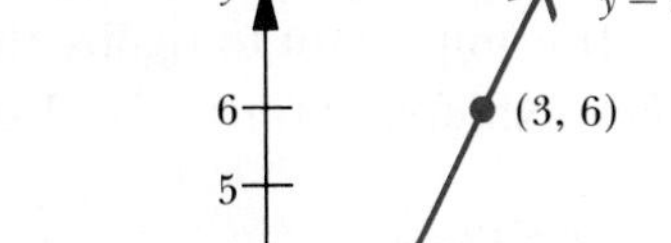

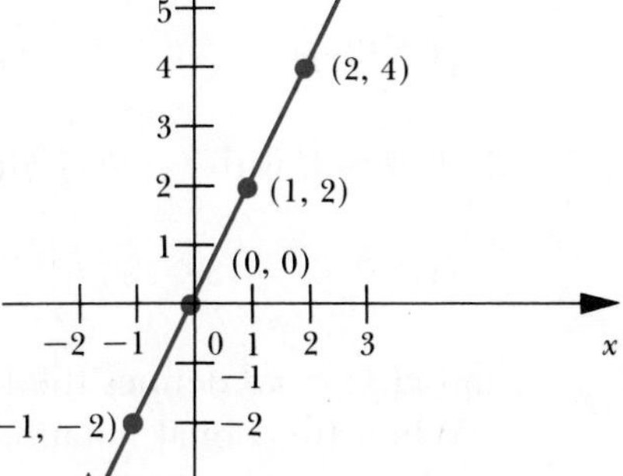

2 Examine the graphs of each of the relations given in Figures 10*a* and *b* to decide whether or not the relation is a function.

Figure 10

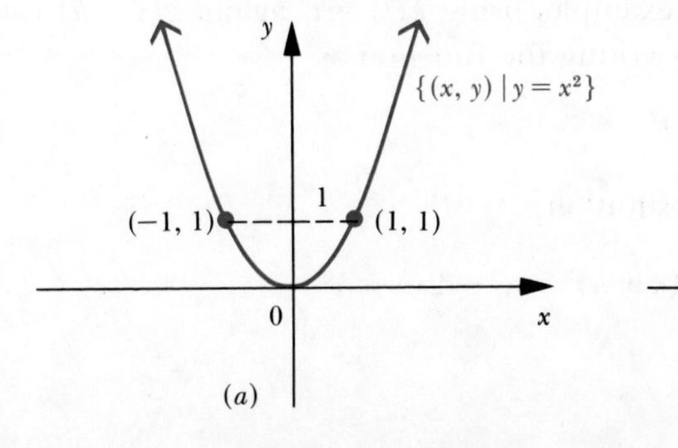

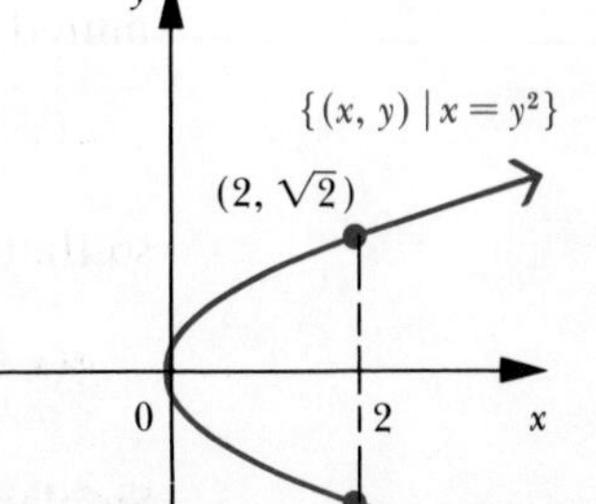

SOLUTION. Since the graph in Figure 10a does not have two different points with equal abscissas, no two different ordered pairs of the relation have equal first members, so the relation $\{(x, y) \mid y = x^2\}$ is a function. It can be seen from the graph in Figure 10b that there are two different points with abscissa 2, so the relation $\{(x, y) \mid x = y^2\}$ is not a function.

6.2 Function Notation

If x represents members of the domain of $y = x^2$, then y is considered to be a function of x and we indicate this by writing $f(x) = x^2$, which reads "f of x equals x squared." and means "the value of the function f at the number x is the number x^2." Therefore, $f(x) = x^2$ is another way of defining the function $\{(x, y) \mid y = x^2\}$.

In other words, $y = f(x)$ means that (x, y) is a member of the function f. For example, the equation $4x - 2y = 1$ determines y as a function of x, which we may see by solving for y in terms of x, getting $y = 2x - \frac{1}{2}$. Thus, we could write the function either as $f = \{(x, y) \mid y = 2x - \frac{1}{2}\}$ or as $f(x) = 2x - \frac{1}{2}$.

It is important to realize that letters other than x, y, or f can be used. For example, $h(r) = r^2 - 1$ defines the function given by

$$\{(r, h(r)) \mid h(r) = r^2 - 1\}$$

$3r + 5t = 3$, with $t = g(r)$, defines the function given by

$$\{(r, t) \mid t = (3 - 3r)/5\}$$

and $c(d) = \pi d$ defines the function given by $\{(d, c(d)) \mid c(d) = \pi d\}$.

When functional notation is used, it is helpful sometimes to think of the variable that represents the members of the domain as a "blank." For example, $g(t) = t^2$ can be thought of as $g(\quad) = (\quad)^2$; hence, if any expression (representing a real number) is to be used to represent a member of the domain, it is easy to see where this same expression is to be substituted in the equation describing the function. For example, using $g(t) = t^2$ again, $g(x + h)$ can be determined by first writing the function as

$$g(\quad) = (\quad)^2$$

so that, by substitution,

$$g(x + h) = (x + h)^2 = x^2 + 2xh + h^2$$

Similarly,

$$g(3-5x) = (3-5x)^2 = 9 - 30x + 25x^2$$

EXAMPLES

1 Let $f(x) = x + 1$. Find $f(2), f(3), f(5), f(a-6)$, and $f(a) - f(6)$. What is the domain of f?

SOLUTION

$$f(2) = (2) + 1 = 3$$
$$f(3) = (3) + 1 = 4$$
$$f(5) = (5) + 1 = 6$$
$$f(a-6) = (a-6) + 1 = a - 5$$
$$f(a) - f(6) = [(a) + 1] - [(6) + 1] = a - 6$$

The domain is the set of real numbers R.

2 Let $f(x) = \sqrt{25 - x^2}$. Find $f(0), f(3), f(4), f(5)$, and $f(-3)$.

SOLUTION

$$f(0) = \sqrt{25 - (0)^2} = 5$$
$$f(3) = \sqrt{25 - (3)^2} = \sqrt{25 - 9} = \sqrt{16} = 4$$
$$f(4) = \sqrt{25 - (4)^2} = \sqrt{25 - 16} = \sqrt{9} = 3$$
$$f(5) = \sqrt{25 - (5)^2} = \sqrt{25 - 25} = 0$$
$$f(-3) = \sqrt{25 - (-3)^2} = \sqrt{25 - 9} = \sqrt{16} = 4$$

3 Let $f(x) = x + 1$ and $g(x) = x^2 - 2$. Find $f[g(x)]$ and $g[f(x)]$.

SOLUTION

$$f[g(x)] = f[x^2 - 2] = (x^2 - 2) + 1 = x^2 - 1$$

and

$$g[f(x)] = g[x+1] = (x+1)^2 - 2 = (x^2 + 2x + 1) - 2 = x^2 + 2x - 1$$

Consider the function f defined by the equation $f(x) = 3x$ with domain $\{1, 2, 3\}$ and range $\{3, 6, 9\}$. In this example, 1 can be considered as "mapped" to 3, 2 "mapped" to 6, and 3 "mapped" to 9 (Figure 11).

Figure 11

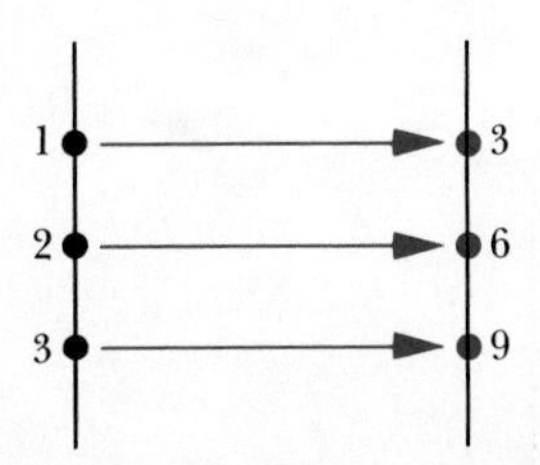

In general, if x is an element of the domain of a function f, then f associates x with $f(x)$, the value of f at x, and we say that f *maps* x *to* $f(x)$ or x *is mapped to* $f(x)$ or, equivalently, $f(x)$ *is the image of* x *under* f. Using arrow notation, we can represent this symbolically by $f: x \rightarrow f(x)$, or $x \xrightarrow{f} f(x)$. Hence, in the above example,

$1 \xrightarrow{f} 3$
$2 \xrightarrow{f} 6$
$3 \xrightarrow{f} 9$

EXAMPLES

1 Interpret $f(x) = 3x - 1$, $x \in \{-1, 1, 2, 3\}$, as a mapping.

SOLUTION. In mapping notation, $-1 \xrightarrow{f} -4$, $1 \xrightarrow{f} 2$, $2 \xrightarrow{f} 5$, and $3 \xrightarrow{f} 8$. Here -4 is the image of -1, 2 is the image of 1, 5 is the image of 2, and 8 is the image of 3.

2 Describe the function f defined by the equation $f(x) = -3x + 5$, $x \in \{-1, 0, 1, 2\}$, using the mapping notation, the ordered pair interpretation and function notation.

SOLUTION

Mapping $x \xrightarrow{f} -3x + 5$	Ordered pair (x, y)	$y = f(x)$
$-1 \rightarrow 8$	$(-1, 8)$	$f(-1) = 8$
$0 \rightarrow 5$	$(0, 5)$	$f(0) = 5$
$1 \rightarrow 2$	$(1, 2)$	$f(1) = 2$
$2 \rightarrow -1$	$(2, -1)$	$f(2) = -1$

3 Given the *absolute value function*, $f(x) = |x|$, find $f(-2)$, $f(-1)$, $f(0)$, $f(1)$, and $f(2)$. Describe the domain and range of f. Sketch the graph of f.

SOLUTION. $f(-2) = 2$, $f(-1) = 1$, $f(0) = 0$, $f(1) = 1$, and $f(2) = 2$. The domain is the set of real numbers R, and the range is the set of all nonnegative real numbers (Figure 12).

Figure 12

x	y
-2	2
-1	1
0	0
1	1
2	2

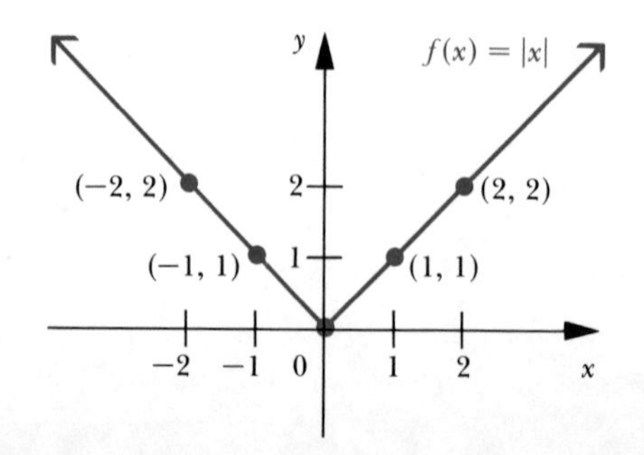

4 Given the function $f(x) = x^3 - 2x^2 - x + 2$, find $f(-2), f(-1), f(0), f(1), f(2)$ and $f(3)$. Also sketch the graph of f.

SOLUTION. The requested values of the function f are listed in the following table.

Figure 13

x	$f(x)$
-2	-12
-1	0
0	2
1	0
2	0
3	8

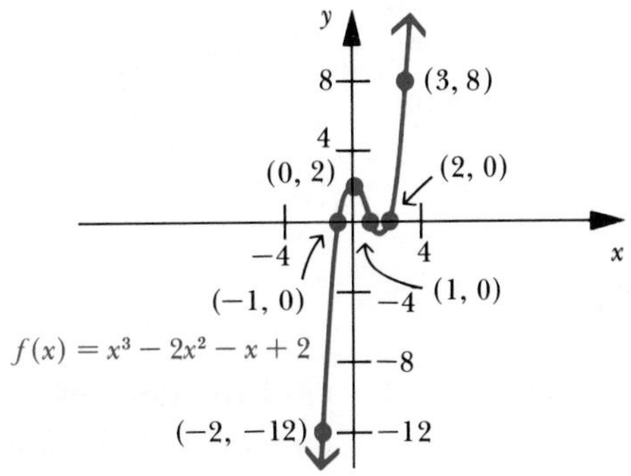

From the table we observe that the points $(-2, -12)$, $(-1, 0)$, $(0, 2)$, $(1, 0)$, $(2, 0)$, and $(3, 8)$ help determine the pattern of the graph (Figure 13).

It should be noticed that the graph of f crosses the x axis at $x = -1$, $x = 1$, and $x = 2$. These values of x are called the *zeros* of the function f. In general, $\{x \mid f(x) = 0\}$ is the set of all zeros of f.

PROBLEM SET 5

1 In each of the following parts, identify which of the relations is a function and which is not a function. If it is not a function, state the reason. Also identify the domain and range and graph the relation.

a) $\{(2, 4), (3, 6), (7, 2), (9, -3)\}$

b) $\{(-8, 0), (-7, 0), (-6, 2), (-7, 3), (5, 1)\}$

c) $\{(x, y) \mid y = 2x - 1\}$ d) $\{(x, y) \mid y \leq 3x + 2\}$

e) $\{(x, y) \mid y = |3x|\}$ f) $\{(x, y) \mid y = 2^x\}$

g) $\{(x, y) \mid y = \log_3 x\}$ h) $y = -5x^2$

2 What numbers in the domain of each of the following functions have 9 as an image?

a) $f(x) = 2x + 1$ b) $f(x) = -2x$

c) $f(x) = 2x^2$ d) $f(x) = 7$

3 Which of the four graphs in Figure 14 could represent functions?

Figure 14

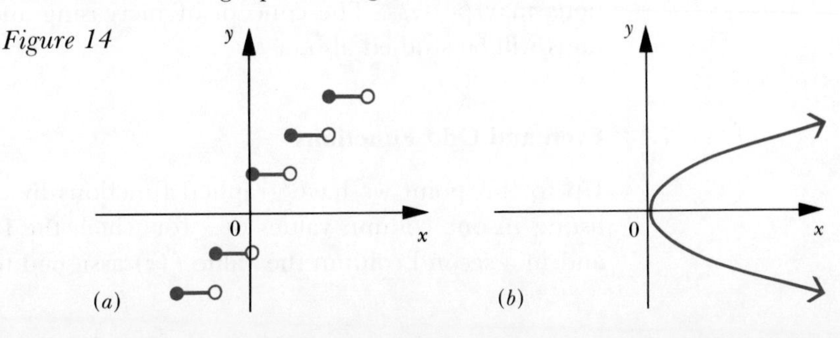

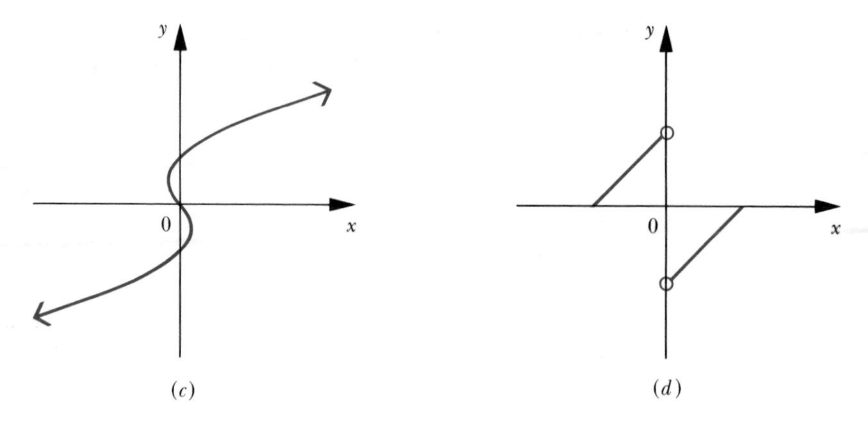

(c) (d)

4 Find the domain and the range of each of the following functions; find $f(1)$, $f(3)$, and $f(5)$; also sketch the graph of f. Identify the zeros of f.

a) $f(x) = 3x + 1$
b) $f = \{(1, 2), (3, 4), (5, 7)\}$
c) $f(x) = |x| - 1$
d) $f(x) = 2x - 3$
e) $f(x) = 5^x$
f) $f(x) = x^2 - 4$
g) $f(x) = 2/x$
h) $f(x) = x^3 - 1$
i) $f(x) = \log_{10} x$

5 Describe each of the functions in Problem 2 using the mapping notation, the ordered pair interpretation, and the function notation.

6 Let $f(x) = |x| - x$.

a) Determine which of the following points lie on the graph of f: $(1, 0)$, $(-1, 0)$, $(-2, 4)$, $(-3, 6)$.
b) What is the domain and the range of f?
c) Evaluate $f(3)$, $2f(-3)$, $f(2)$, $f(-2)$, and $f(a + 1)$.
d) Sketch the graph of f.

7 Let $f(x) = 3x - 1$ and $g(x) = -2x^2$. Find $f[g(x)]$ and $g[f(x)]$.

7 Some Properties of Functions

Next, we shall investigate two types of symmetry that graphs of functions may possess. The concept of increasing and decreasing functions will be studied also.

7.1 Even and Odd Functions

Up to this point we have graphed functions by constructing tables, listing in one column values of x for which the function is defined, and in a second column the value $f(x)$ assigned to each x. This procedure involves selecting enough values of x that the pattern suggested by the corresponding points $(x, f(x))$ enables us to sketch the complete graph. At times we can reduce the number of points needed to graph some functions by employing special features of the emerging pattern, such as symmetry.

The graph of a function f is *symmetric with respect to the y axis* if whenever the point (x, y) is on the graph of f, the point $(-x, y)$ is also on the graph of f, that is, if $f(-x) = f(x)$ (Figure 1).

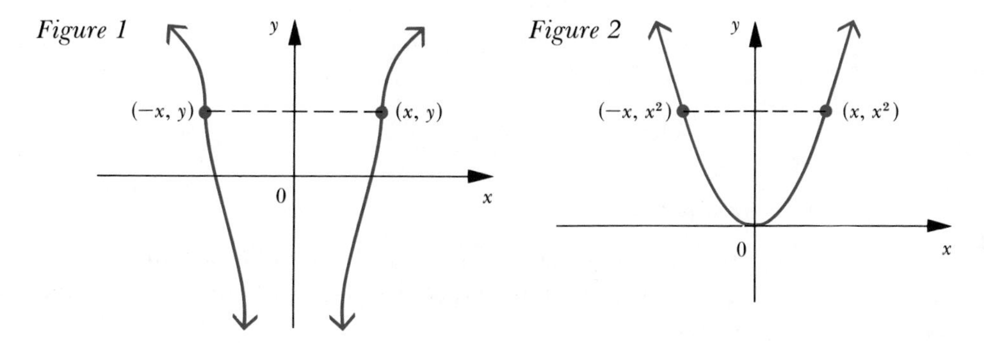

Figure 1

Figure 2

A function f whose graph possesses the property $f(-x) = f(x)$ is called an *even function.* For example, the graph of $f(x) = x^2$ is symmetric with respect to the y axis, since $f(-x) = (-x)^2 = x^2 = f(x)$ (Figure 2) and $f(x) = x^2$ is said to be an even function. In graphing such a function we need only to plot points to the right of the y axis and then "copy" the graph onto the left side as though the y axis were a mirror.

The graph of a function f is *symmetric with respect to the origin* if whenever the point (x, y) is on the graph of f, the point $(-x, -y)$ is also on the graph of f, that is, if $f(-x) = -f(x)$ (Figure 3). In this

Figure 3

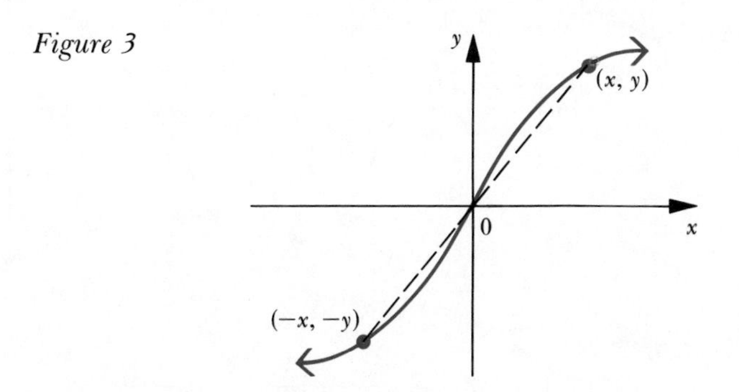

case, the function f is said to be an *odd function.* For example, the graph of $f(x) = x^3$ is symmetric with respect to the origin, since $f(-x) = (-x)^3 = -x^3 = -f(x)$ (Figure 4) and $f(x) = x^3$ is said to be an odd function.

Figure 4

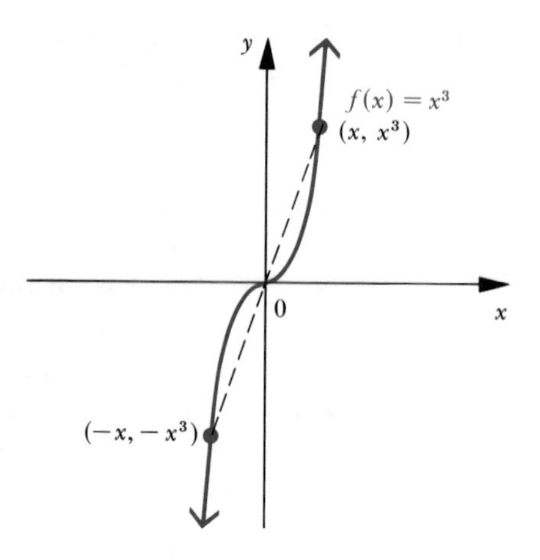

EXAMPLES

Discuss the symmetry of each of the following functions, sketch the graphs using symmetry, and indicate whether the function is even or odd.

1 $f(x) = -3x^2$

SOLUTION. $f(-x) = -3(-x)^2 = -3x^2 = f(x)$, so the graph of f is symmetric with respect to the y axis; hence, it is only necessary to graph the function f for nonnegative values of x (Figure 5a). The remainder ot the graph is determined by a "reflection" across the y axis (Figure 5b). f is an even function.

Figure 5

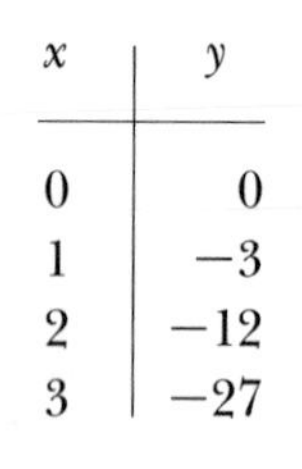

x	y
0	0
1	−3
2	−12
3	−27

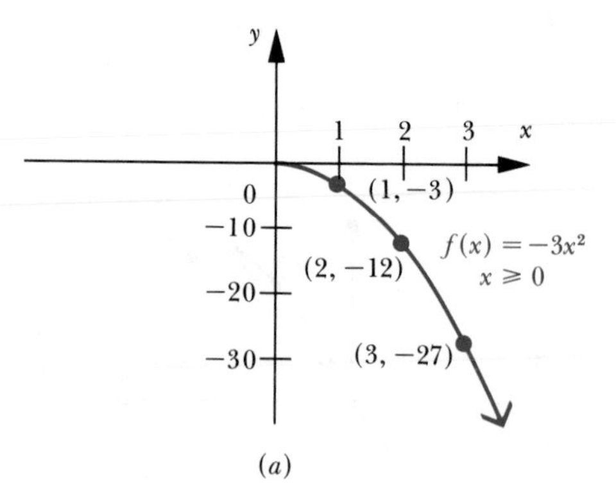

(a)

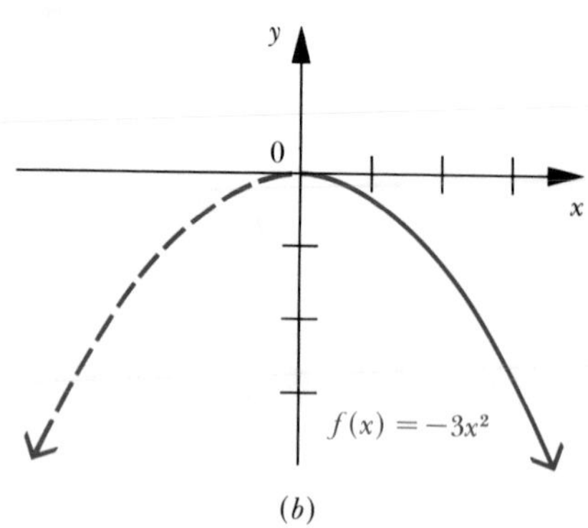

(b)

2 $f(x) = \sqrt{1 - x^2}$

SOLUTION. $f(-x) = \sqrt{1 - (-x)^2} = \sqrt{1 - x^2} = f(x)$, so the graph of f is symmetric with respect to the y axis. Hence, we can get the graph of the function f by first graphing points from the above table (Figure 6a), and then reflecting this graph across the y axis (Figure 6b). f is an even function.

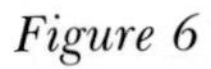

Figure 6

x	y
0	1
$\frac{1}{2}$	$\frac{\sqrt{3}}{2}$
1	0

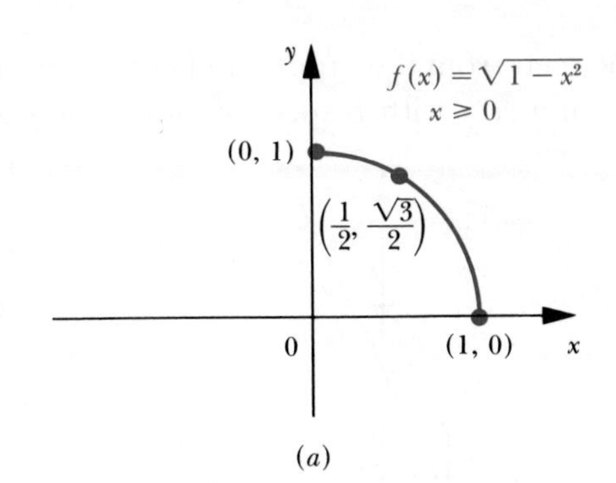

(*a*)

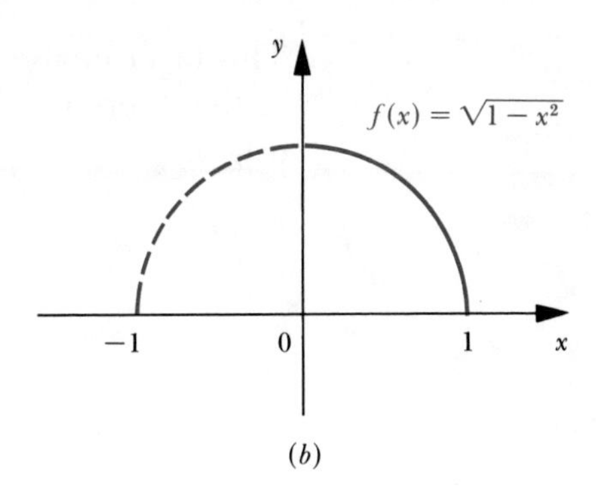

(*b*)

3 $f(x) = -5x^3$

SOLUTION. $f(-x) = -5(-x)^3 = -(-5x^3) = -f(x)$, so the graph of f is symmetric with respect to the origin. Hence, we can graph f by first locating points from the table (Figure 7*a*), and then reflecting this graph across the origin (Figure 7*b*). f is an odd function.

Figure 7

x	y
0	0
−1	5
−2	40

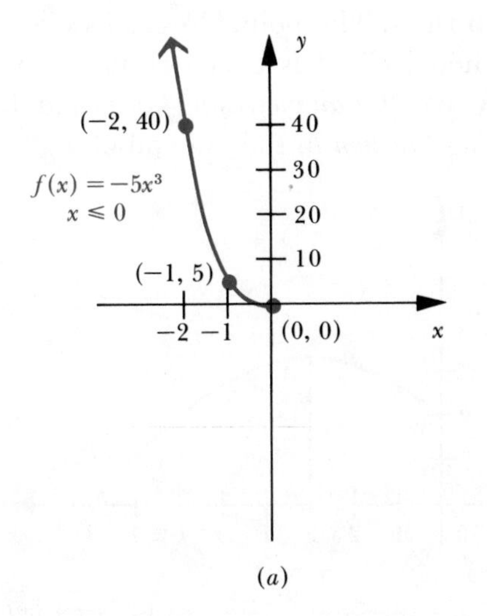

(*a*)

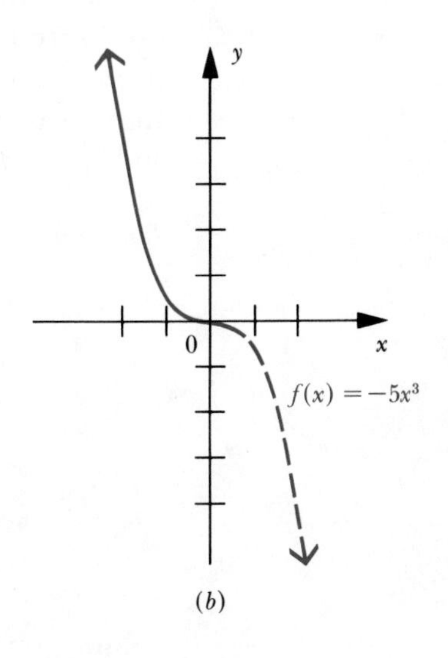

(*b*)

Some functions are neither even nor odd. For example, $f(x) = 2x + 1$ is neither even nor odd because

$$f(-x) = 2(-x) + 1 = -2x + 1$$

and

$$-f(x) = -2x - 1$$

so that

$$f(x) \neq f(-x) \qquad \text{and} \qquad -f(x) \neq f(-x)$$

This fact can also be seen from the graph of $f(x) = 2x + 1$ (Figure 8) in that there is no symmetry with respect to either the y axis or the origin.

Figure 8

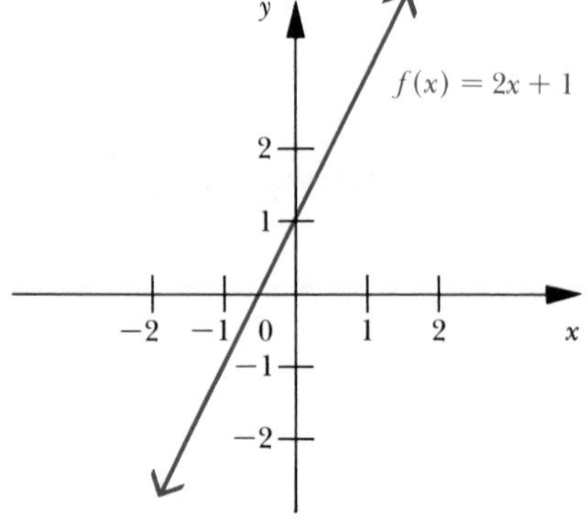

7.2 Increasing and Decreasing Functions

Consider the motion of a point traveling along the graph of f (Figure 9) from left to right. The point rises as its x coordinate increases from 0 to 3, and then declines as its x coordinate increases from 3 to 8. In this case we say that "f *is an increasing function* in the interval $(0, 3)$," and "f *is a decreasing function* in the interval $(3, 8)$."

Figure 9

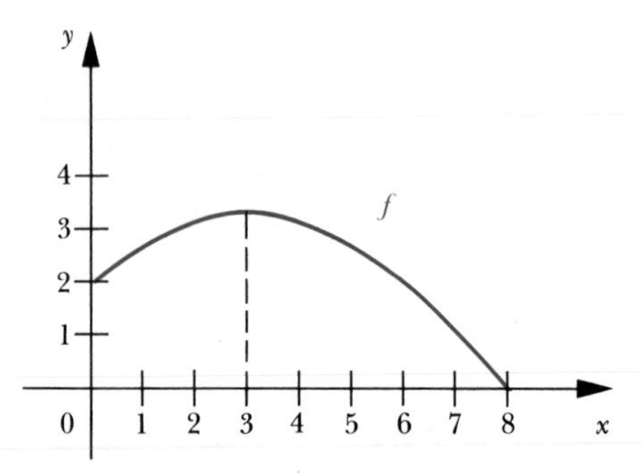

More formally, a function f is said to be a *(strictly) increasing function* in an interval I if whenever a and b are two numbers in I such that $a < b$, we have $f(a) < f(b)$ (Figure 10).

Figure 10

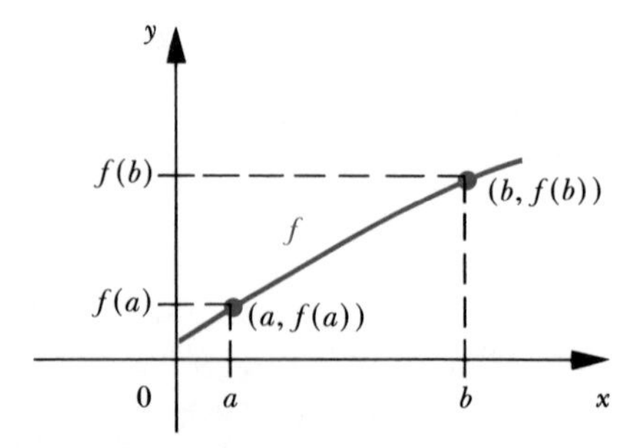

A function f is said to be a (*strictly*) *decreasing function* in an interval I if whenever a and b are two numbers in I such that $a < b$, we have $f(a) > f(b)$ (Figure 11).

Figure 11

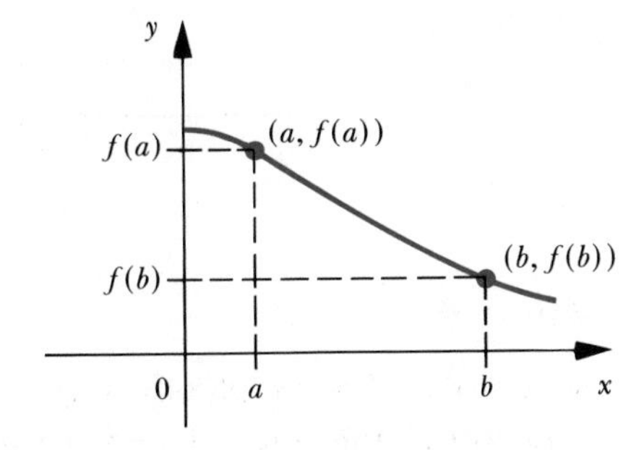

EXAMPLES

Determine where each of the following functions are increasing or decreasing. Also sketch the graph of the functions.

1 $f(x) = 3x + 2$

SOLUTION. The graph of f (Figure 12) indicates that f is an increasing function on R. For example, for -2 and 7, we have $-2 < 7$ and $f(-2) < f(7)$ since $3(-2) + 2 < 3(7) + 2$ or $-4 < 23$.

Figure 12

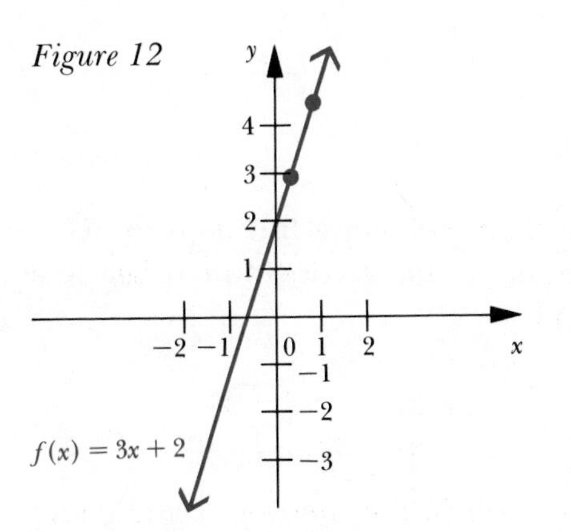

Figure 13

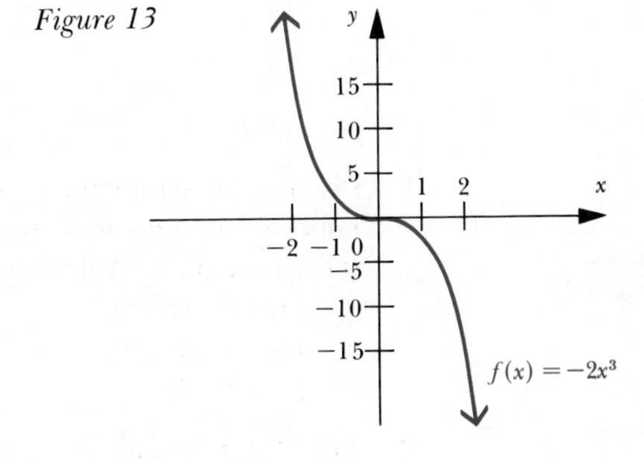

2 $f(x) = -2x^3$

SOLUTION. The graph of f (Figure 13) indicates that f is a decreasing function on R. For any two real numbers $a < b$, we have $f(a) > f(b)$. For example, for 1 and 2, $1 < 2$ and $f(1) > f(2)$ since $-2 > -16$.

3 $f(x) = 3$

SOLUTION. The graph of f (Figure 14) indicates that f is neither an increasing nor a decreasing function on R. For example, for 1 and 5, $1 < 5$ but $f(1) = f(5) = 3$.

Figure 14

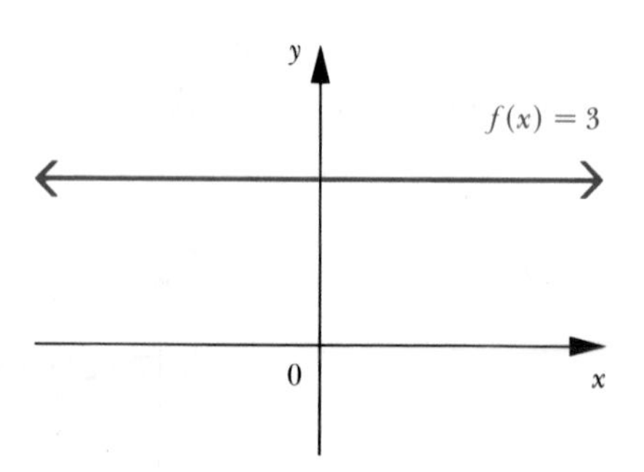

4 $f(x) = x^2$

SOLUTION. The graph of f (Figure 15) indicates that f is a decreasing function in the interval $(-\infty, 0)$ and f is increasing in the interval $(0, \infty)$.

Figure 15

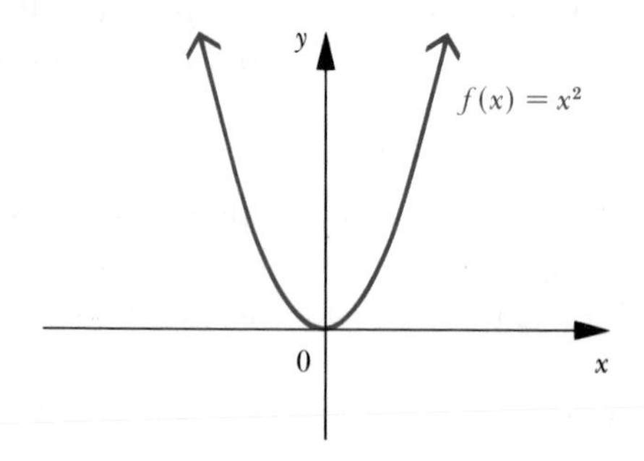

PROBLEM SET 6

1 Discuss the symmetry of each of the following functions to determine if f is symmetric with respect to the y axis or the origin. Use the symmetry to sketch the graph.

a) $f(x) = -2x$ b) $f(x) = |x|$

c) $f(x) = 5x^3$ d) $f(x) = -\sqrt{4 - x^2}$

e) $f(x) = -2x^4$ f) $f(x) = 3x^2 + 1$

2 Let $f(x) = |x|$ and $g(x) = 3x$. Is each of the functions f and g symmetric with respect to the y axis? Indicate which of the following functions is symmetric with respect to the y axis.

a) $f[g(x)]$ b) $g[f(x)]$

3 Indicate whether each of the given functions is even or odd or neither.

a) $f(x) = x^3 + 1/x$ b) $f(x) = x^4 + 1$

c) $f(x) = x^3 + x^2$ d) $f(x) = \sqrt{x} + 3$

e) $f(x) = (x^2 + 1)^3$ f) $f(x) = x^2 - 2x + 3$

4 Can a function be both even and odd? If so, give an example.

5 Let $f(x) = -x^3$. Sketch the graph of f for $x \in [1, 4]$. Is f increasing or decreasing in the interval $[1, 4]$?

6 Indicate the interval where each of the given functions is increasing or decreasing. Sketch the graph of f.

a) $f(x) = -5x + 7$
b) $f(x) = |x| + x$
c) $f(x) = 2x$
d) $f(x) = |x|/x$
e) $f(x) = x^3 + 1$
f) $f(x) = \sqrt[3]{x}$
g) $f(x) = \log_5 x$

7 Given the two functions f and g defined by $f(x) = x^2 + 4x + 1$ and $g(x) = 2x + 4$.

a) Is f even or odd? Is g even or odd?
b) Is f increasing or decreasing? Is g increasing or decreasing?
c) Show that $f(x) - 2g(x) = x^2 - 7$.

REVIEW PROBLEM SET

1 Let $A = \{1, 2, 3\}$, $B = \{3, 4, 5, 6\}$, and $C = \{1, 3, 4, 6\}$. Find each of the following sets.

a) $A \cap B$
b) $B \cup A$
c) $B \cap C$
d) $B \cup C$
e) $A \cap C$
f) $C \cup B$
g) $C \cap A$
h) $A \cap \emptyset$
i) $C \cup A$
j) $B \cup \emptyset$

2 Express each of the given sets as intervals and illustrate each on a number line.

a) $\{x \mid x \leqslant 1\}$
b) $\{x \mid x \geqslant 1\}$
c) $\{x \mid x < 2\} \cup \{x \mid x \geqslant -2\}$
d) $\{x \mid x \leqslant 1\} \cap \{x \mid x \geqslant -1\}$
e) $\{x \mid x \geqslant 0\} \cap \{x \mid x \leqslant 2\}$

3 Suppose that a and b are real numbers, with $a < b$. Indicate which of the following are true and which are false.

a) $a < 3b$
b) $a - 4 < b - 4$
c) $2a > -(-2)b$
d) $a - 2 < b - 3$
e) $a + c < b + c,\ c \in R$
f) $1/a < 1/b$
g) $a + 3 < b + 4$
h) $|a| < b$

4 Let x, y, and z be real numbers. Which of the following are true?

a) If $0 < x < \frac{1}{2}$, then $-\frac{1}{3} < x < \frac{1}{3}$.
b) If $-\frac{1}{2} < x < \frac{1}{2}$, then $-1 < x < 1$.
c) If $x > y$, then $x - z > y - z$.
d) If $x > y$ and $z > 0$, then $x/z > y/z$.
e) If $x > y$, with $x > 0$, $y > 0$, then $x^2 > y^2$ and $1/x < 1/y$.
f) If $x^3 > y^3$, and $x > 0$, $y > 0$, then $x < y$.

5 Solve each of the following inequalities. Illustrate your solution on the real line.

a) $3x + 5 \leqslant 2x + 1$
b) $3x - 5 > 2x + 7$
c) $3x - 2 < 5 - x$
d) $1 - 2x \geqslant 5 + x$
e) $1 - 3x \geqslant 1 + 2x$
f) $3x + 4 < 8x + 7$
g) $3(x - 4) < 2(4 - x)$
h) $5x + 2 \geqslant 3x + 1$
i) $3(x + 2) \geqslant 2(x + 3)$
j) $5x + 2(3x - 1) < -1 - 5(x - 3)$

6 Solve each of the following absolute value equations. Illustrate your solution on the real line.

a) $|2x| = 5$
b) $|7x| = 6$
c) $|3x + 4| = 12$
d) $|1 - 2x| = 2$
e) $|7x - 3| = 5$
f) $|3 - x| = |-3|$
g) $|5x - 8| = 3$
h) $|3(x - 4)| = |2(4 - x)|$

7 Solve each of the following absolute value inequalities. Illustrate your solution on the real line.

a) $|x| < 7$
b) $|x| > \frac{5}{2}$
c) $|x - 5| < \frac{3}{2}$
d) $|x - 2| \leqslant |2 - x|$
e) $|3x + 1| > 8$
f) $|2x - 5| < 9$
g) $|x - 3| > 0$
h) $|x - 2| \geqslant 5$
i) $|x - 2| > 3$
j) $|2 - 3x| < 1$

8 Plot each of the following points and indicate which quadrant, if any, contains the points.

a) $(1, 3)$
b) $(3.1, -1.7)$
c) $(-3, 2)$
d) $(-3, -2)$
e) $(-3, -9)$
f) $(-2, -4)$
g) $(3, 0)$
h) $(-3, 7)$

9 Let $A = \{3, -3, 4\}$ and $B = \{-1, 2, 4\}$. List all the ordered pairs of the following and graph each set.

a) $\{(a, b) \mid a \in A, b \in A\}$
b) $\{(a, b) \mid a \in B, b \in B\}$
c) $\{(a, b) \mid a \in A, b \in B\}$
d) $\{(a, b) \mid a \in B, b \in A\}$

10 Use the distance formula to show that each of the following triangles with the given vertices is an isosceles triangle.

a) $(2, 1)$, $(9, 3)$, and $(4, -6)$
b) $(0, 2)$, $(-1, 4)$, and $(-3, 3)$
c) $(5, -2)$, $(6, 5)$, and $(2, 2)$

11 Which of the following relations are functions? Indicate the domain and range of each.

a) $\{(4, 1), (3, -1), (5, 1), (-1, 1)\}$
b) $\{(1, 1), (2, 2), (3, 3), (4, 4)\}$

c) $\{(0, 1), (0, -1), (1, 3), (2, 4)\}$
d) $\{(x, y), y \leq 3x + 1\}$

12 Which of the graphs in Figure 16 represent a function?

Figure 16

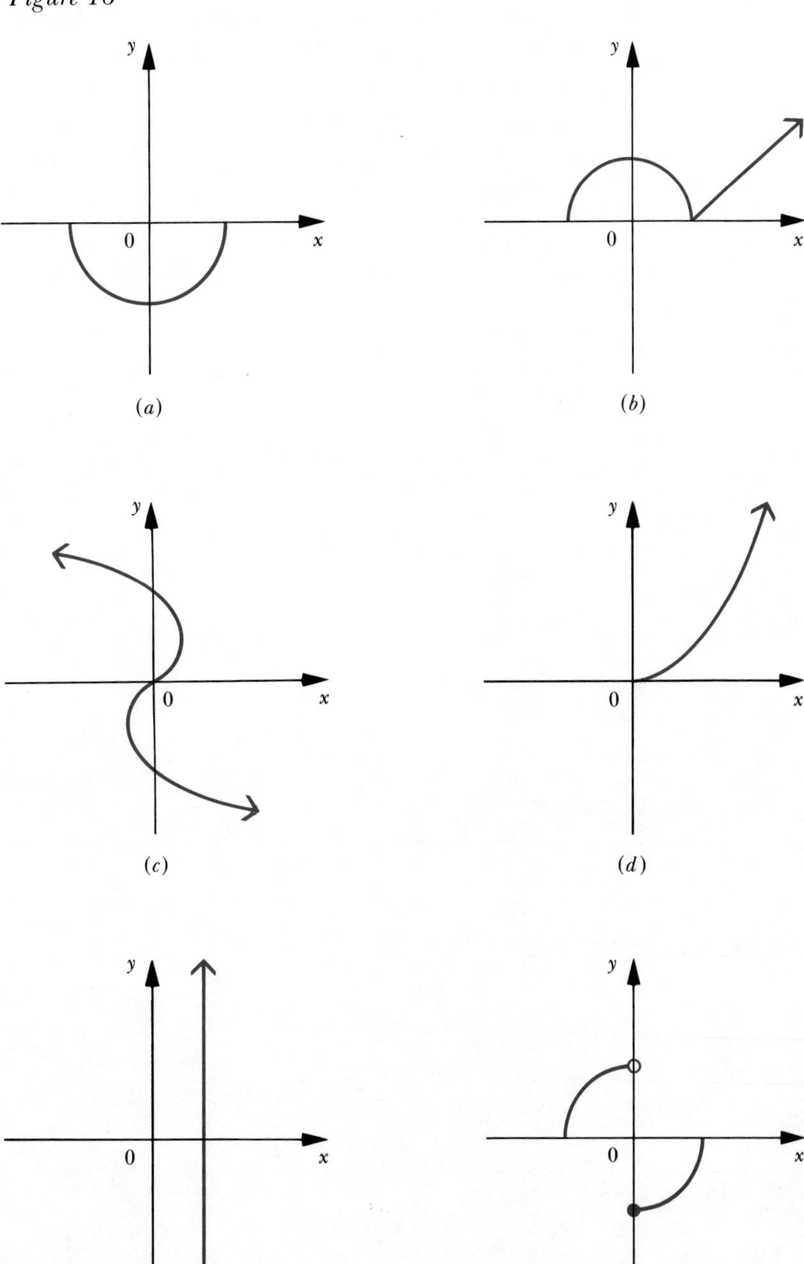

13 Discuss the symmetry for the following functions. Use your results to graph the functions.

a) $f(x) = -5x^2$
b) $f(x) = -x - 2$
c) $f(x) = -7/x$
d) $f(x) = |x - 2|$
e) $f(x) = -4x^3$
f) $F(x) = \sqrt{25 - x^2}$

14 Indicate whether each of the following functions is symmetric with respect to the y axis or the origin. Indicate the domain and range of f. Are the functions increasing or decreasing? Are the functions even or odd? Graph the functions.

a) $f = \{(x, y) \mid y = 2x - 1\}$
b) $f(x) = |x|^2$
c) $f(x) = 10^x$
d) $f = \{(x, y) \mid y = -\frac{1}{2}x + 1\}$
e) $f(x) = \log_7 x$
f) $f(x) = 3/x$

15 Let $f(x) = \frac{1}{2}(x + 1)$ and $g(x) = 2x - 1$. Find $f[g(x)]$ and $g[f(x)]$.

CHAPTER 2

Trigonometric Functions

CHAPTER 2

Trigonometric Functions

1 Introduction

This chapter is devoted to a special class of functions—the trigonometric functions. Trigonometry was originally developed to solve problems in astronomy and land measurement. The word "trigonometry" was derived from Greek words meaning "triangle measurement." Our objective will be to study trigonometric functions of angles, but before we begin we shall review elementary notions about angles and circles.

2 Angles

We know from plane geometry that two distinct points determine a (straight) *line* (Figure 1*a*). A line has no end points. A portion of a line joining any two of its points is called a line segment (Figure 1*b*). The two points A and B are called the *end points* of the segment joining A and B. Any point P on a line separates the line into two parts, called *rays* (Figure 1*c*). For example, ray PB means that the ray starts at P, goes through B, and continues indefinitely.

Figure 1

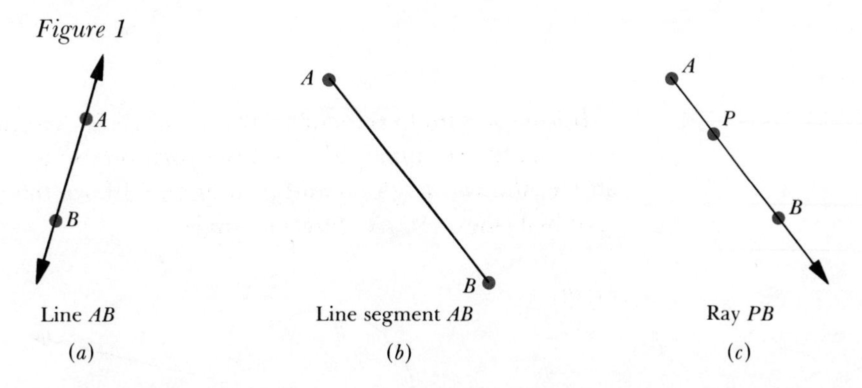

An *angle* is determined by rotating a ray about its end point, the *vertex* of the angle, from an initial position, the *initial side* of the angle, to a terminal position, the *terminal side* of the angle (Figure 2). If the angle is formed by a counterclockwise rotation, the angle is said to be an angle of *positive* sense, whereas if the angle is formed by a clockwise rotation, the angle is of *negative* sense. In Figure 2, the angle determined by Q, P, and R, $\angle QPR$, is a positive angle, whereas $\angle CAB$ is negative.

Figure 2

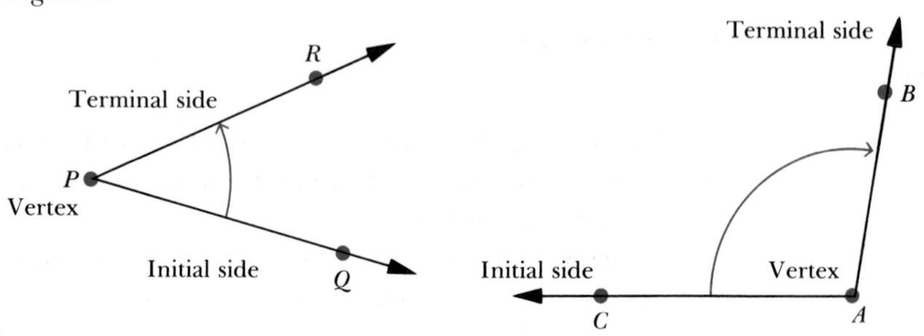

For convenience, Greek letters are often used to label angles. In Figure 3, α is the positive angle $\angle DBC$ and θ is the negative angle $\angle ABC$.

Figure 3

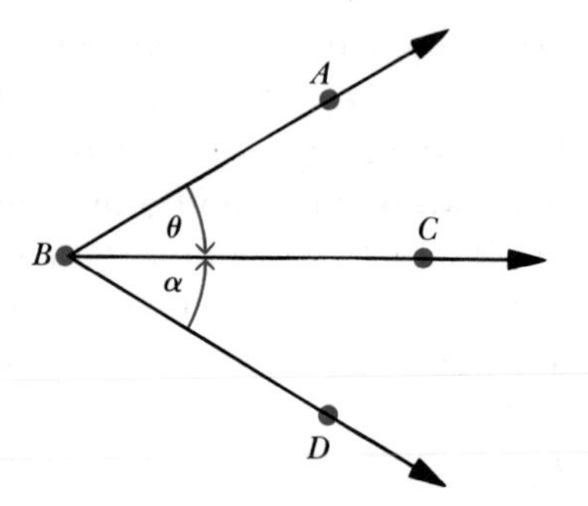

It is important to remember that an angle is determined by the initial side, the terminal side, and the rotation used to form it. For example, the two angles α and β in Figure 4 have the same initial and terminal sides, yet are different angles.

Figure 4

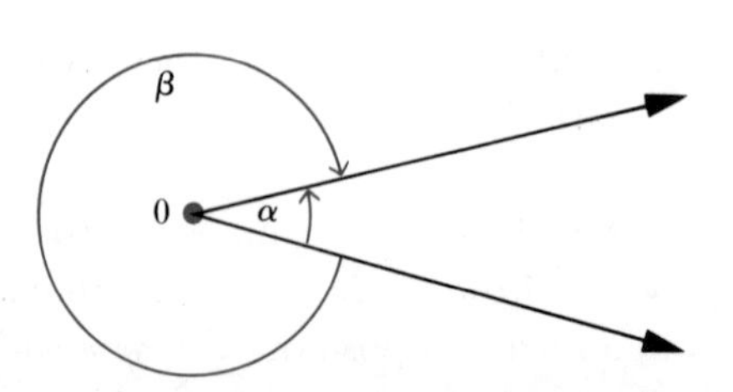

2.1 Angle Measure

Angles are measured by using either *degrees* or *radians*. *One degree* ($1°$) is the measure of a positive angle formed by $\frac{1}{360}$ of one complete revolution (Figure 5*a*). Furthermore, a degree can be divided into 60 equal parts called *minutes* ($'$); a minute can be divided into 60 equal parts called *seconds* ($''$). On the other hand, one *radian* is the measure of a positive angle that intercepts an arc of length 1 on a circle of radius 1 (Figure 5*b*). Note that we indicate degree measurement by use of the symbol °; the absence of any symbol indicates radian measure. The angle measure is positive or negative according to whether the angle is formed by a counterclockwise or clockwise rotation. Hence, the radian measure of an angle is the "directed" length of its subtended arc on a circle of radius 1.

Figure 5

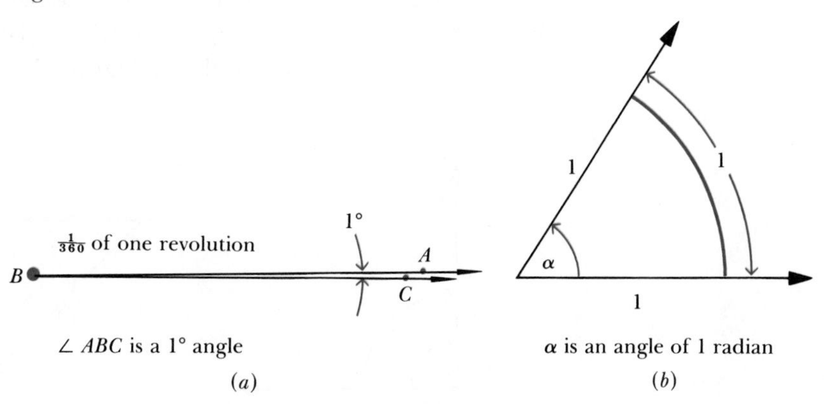

EXAMPLES

1 Indicate both the degree measure and the radian measure of an angle α formed by

a) 1 complete counterclockwise rotation
b) $\frac{1}{4}$ counterclockwise rotation
c) $\frac{1}{8}$ clockwise rotation

SOLUTION

a) Since $1°$ is the measure of an angle formed by $\frac{1}{360}$ of one complete revolution, the degree measure of α is $(360)(1°) = 360°$. The radian measure of α is 2π, the length of the circumference of a circle of radius 1 (Figure 6).

Figure 6

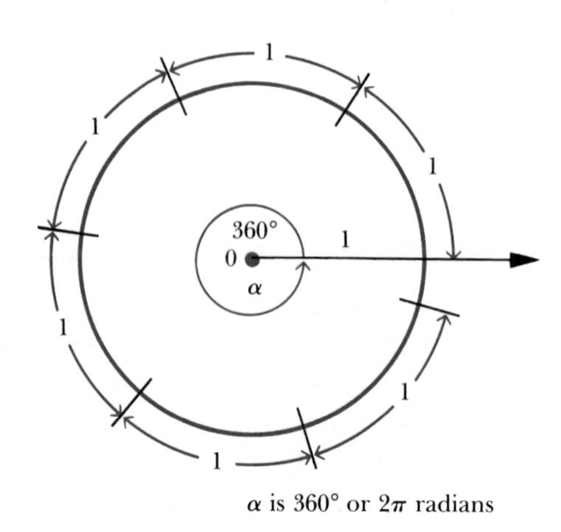

α is 360° or 2π radians

b) α is either $\frac{1}{4}(360°) = 90°$ or $\frac{1}{4}(2\pi) = \pi/2$ radians (Figure 7*a*).
c) $\frac{1}{8}(360°) = 45°$ and $\frac{1}{8}(2\pi) = \pi/4$, however, α is a negative angle; hence, α is $-45°$ or $-\pi/4$ radians (Figure 7b).

Figure 7

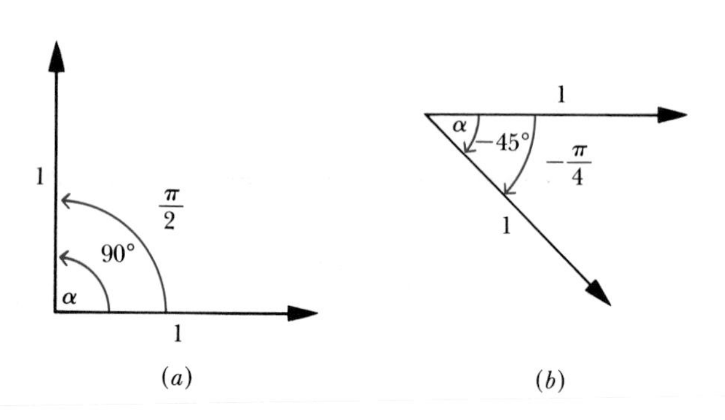

2 Express 37.45° in terms of degrees, minutes, and seconds.

SOLUTION. $37.45° = 37° + 0.45°$; however, $0.45° = (0.45)(60') = 27.00'$ so that $37.45° = 37°27'0''$.

3 Express $-84.32°$ in terms of degrees, minutes, and seconds.

SOLUTION

$-84.32° = -(84° + 0.32°)$. However, $0.32° = (0.32)(60') = 19.20'$. Also, $0.20' = (0.20)(60'') = 12''$, so that $-84.32° = -84°19'12''$.

2.2 Conversion of Angle Measurements

Quite often an angle is expressed in degrees, and we need to express its measure in radians or vice versa. Let us consider a specific angle of 1° (an angle of 1° subtends an arc of the unit circle of length $\frac{1}{360}$ of the circumference of the circle). In establishing the radian measure of this angle, we use the fact that R_1 is the length of the subtended arc, together with the fact that the circumference of the circle is 2π. We have $R_1 = \frac{1}{360}(2\pi) = \pi/180$ (Figure 8). Hence, 1° corresponds to $\pi/180$ radians. This means that an angle of D degrees is $(\pi/180)D$ radians.

This fact can be generalized by the formula

$$R_1 = \left(\frac{\pi}{180}\right) D$$

where R_1 represents the radian measure of any angle and D represents the degree measure of the same angle. This relationship between R_1 and D can also be expressed by the formula

$$D = \left(\frac{180}{\pi}\right) R_1$$

Figure 8

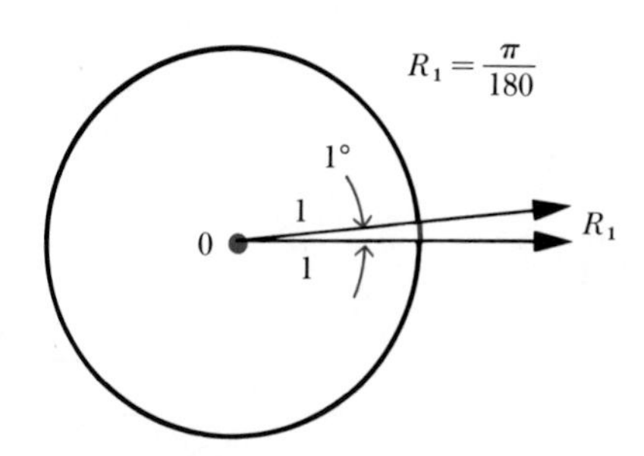

EXAMPLES

1 In each case, sketch the angle with the given radian measure and find its degree measure.

a) $\pi/3$ b) $3\pi/4$ c) $-5\pi/12$
d) $17\pi/10$ e) $23\pi/10$

SOLUTION. Using the fact that the degree measure D corresponds to $(180/\pi)R_1$, where R_1 is the radian measure of the angle we have

a) $\pi/3$ radians corresponds to $(180/\pi)(\pi/3) = 60°$ (Figure 9*a*).
b) $3\pi/4$ radians corresponds to $(180/\pi)(3\pi/4) = 135°$ (Figure 9*b*).
c) $-5\pi/12$ radians corresponds to $(180/\pi)(-5\pi/12) = -75°$ (Figure 9*c*).
d) $17\pi/10$ radians corresponds to $(180/\pi)(17\pi/10) = 306°$ (Figure 9*d*).

e) $23\pi/10$ radians corresponds to $(180/\pi)(23\pi/10) = 414°$ (Figure 9*e*).

Figure 9

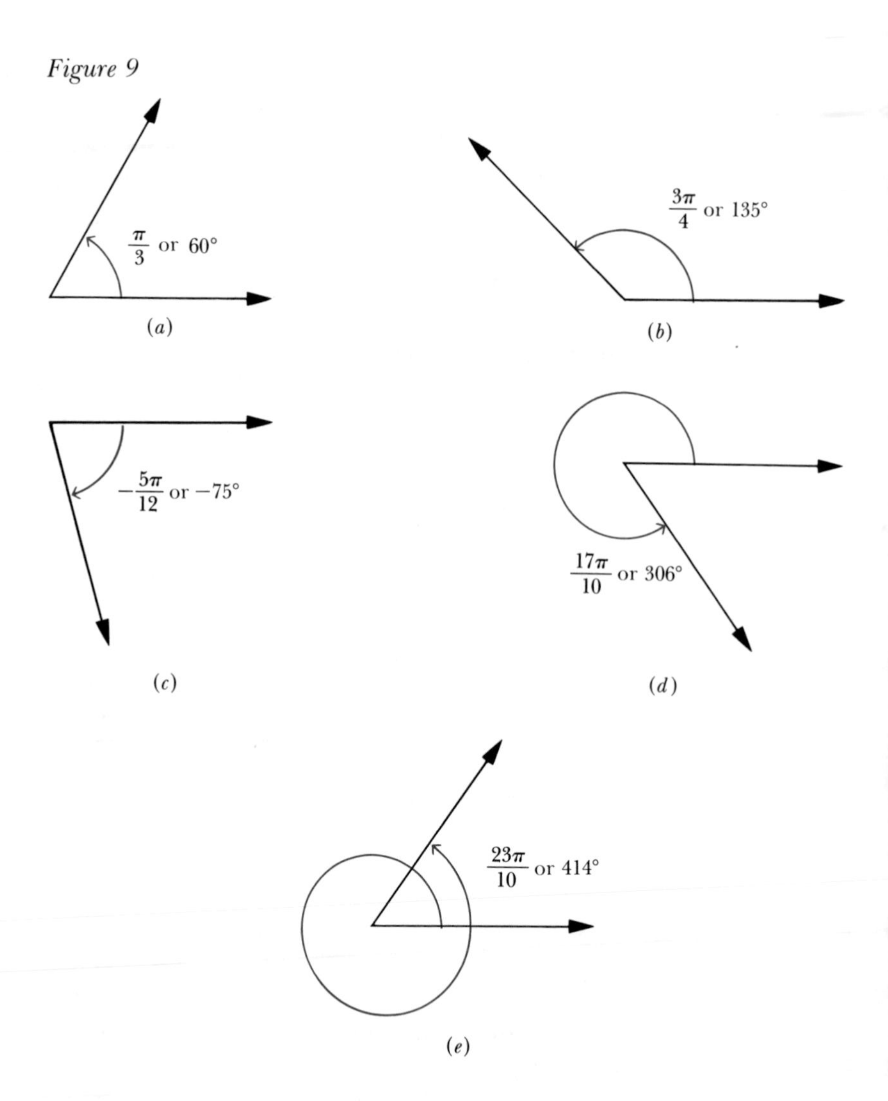

2 In each case, sketch the angle with the given degree measure and find its radian measure.
a) 45° b) 150° c) 215°
d) −15° e) 810°

SOLUTION. Using the fact that the radian measure R_1 corresponds to $(\pi/180)D$, where D is the degree measure of the angle, we have
a) 45° corresponds to $(\pi/180)(45°) = \pi/4$ radians (Figure 10*a*).
b) 150° corresponds to $(\pi/180)(150°) = 5\pi/6$ radians (Figure 10*b*).
c) 215° corresponds to $(\pi/180)(215°) = 43\pi/36$ radians (Figure 10*c*).

d) $-15°$ corresponds to $(\pi/180)(-15°) = -\pi/12$ radians (Figure 10*d*).

e) $810°$ corresponds to $(\pi/180)(810) = 9\pi/2$ radians (Figure 10*e*).

Figure 10

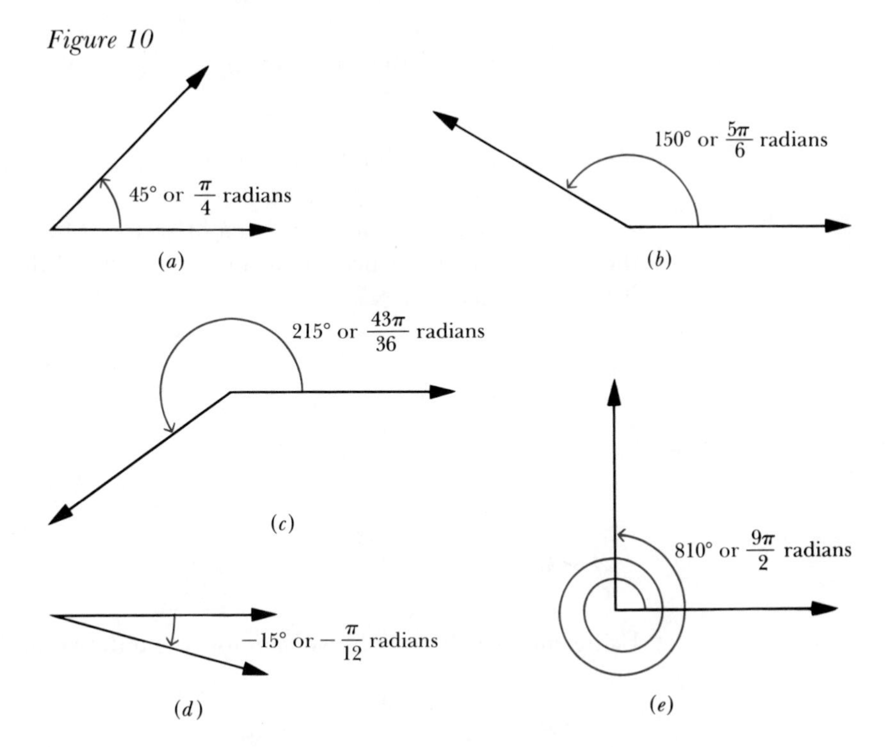

2.3 Arcs and Sectors of Circles

The radian measure of a central angle of a circle can be used to determine both the length of the arc which the angle subtends and the area of the sector determined by the central angle.

Suppose that θ is the central angle of a circle of radius r (Figure 11). Also assume that the angle θ has a radian measure t. (If θ is measured in degrees, we can always convert to radians.)

Figure 11

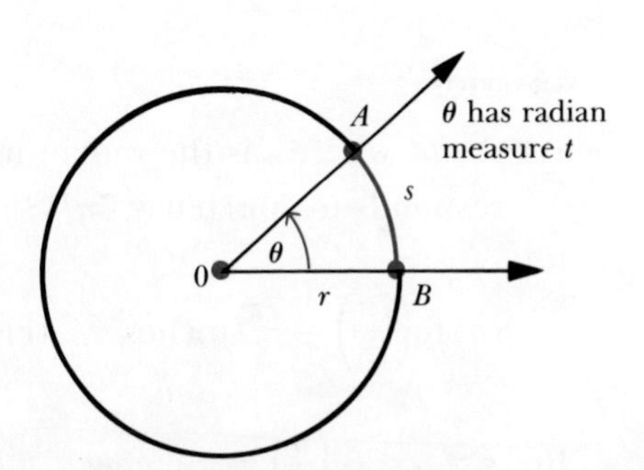

Our task is to determine s, the length of arc AB. The length of the circumference of the circle is $2\pi r$; but, since θ determines an arc that is "$(t/2\pi)$th" of the circumference,

$$s = \frac{t}{2\pi}(2\pi r) \qquad \text{that is} \qquad s = tr$$

Hence, we have a formula for determining the length of an arc subtended by a central angle of t radians.

Next, we can determine the area of sector OAB. Since the area of the circle is πr^2 and since θ determines $(t/2\pi)$th of the circle, the area A is determined by

$$A = \frac{t}{2\pi}\pi r^2$$

so

$$A = \tfrac{1}{2}r^2 t$$

Using the fact that $s = tr$, we can also write the latter formula as

$$A = \tfrac{1}{2}sr$$

EXAMPLES

1 Find the length of an arc of a circle of radius 6 inches and with a central angle of

a) $70°$ b) $\pi/6$

SOLUTION

a) $s = tr$, where t is the radian measure of the angle. Since $70°$ corresponds to $70\pi/180 = 7\pi/18$ radians, so that

$$s = 6\left(\frac{7\pi}{18}\right) = \frac{7\pi}{3} \text{ inches} \qquad \text{(Figure 12}a\text{)}$$

b) $s = tr = 6\left(\frac{\pi}{6}\right) = \pi$ inches (Figure 12b)

Figure 12

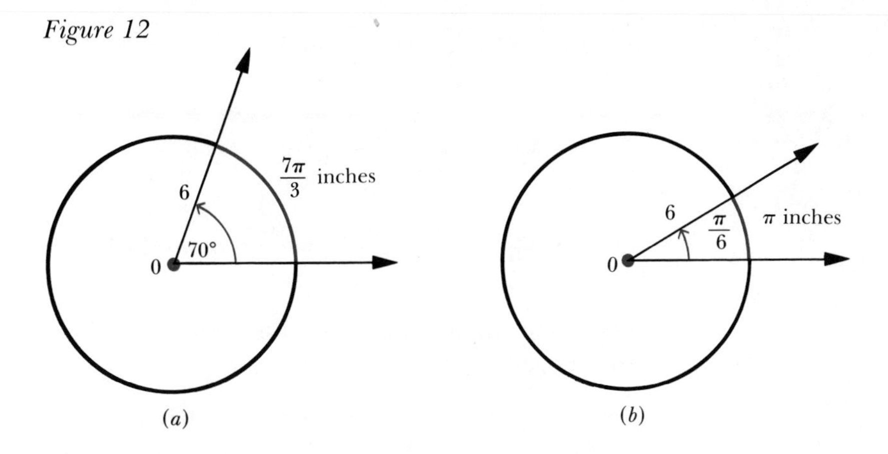

2 Find the area of the sector of a circle of radius 6 inches and subtended by a central angle of

a) 65° b) $\pi/6$

SOLUTION

a) $A = \frac{1}{2}r^2t$. Since 65° corresponds to $65\pi/180 = 13\pi/36$ radians,

$$A = \frac{1}{2}\,(36)\left(\frac{13\pi}{36}\right) = \frac{13\pi}{2} \text{ square inches}$$

b) $A = \frac{1}{2}\,(36)\left(\frac{\pi}{6}\right) = 3\pi$ square inches

3 Find the area and the arc length of the sector of a circle of radius 8 inches subtended by a central angle of

a) $\pi/4$ b) $3\pi/8$ c) 85°

SOLUTION

a) $s = rt = 8\left(\frac{\pi}{4}\right) = 2\pi$ inches

$$A = \frac{1}{2}\,r^2t = \frac{1}{2}\,(64)\left(\frac{\pi}{4}\right) = 8\pi \text{ square inches}$$

b) $s = rt = 8\left(\frac{3\pi}{8}\right) = 3\pi$ inches

$$A = \frac{1}{2}\,r^2t = \frac{1}{2}\,(64)\left(\frac{3\pi}{8}\right) = 12\pi \text{ square inches}$$

c) $85° = 85\left(\dfrac{\pi}{180}\right) = \dfrac{17\pi}{36}$, so that

$$s = rt = 8\left(\frac{17\pi}{36}\right) = \frac{34\pi}{9} \text{ inches}$$

$$A = \frac{1}{2}sr = \frac{1}{2}\left(\frac{34\pi}{9}\right)8 = \frac{136\pi}{9} \text{ square inches}$$

4 The tip of the minute hand of a clock travels $7\pi/10$ inches in 3 minutes. How long is the minute hand?

SOLUTION. In 3 minutes the minute hand generates a central angle of $(3)(\frac{1}{60})(360°) = 18°$. (Why?) But an angle of 18° has a radian measure of $t = (\pi/180)(18) = \pi/10$ (Figure 13). Hence, $7\pi/10 = r(\pi/10)$, so that $r = 7$ inches.

Figure 13

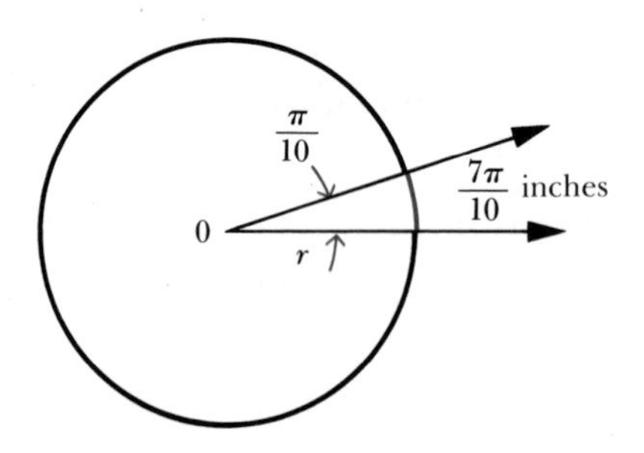

PROBLEM SET 1

1 Fill in the missing blanks in each of the following statements.

a) An angle is determined by rotating a ray about its end point, the ______ of the angle, from some initial position, the ______ of the angle, to the terminal position, the ______ of the angle.

b) There are 360° in a complete counterclockwise revolution and ______ radians in a complete counterclockwise revolution.

c) Radian measure is related to a degree measure by 1 radian = ______.

2 In each case, find the radian measure for the angle with the given degree measures.

a) 30°	b) 90°	c) −45°
d) −90°	e) 240°	f) 855°
g) −315°	h) 405°	i) 175°
j) 210°	k) 1230°	l) −300°

m) $540°$ n) $-420°$ o) $-220°$
p) $-100°$ q) $72°$ r) $150°$

3 In each case, find the degree measure for the angle with the given radian measures.

a) $-\pi/3$ b) $2\pi/9$ c) $-5\pi/4$
d) $3\pi/2$ e) $\pi/4$ f) $11\pi/3$
g) $-7\pi/12$ h) $-5\pi/8$ i) $-7\pi/8$
j) $5\pi/2$ k) $7\pi/6$ l) $91\pi/4$
m) $31\pi/3$ n) $43\pi/6$ o) $2\pi/5$

4 Express each of the following numbers in terms of degrees, minutes, and seconds.

a) $32.12°$ b) $-45.35°$ c) $110.25°$
d) $60.20°$ e) $35.16°$ f) $75.35°$
g) $72.31°$ h) $67.45°$

5 Find the area and the arc length of the sector of the circle with the given radius r and with the given subtended central angle whose radian measure is t.

a) $r = 6$ inches and $t = \pi/7$ b) $r = 4$ inches and $t = 2\pi/9$
c) $r = 8$ inches and $t = 75°$ d) $r = 20$ feet and $t = 35°$
e) $r = 3$ inches and $t = 7\pi/12$ f) $r = 6$ inches and $t = 5\pi/6$

6 Find the measure in radians of the angles that subtend the given arc s of the circle with radius r.

a) $r = 3$ inches and $s = 6.73$ inches
b) $r = 4$ inches and $s = 7\pi/3$ inches
c) $r = 5$ inches and $s = 65.32$ inches
d) $r = 5$ feet and $s = 6$ inches
e) $r = 5$ feet and $s = 13\pi$ feet
f) $r = 2.5$ feet and $s = 8.5$ feet

7 Find the measure in radians of the central angles which determine areas A of each circle of radius r.

a) $r = 2$ inches and $A = 12.45$ square inches.
b) $r = 6$ inches and $A = 13\pi/2$ square inches.

8 A central angle of $100.35°$ intercepts an arc of 20 inches on the circumference of a circle. Find the radius of the circle.

3 Functions Defined on Angles

In this section we shall discuss and formulate the definition of trigonometric functions of angles in standard positions. An angle is in *standard position* if it is placed on a Cartesian coordinate system so that the vertex corresponds to the origin and the initial side coincides with the positive x axis. For example, an 80° angle in standard position would have its terminal side in quadrant I (Figure 1*a*), whereas an angle of radian measure $-3\pi/5$ in standard position has its terminal side in quadrant III (Figure 1*b*). [Notice that $-3\pi/5$ corresponds to $(180/\pi)(-3\pi/5) = -108°$.]

Figure 1

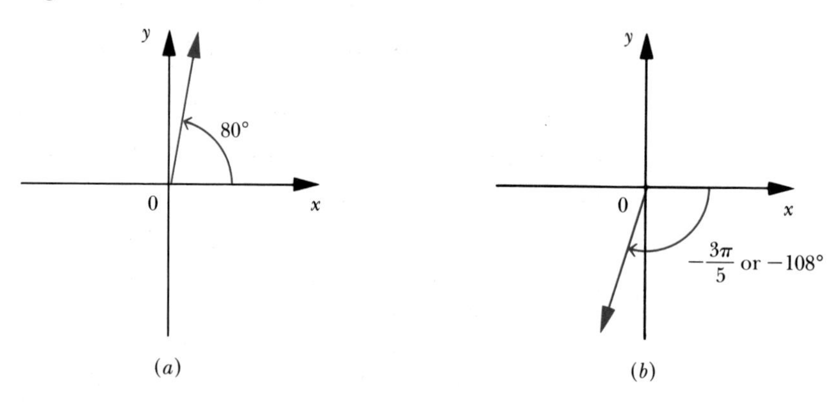

EXAMPLES

1 In what quadrants do the terminal sides of the following angles in standard position lie? Illustrate each case by a diagram.
a) 50° b) −25° c) −230°

SOLUTION. We notice in Figure 2*a*, *b*, and *c* that
a) 50° lies in quadrant I
b) −25° lies in quadrant IV
c) −230° lies in quadrant II

Figure 2

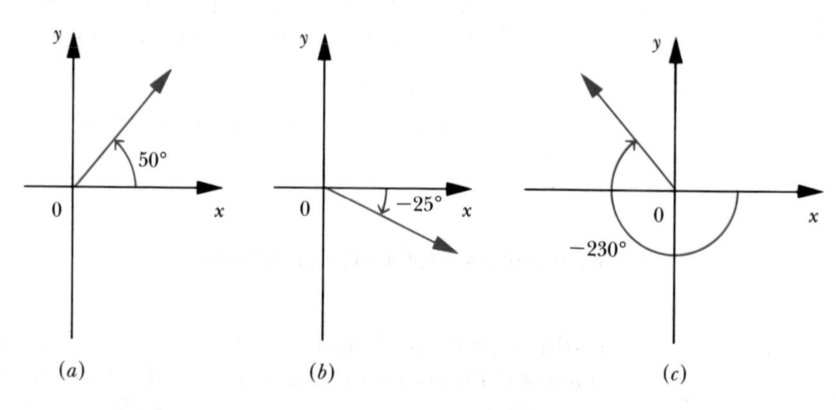

2 Sketch an angle in standard position whose terminal side contains the given point. Indicate by θ one of the positive angles and by ϕ one of the negative angles thus formed. Indicate the quadrant in which the terminal side of θ lies.

a) $(3, 4)$ b) $(-4, 3)$ c) $(-2, -3)$ d) $(4, -1)$

SOLUTION

a) The terminal side of any angle θ that contains the point $(3, 4)$ lies in quadrant I (Figure 3*a*).
b) The terminal side of any angle θ that contains the point $(-4, 3)$ lies in quadrant II (Figure 3*b*).
c) The terminal side of any angle θ that contains the point $(-2, -3)$ lies in quadrant III (Figure 3*c*).
d) The terminal side of any angle θ that contains the point $(4, -1)$ lies in quadrant IV (Figure 3*d*).

Figure 3

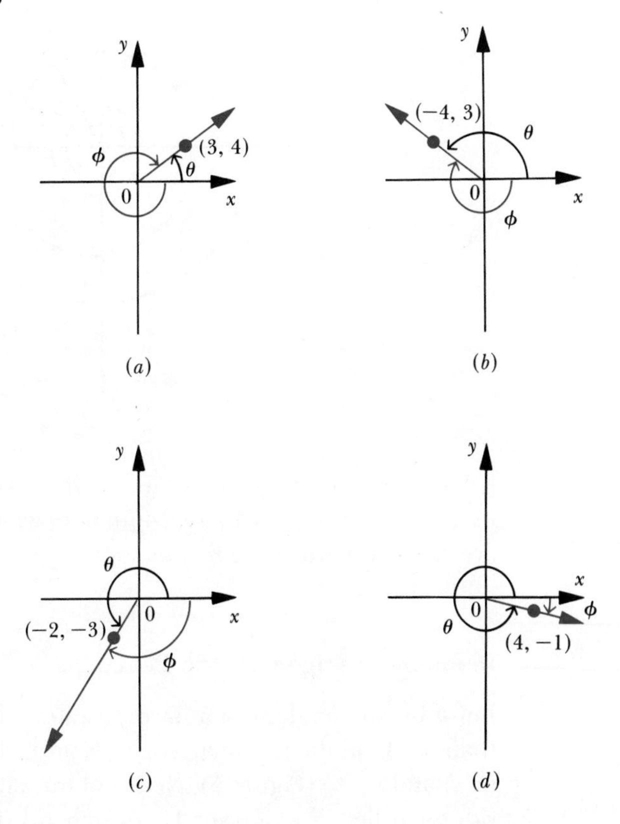

We should note that no limit is placed on the possible magnitude (size) of an angle. If a ray makes one complete revolution in a counterclockwise sense, it will generate one angle of 360°, and so two complete revolutions in a counterclockwise sense is one angle of 720°, and so on. Angles having the same initial sides and the same terminal sides are called *coterminal angles.*

EXAMPLE

List three angles in standard position coterminal with the angle of 60°.

SOLUTION. Three angles in standard position with the same terminal sides as 60° are 420°, 780°, and −300° (Figure 4). Note that other possibilities exist.

Figure 4

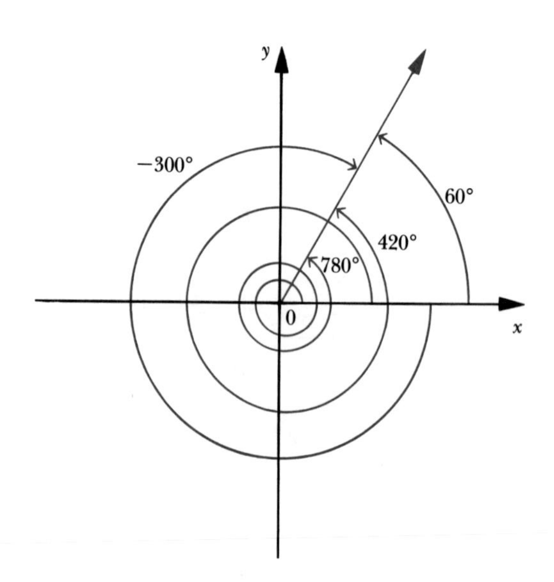

We now define six functions, each of which has a set of angles as its domain and a set of real numbers as its range. These functions are called the *trigonometric functions.*

Definition Trigonometric Functions

Let θ be any angle in standard position, let (x, y) be any point other than $(0, 0)$ on the terminal side of θ, and let r be the distance between (x, y) and $(0, 0)$ (Figure 5). Notice in this illustration that the terminal side of θ lies in quadrant II. In general the quadrant in which the terminal side of angle θ lies depends, of course, upon the measure of the angle. The following definition applies to all cases.

Figure 5

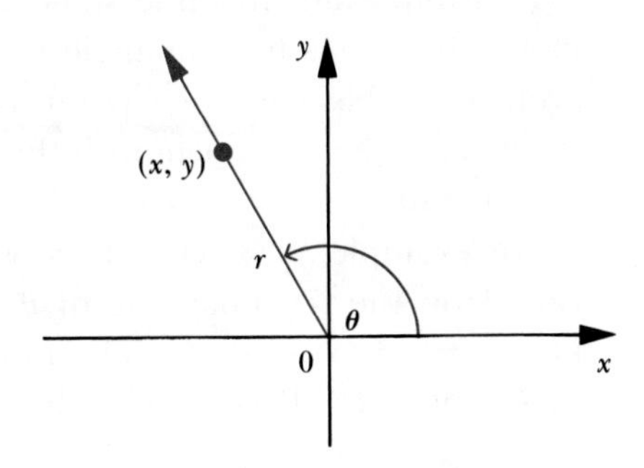

The six *trigonometric* functions are defined in the following manner, using ordered pair notation:

$$\text{sine} = \left\{(\theta, \sin\theta) \mid \sin\theta = \frac{y}{r}\right\}$$

$$\text{cosine} = \left\{(\theta, \cos\theta) \mid \cos\theta = \frac{x}{r}\right\}$$

$$\text{tangent} = \left\{(\theta, \tan\theta) \mid \tan\theta = \frac{y}{x},\ x \neq 0\right\}$$

$$\text{cotangent} = \left\{(\theta, \cot\theta) \mid \cot\theta = \frac{x}{y},\ y \neq 0\right\}$$

$$\text{secant} = \left\{(\theta, \sec\theta) \mid \sec\theta = \frac{r}{x},\ x \neq 0\right\}$$

$$\text{cosecant} = \left\{(\theta, \csc\theta) \mid \csc\theta = \frac{r}{y},\ y \neq 0\right\}$$

where $r = \sqrt{x^2 + y^2}$ and $r > 0$. These functions are usually written in the following abbreviated forms.

$$\sin\theta = \frac{y}{r}$$

$$\cos\theta = \frac{x}{r}$$

$$\tan\theta = \frac{y}{x},\ x \neq 0$$

$$\cot\theta = \frac{x}{y},\ y \neq 0$$

$$\sec\theta = \frac{r}{x},\ x \neq 0$$

$$\csc\theta = \frac{r}{y},\ y \neq 0$$

It is important to understand that the values of the six trigonometric functions for the angle θ depend only on the angle. That is, no matter what particular point may be selected on the terminal side to evaluate the six functions [other than (0, 0) of course], the results are the same.

For example, if (x, y) and (x_1, y_1) are two different points in quadrant I on the terminal side of θ other than (0, 0) (Figure 6), then $r = \sqrt{x^2 + y^2}$, $r_1 = \sqrt{x_1{}^2 + y_1{}^2}$, and, because of the similar triangles OBP_1 and OAP (Figure 6), we have

$$\frac{y_1}{r_1} = \frac{y}{r}, \quad \frac{x_1}{r_1} = \frac{x}{r}, \quad \frac{y_1}{x_1} = \frac{y}{x}, \quad \frac{x_1}{y_1} = \frac{x}{y}, \quad \frac{r_1}{x_1} = \frac{r}{x}, \quad \frac{r_1}{y_1} = \frac{r}{y}$$

Figure 6

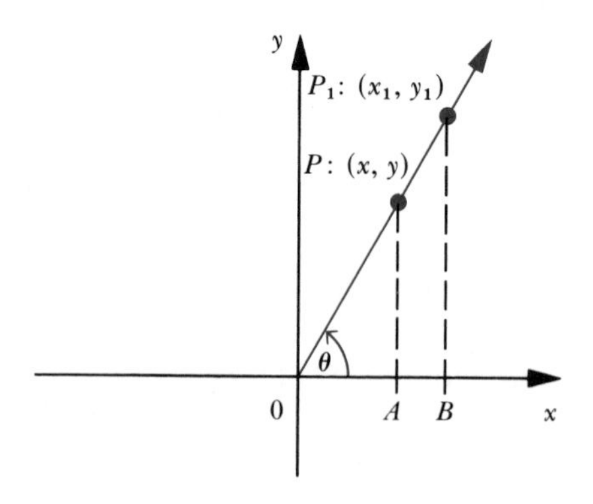

Hence, the values of the trigonometric functions are the same no matter what two points are selected on the terminal side of the angle other than (0, 0). Furthermore, by a similar construction, we can show that the above results are true for angles whose terminal sides lie in quadrant II, III, or IV.

EXAMPLES

Evaluate the trigonometric functions of an angle θ whose terminal side contains the given point. Is θ unique?

1 (3, 4)

SOLUTION. Figure 7 suggests the fact that there are infinitely many different angles whose terminal sides contain the point (3, 4). Hence, θ is not unique. Using the fact that $r = \sqrt{3^2 + 4^2} = 5$, and the definition of trigonometric functions, we have

$$\sin \theta = \frac{y}{r} = \frac{4}{5} \qquad \csc \theta = \frac{r}{y} = \frac{5}{4}$$

$$\cos\theta = \frac{x}{r} = \frac{3}{5} \qquad \sec\theta = \frac{r}{x} = \frac{5}{3}$$

$$\tan\theta = \frac{y}{x} = \frac{4}{3} \qquad \cot\theta = \frac{x}{y} = \frac{3}{4}$$

Figure 7

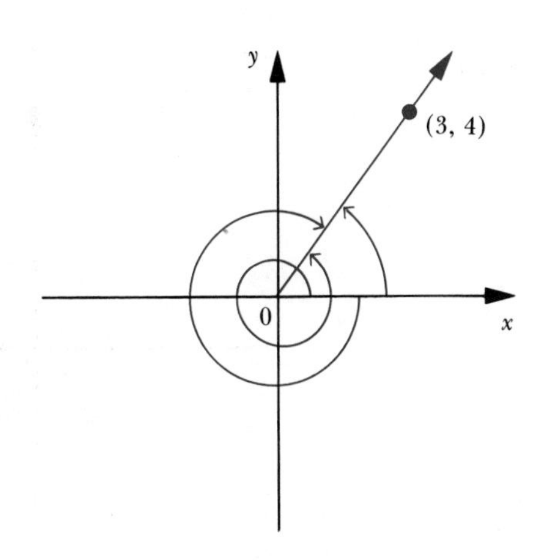

2 (−4, 3)

SOLUTION. Figure 8 suggests the fact that there are infinitely many different angles whose terminal side contains the point (−4, 3). Hence, θ is not unique. Using the fact that $r = \sqrt{(-4)^2 + 3^2} = 5$ and the definition of trigonometric functions, we have

$$\sin\theta = \frac{y}{r} = \frac{3}{5} \qquad \csc\theta = \frac{r}{y} = \frac{5}{3}$$

$$\cos\theta = \frac{x}{r} = \frac{-4}{5} \qquad \sec\theta = \frac{r}{x} = \frac{5}{-4}$$

$$\tan\theta = \frac{y}{x} = \frac{3}{-4} \qquad \cot\theta = \frac{x}{y} = \frac{-4}{3}$$

Figure 8

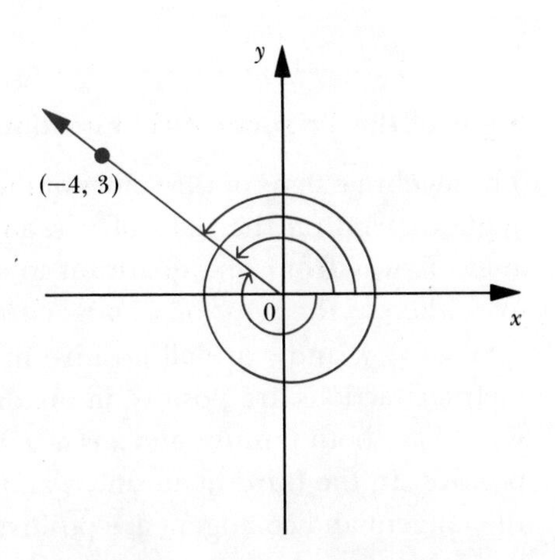

3 $(3, -5)$

SOLUTION. Figure 9 suggests the fact that there are infinitely many different angles whose terminal side contains the point $(3, -5)$. Therefore, θ is not unique.

Figure 9

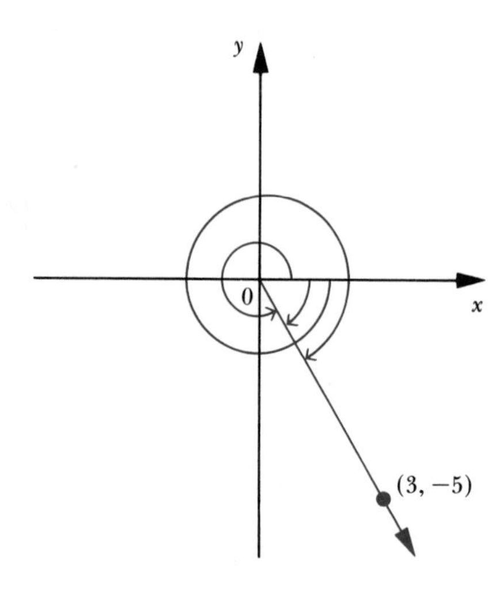

Using the fact that $r = \sqrt{3^2 + (-5)^2} = \sqrt{34}$, and the definition of trigonometric functions, we have

$$\sin\theta = \frac{y}{r} = \frac{-5}{\sqrt{34}} \qquad \csc\theta = \frac{r}{y} = \frac{\sqrt{34}}{-5}$$

$$\cos\theta = \frac{x}{r} = \frac{3}{\sqrt{34}} \qquad \sec\theta = \frac{r}{x} = \frac{\sqrt{34}}{3}$$

$$\tan\theta = \frac{y}{x} = -\frac{5}{3} \qquad \cot\theta = \frac{x}{y} = -\frac{3}{5}$$

3.1 Signs of the Trigonometric Functions

The algebraic signs of the values of the trigonometric functions of any angle depend on the signs of x, y, and r. As the terminal side of the angle moves from one quadrant to another, r always remains positive, whereas the signs of x and y change alternately.

Since x, y, and r are all positive in the first quadrant, all trigonometric functions are positive in quadrant I. In the second quadrant, y and r are both positive and x is negative, so the sine and cosecant are positive. In the third quadrant, x and y are both negative, and hence the tangent and cotangent are positive. In the fourth quadrant, x and

r are both positive and y is negative; consequently, the cosine and secant are positive (Figure 10).

Figure 10

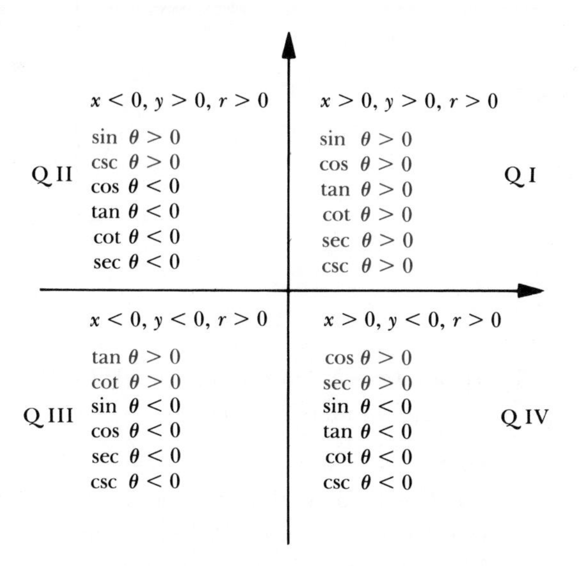

EXAMPLES

1 Indicate the quadrant in which the terminal side of θ lies if

a) $\sin\theta > 0$ and $\tan\theta > 0$ b) $\cot\theta < 0$ and $\sin\theta > 0$

c) $\sin\theta > 0$ and $\sec\theta < 0$ d) $\tan\theta > 0$ and $\csc\theta < 0$

SOLUTION. Using Figure 10 we see that

a) If $\sin\theta > 0$ and $\tan\theta > 0$, then the terminal side of θ lies in quadrant I.

b) If $\cot\theta < 0$ and $\sin\theta > 0$, then the terminal side of θ lies in quadrant II.

c) If $\sin\theta > 0$ and $\sec\theta < 0$, then the terminal side of θ lies in quadrant II.

d) If $\tan\theta > 0$ and $\csc\theta < 0$, then the terminal side of θ lies in quadrant III.

2 If $\sin\theta = \frac{12}{13}$, find $\cos\theta$ and indicate which quadrant contains the terminal side of θ.

SOLUTION. Since $\sin\theta = y/r = \frac{12}{13}$, we have two possibilities (Figure 11). Since $13 = \sqrt{(12)^2 + x^2}$ or $169 = 144 + x^2$, then $x^2 = 25$ or $x = \pm 5$. Hence, $\cos\theta = \frac{5}{13}$ if the terminal side of θ is in quadrant I and $\cos\theta = \frac{-5}{13}$ if the terminal side of θ is in quadrant II (Figure 11).

Notice that θ may be one of infinitely many possible angles; however, any such θ must have one of the two terminal sides (Figure 11).

Figure 11

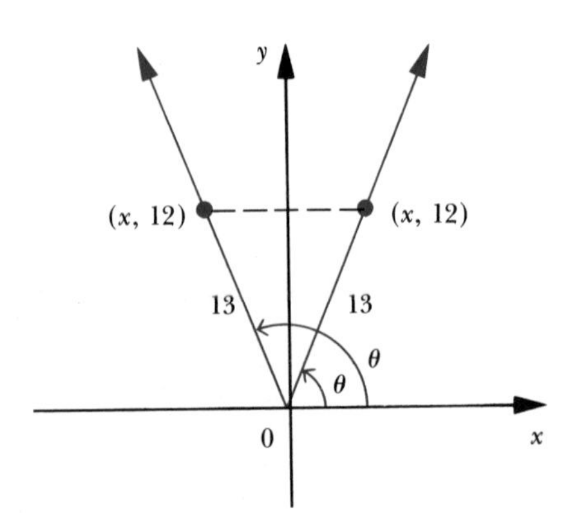

3 If $\tan \theta = -4/3$ and θ is in quadrant II, find the values of the remaining trigonometric functions.

SOLUTION. Since $\tan \theta = y/x = -4/3$ and θ is in quadrant II, we have $\tan \theta = 4/-3$ so that one point on the terminal side of θ is $(-3, 4)$ (Figure 12). Also, $r = \sqrt{(-3)^2 + 4^2} = 5$ and the values of the remaining trigonometric functions are

$$\sin \theta = \frac{4}{5}$$

$$\cos \theta = \frac{-3}{5}$$

$$\cot \theta = \frac{-3}{4}$$

$$\sec \theta = \frac{5}{-3}$$

$$\csc \theta = \frac{5}{4}$$

Figure 12

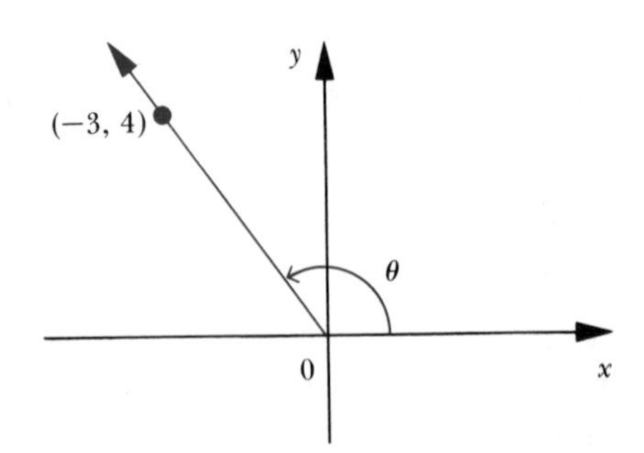

4 In what quadrant is the terminal side of θ if $\sin \theta = -\frac{7}{9}$? Evaluate $\cos \theta$.

SOLUTION. $\sin \theta = y/r = -\frac{7}{9}$, from which we can deduce the possibilities shown in Figure 13. Since $9 = \sqrt{x^2 + 49}$, $x^2 = 32$. Hence, $x = \pm 4\sqrt{2}$, so $\cos \theta = 4\sqrt{2}/9$ if θ has a terminal side in quadrant IV or $\cos \theta = -4\sqrt{2}/9$ if θ has a terminal side in quadrant III. Notice that there are actually infinitely many values possible for θ; however, any θ must have one of the two terminal sides (Figure 13).

Figure 13

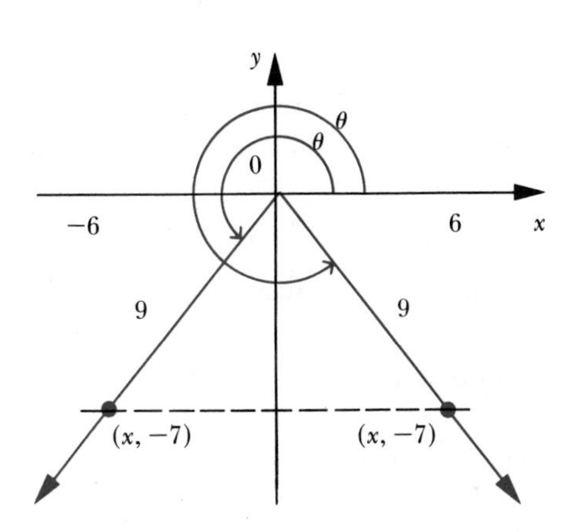

PROBLEM SET 2

1 Sketch the following angles in standard position and identify the quadrant that contains the terminal side of each angle.

a) 125° b) 260° c) −90° d) 210°
e) −720° f) $5\pi/6$ radians g) −3 radians h) 330°
i) 682° j) −342° k) −665° l) 704°
m) $15\pi/7$ n) 585° o) 858° p) 682°
q) −298°

2 Sketch three angles in standard position coterminal with each angle θ whose measure is given.

a) $\theta = 70°$ b) $\theta = -28°$ c) $\theta = \pi/12$
d) $\theta = 147°$ e) $\theta = 298°$ f) $\theta = 513°$

3 Sketch an angle in standard position whose terminal side contains the given point. Indicate by θ one of the positive angles and by ϕ one of the negative angles thus formed. Also, indicate the quadrant that contains the terminal side of θ.

a) (1, 3) b) (−2, −5) c) (−2, 4)
d) (1, −2) e) (1, −3) f) (−1, −2)

4 Suppose that two angles in standard position have the same initial side and the same terminal side. Are the two angles necessarily identical? Sketch a figure to illustrate this situation.

5. Evaluate the trigonometric functions of θ in standard position if the terminal side contains the following points.
   a) $(-1, \sqrt{3})$ b) $(-12, -5)$ c) $(-3, 4)$
   d) $(1, -1)$ e) $(\sqrt{3}, 1)$ f) $(-5, -5)$

6. Determine the values of the trigonometric functions on θ in each of the following cases.
   a) θ has $(-5, 0)$ on its terminal side.
   b) θ has $(-3, -4)$ on its terminal side.
   c) θ has $(7, -10)$ on its terminal side.
   d) θ has $(0, 1)$ on its terminal side.
   e) θ has $(1, -\sqrt{3})$ on its terminal side.
   f) $(x, -4)$ is 11 units from $(0, 0)$ and is on the terminal side of θ.

7. Construct the following angles in standard positions whose terminal side is in the indicated quadrant, and evaluate the other five trigonometric functions of θ.
   a) $\sin\theta = 15/17$, quadrant II
   b) $\tan\theta = -1$, quadrant IV
   c) $\cos\theta = \sqrt{3}/2$, quadrant IV
   d) $\cot\theta = \sqrt{3}$, quadrant III
   e) $\sec\theta = -13/5$, quadrant II
   f) $\csc\theta = \sqrt{2}$, quadrant I

4 Trigonometric Functions of Special Angles

In this section we shall consider the values of the trigonometric functions of certain special angles. These angles are special in that it is possible to apply elementary properties from plane geometry to evaluate the trigonometric functions.

4.1 Trigonometric Functions of Quadrantal Angles

Angles with measures such as 0°, 90°, 180°, 270°, and 360°, whose terminal side lies along either the x axis or the y axis, are called *quadrantal angles.* The values of the trigonometric functions of these angles can easily be found by inspection from a figure. This can be illustrated by the following examples.

EXAMPLES

In each of the following, sketch the angle θ in standard position and determine the six trigonometric functions of θ.

1 $\theta = 0°$

SOLUTION. Construct the angle 0° in standard position (Figure 1). For convenience, select the point (1, 0) on the terminal side of θ so that $x = 1$, $y = 0$, and $r = 1$. Notice that the selection of other points such as (2, 0), (4, 0), and (5, 0) will yield the same results.

Figure 1

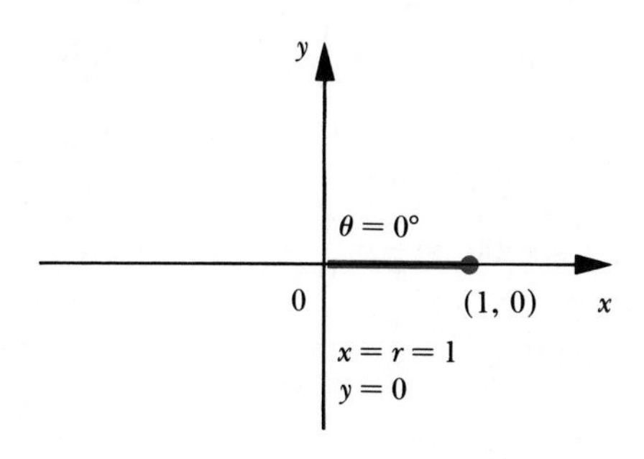

From the definition of trigonometric functions, we have

$$\sin 0° = \frac{y}{r} = \frac{0}{1} = 0$$

$$\cos 0° = \frac{x}{r} = \frac{1}{1} = 1$$

$$\tan 0° = \frac{y}{x} = \frac{0}{1} = 0$$

$$\cot 0° = \frac{x}{y} = \frac{1}{0} \quad \text{(undefined)}$$

$$\sec 0° = \frac{r}{x} = \frac{1}{1} = 1$$

$$\csc 0° = \frac{r}{y} = \frac{1}{0} \quad \text{(undefined)}$$

2 $\theta = 90°$

SOLUTION. Select the point (0, 1) on the terminal side of the 90° angle, constructed in standard position so that $x = 0$, $y = 1$, and $r = 1$ (Figure 2).

Figure 2

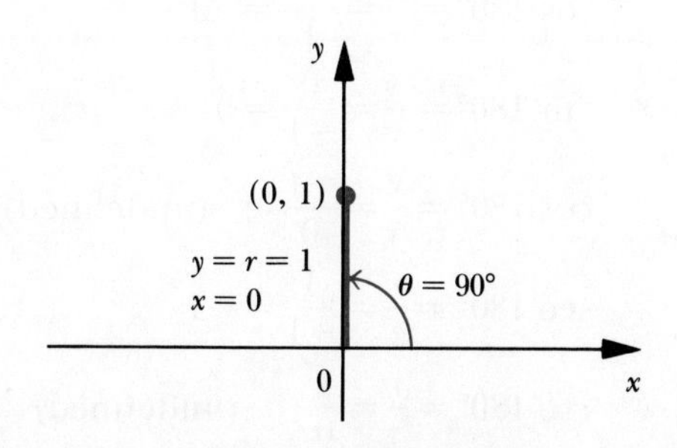

Then by the definition of trigonometric functions, we obtain the values of the trigonometric functions for a 90° angle.

$$\sin 90° = \frac{y}{r} = \frac{1}{1} = 1$$

$$\cos 90° = \frac{x}{r} = \frac{0}{1} = 0$$

$$\tan 90° = \frac{y}{x} = \frac{1}{0} \quad \text{(undefined)}$$

$$\cot 90° = \frac{x}{y} = \frac{0}{1} = 0$$

$$\sec 90° = \frac{r}{x} = \frac{1}{0} \quad \text{(undefined)}$$

$$\csc 90° = \frac{r}{y} = \frac{1}{1} = 1$$

3 $\theta = 180°$

SOLUTION. Select the point $(-1, 0)$ on the terminal side of a 180° angle that is constructed in standard position so that $x = -1$, $y = 0$, and $r = 1$ (Figure 3).

Figure 3

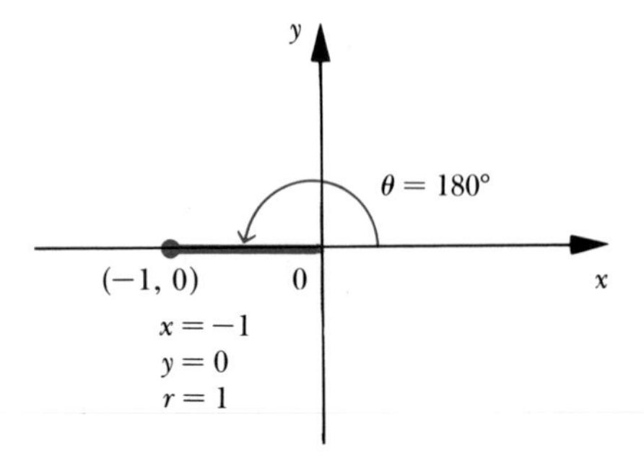

Hence, by the definition of trigonometric functions, we obtain the value of the trigonometric functions for a 180° angle.

$$\sin 180° = \frac{y}{r} = \frac{0}{1} = 0$$

$$\cos 180° = \frac{x}{r} = \frac{-1}{1} = -1$$

$$\tan 180° = \frac{y}{x} = \frac{0}{-1} = 0$$

$$\cot 180° = \frac{x}{y} = \frac{-1}{0} \quad \text{(undefined)}$$

$$\sec 180° = \frac{r}{x} = \frac{1}{-1} = -1$$

$$\csc 180° = \frac{r}{y} = \frac{1}{0} \quad \text{(undefined)}$$

4 $\theta = -90°$

SOLUTION. Select the point $(0, -1)$ on the terminal side of the $-90°$ angle constructed in standard position so that $x = 0$, $y = -1$, and $r = 1$ (Figure 4).

Figure 4

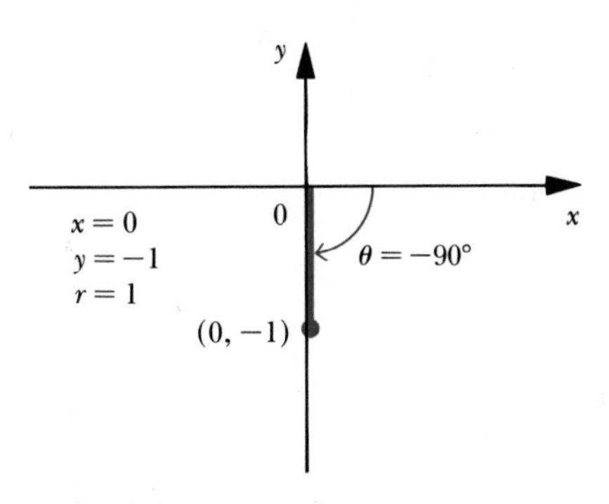

Using the definition of the trigonometric functions, we find that the values of the six trigonometric functions are as follows:

$$\sin(-90°) = \frac{y}{r} = \frac{-1}{1} = -1$$

$$\cos(-90°) = \frac{x}{r} = \frac{0}{1} = 0$$

$$\tan(-90°) = \frac{y}{x} = \frac{-1}{0} \quad \text{(undefined)}$$

$$\cot(-90°) = \frac{x}{y} = \frac{0}{-1} = 0$$

$$\sec(-90°) = \frac{r}{x} = \frac{1}{0} \quad \text{(undefined)}$$

$$\csc(-90°) = \frac{r}{y} = \frac{1}{-1} = -1$$

4.2 Trigonometric Functions of 45°, 30° and 60° Angles

In order to find the values of the trigonometric functions of a 45° angle, we recall from plane geometry that in a right triangle with an acute angle of 45°, the two sides are equal. Thus, we can construct an angle of 45° in standard position and choose the point (1, 1) on its terminal side (Figure 5). Hence, we have $x = 1$, $y = 1$, and $r = \sqrt{1^2 + 1^2} = \sqrt{2}$. Applying the definition of trigonometric functions for $\theta = 45°$, we have

$$\sin 45° = \frac{y}{r} = \frac{1}{\sqrt{2}} = \frac{\sqrt{2}}{2}$$

$$\cos 45° = \frac{x}{r} = \frac{1}{\sqrt{2}} = \frac{\sqrt{2}}{2}$$

$$\tan 45^\circ = \frac{y}{x} = \frac{1}{1} = 1$$

$$\cot 45^\circ = \frac{x}{y} = \frac{1}{1} = 1$$

$$\sec 45^\circ = \frac{r}{x} = \frac{\sqrt{2}}{1} = \sqrt{2}$$

$$\csc 45^\circ = \frac{r}{y} = \frac{\sqrt{2}}{1} = \sqrt{2}$$

Figure 5

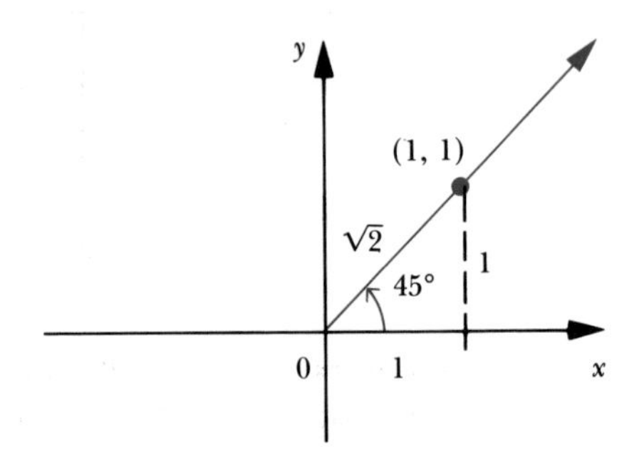

In order to find the values of the trigonometric functions of a 30° angle, we recall from plane geometry that in a right triangle with acute angles of 30° and 60°, the hypotenuse is twice the length of the side opposite the 30° angle. Thus, we can construct an angle of 30° in standard position and choose the point $(x, 1)$ on its terminal side so that the hypotenuse r is 2 (Figure 6). By this construction, we have $y = 1$ and $r = 2$, so that,

$$2 = \sqrt{x^2 + 1^2} \quad \text{or}$$
$$4 = x^2 + 1 \quad \text{or}$$
$$x^2 = 3$$

Figure 6

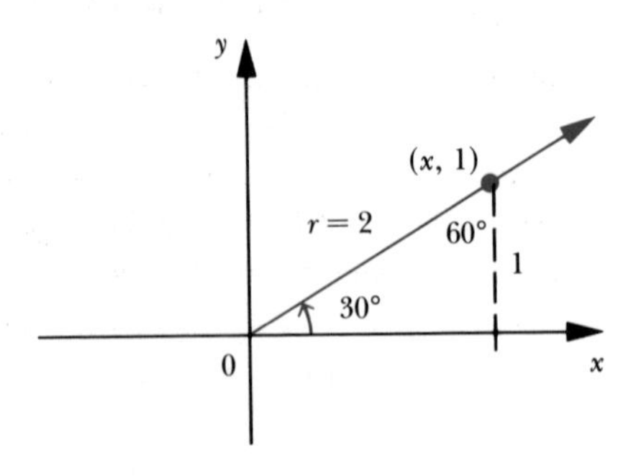

Hence, $x = \sqrt{3}$, since the value of x is positive in quadrant I. Thus the coordinates of a point on the terminal side of the 30° angle are

$(\sqrt{3}, 1)$, and r is 2. Applying the definition of trigonometric functions for $\theta = 30°$, we have

$$\sin 30° = \frac{y}{r} = \frac{1}{2}$$

$$\cos 30° = \frac{x}{r} = \frac{\sqrt{3}}{2}$$

$$\tan 30° = \frac{y}{x} = \frac{1}{\sqrt{3}} = \frac{\sqrt{3}}{3}$$

$$\cot 30° = \frac{x}{y} = \frac{\sqrt{3}}{1} = \sqrt{3}$$

$$\sec 30° = \frac{r}{x} = \frac{2}{\sqrt{3}} = \frac{2\sqrt{3}}{3}$$

$$\csc 30° = \frac{r}{y} = \frac{2}{1} = 2$$

By constructing the right triangle from Figure 6 with the 60° angle in standard position (Figure 7), we have $r = 2$, $x = 1$, and $y = \sqrt{3}$. Applying definition of trigonometric functions for $\theta = 60°$, we have

$$\sin 60° = \frac{y}{r} = \frac{\sqrt{3}}{2}$$

$$\cos 60° = \frac{x}{r} = \frac{1}{2}$$

$$\tan 60° = \frac{y}{x} = \frac{\sqrt{3}}{1} = \sqrt{3}$$

$$\cot 60° = \frac{x}{y} = \frac{1}{\sqrt{3}} = \frac{\sqrt{3}}{3}$$

$$\sec 60° = \frac{r}{x} = \frac{2}{1} = 2$$

$$\csc 60° = \frac{r}{y} = \frac{2}{\sqrt{3}} = \frac{2\sqrt{3}}{3}$$

Figure 7

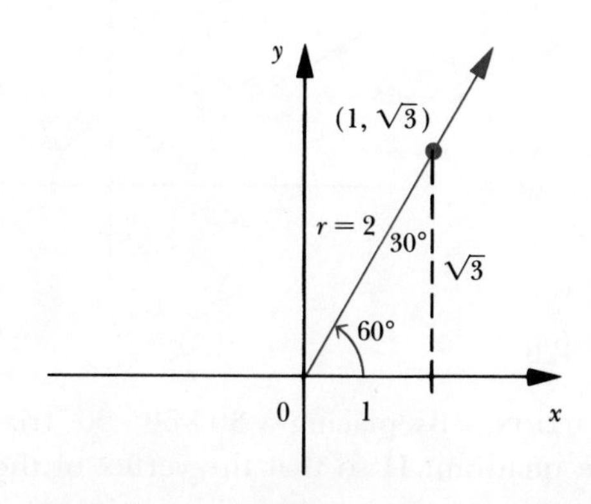

The trigonometric functions of angles that are multiples of either 30° or 45° (excluding quadrantal angles) can be determined by the values of the functions of 30° or 45° from the 30°–60°–90° triangle or the right triangle with 45° angle. This procedure is illustrated by the following examples.

EXAMPLES

Find the values of the six trigonometric functions of the angle θ.

1 $\theta = 150°$

SOLUTION. By placing a 30°–60°–90° triangle with sides 1, $\sqrt{3}$, and 2 in quadrant II so that the vertex of the 30° angle is at the origin, we obtain an angle of 150° in standard position whose terminal side contains the point $(-\sqrt{3}, 1)$ and whose corresponding value of r is 2 (Figure 9). Applying the definition of trigonometric functions for $\theta = 120°$, we obtain

$$\sin 150° = \frac{y}{r} = \frac{1}{2}$$

$$\cos 150° = \frac{x}{r} = \frac{-\sqrt{3}}{2}$$

$$\tan 150° = \frac{y}{x} = \frac{1}{-\sqrt{3}} = -\frac{\sqrt{3}}{3}$$

$$\cot 150° = \frac{x}{y} = \frac{-\sqrt{3}}{1} = -\sqrt{3}$$

$$\sec 150° = \frac{r}{x} = \frac{2}{-\sqrt{3}} = -\frac{2\sqrt{3}}{3}$$

$$\csc 150° = \frac{r}{y} = \frac{2}{1} = 2$$

Figure 8

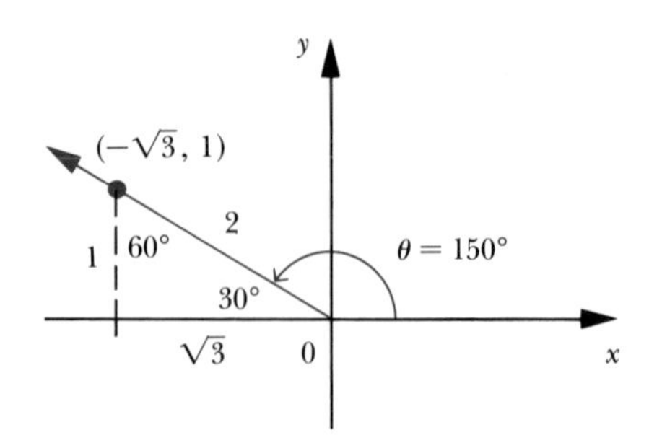

2 $\theta = 120°$

SOLUTION. By placing a 30°–60°–90° triangle with sides 1, $\sqrt{3}$, and 2 in quadrant II so that the vertex of the 60° angle is at the origin, we obtain an angle of 120° in standard position whose terminal side contains the point $(-1, \sqrt{3})$ and whose corresponding value of r is (Figure 9). Applying the definition of trigonometric functions for $\theta = 120°$, we obtain

$$\sin 120° = \frac{y}{r} = \frac{\sqrt{3}}{2}$$

$$\cos 120° = \frac{x}{r} = \frac{-1}{2}$$

$$\tan 120° = \frac{y}{x} = \frac{\sqrt{3}}{-1} = -\sqrt{3}$$

$$\cot 120° = \frac{x}{y} = \frac{-1}{\sqrt{3}} = -\frac{\sqrt{3}}{3}$$

$$\sec 120° = \frac{r}{x} = \frac{2}{-1} = -2$$

$$\csc 120° = \frac{r}{y} = \frac{2}{\sqrt{3}} = \frac{2\sqrt{3}}{3}$$

Figure 9

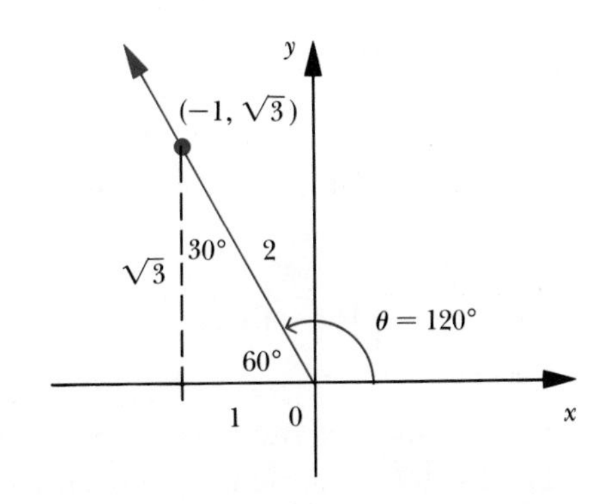

3 $\theta = 240°$

SOLUTION. Construct the 240° angle in standard position by placing the 30°–60°–90° triangle in quadrant III so that the point $(-1, -\sqrt{3})$ lies on the terminal side of the 240° angle and whose corresponding value of r is 2 (Figure 10). Applying the definition of trigonometric functions for $\theta = 240°$, we obtain

$$\sin 240° = \frac{y}{r} = \frac{-\sqrt{3}}{2}$$

$$\cos 240° = \frac{x}{r} = \frac{-1}{2}$$

$$\tan 240° = \frac{y}{x} = \frac{-\sqrt{3}}{-1} = \sqrt{3}$$

$$\cot 240° = \frac{x}{y} = \frac{-1}{-\sqrt{3}} = \frac{\sqrt{3}}{3}$$

$$\sec 240° = \frac{r}{x} = \frac{2}{-1} = -2$$

$$\csc 240° = \frac{r}{y} = \frac{1}{-\sqrt{3}} = -\frac{2\sqrt{3}}{3}$$

Figure 10

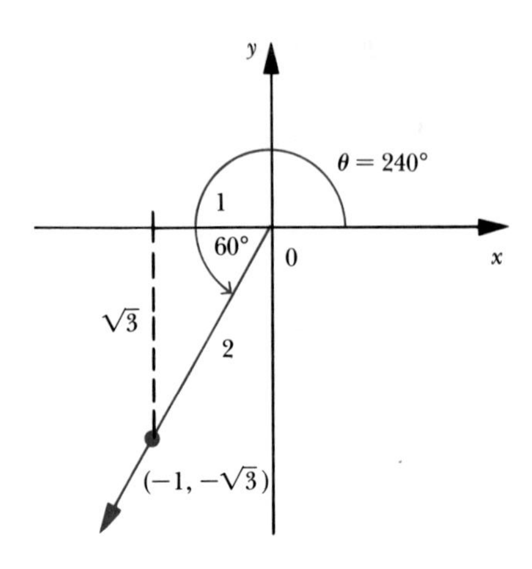

4 $\theta = -\pi/4$

SOLUTION. First, we note that $-\frac{\pi}{4}$ radians corresponds to

$$\left(\frac{180}{\pi}\right)\left(-\frac{\pi}{4}\right) = -45°.$$

Next, construct the $-45°$ angle in standard position by placing the 45°–45°–90° triangle in quadrant IV so that the point $(1, -1)$ lies on the terminal side of the $-45°$ angle (Figure 11). Thus, $x = 1$, $y = -1$, and $r = \sqrt{2}$. Applying the definition of trigonometric functions, for $\theta = -45° = -\pi/4$, we obtain

$$\sin\left(-\frac{\pi}{4}\right) = \sin(-45°) = \frac{y}{r} = \frac{-1}{\sqrt{2}} = -\frac{\sqrt{2}}{2}$$

$$\cos\left(-\frac{\pi}{4}\right) = \cos(-45°) = \frac{x}{r} = \frac{1}{\sqrt{2}} = \frac{\sqrt{2}}{2}$$

$$\tan\left(-\frac{\pi}{4}\right) = \tan(-45°) = \frac{y}{x} = \frac{-1}{1} = -1$$

$$\cot\left(-\frac{\pi}{4}\right) = \cot(-45°) = \frac{x}{y} = \frac{1}{-1} = -1$$

$$\sec\left(-\frac{\pi}{4}\right) = \sec(-45°) = \frac{r}{x} = \frac{\sqrt{2}}{1} = \sqrt{2}$$

$$\csc\left(-\frac{\pi}{4}\right) = \csc(-45°) = \frac{r}{y} = \frac{\sqrt{2}}{-1} = -\sqrt{2}$$

Figure 11

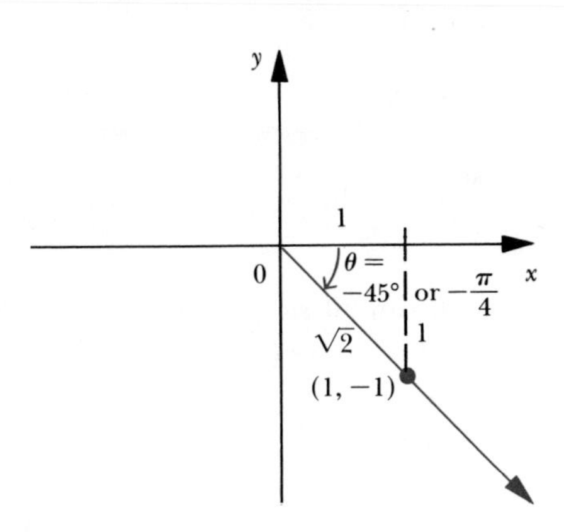

5 $\theta = 390°$

SOLUTION. The 390° angle has the same terminal side as the 30° angle (Figure 12). Therefore, the values of the trigonometric functions of 390° are the same as the values of the functions for 30°. Using the definition of trigonometric functions, we have

$$\sin 390° = \sin 30° = \frac{y}{r} = \frac{1}{2}$$

$$\cos 390° = \cos 30° = \frac{x}{r} = \frac{\sqrt{3}}{2}$$

$$\tan 390° = \tan 30° = \frac{y}{x} = \frac{1}{\sqrt{3}} = \frac{\sqrt{3}}{3}$$

$$\cot 390° = \cot 30° = \frac{x}{y} = \frac{\sqrt{3}}{1} = \sqrt{3}$$

$$\sec 390° = \sec 30° = \frac{r}{x} = \frac{2}{\sqrt{3}} = \frac{2\sqrt{3}}{3}$$

$$\csc 390° = \csc 30° = \frac{r}{y} = \frac{2}{1} = 2$$

Figure 12

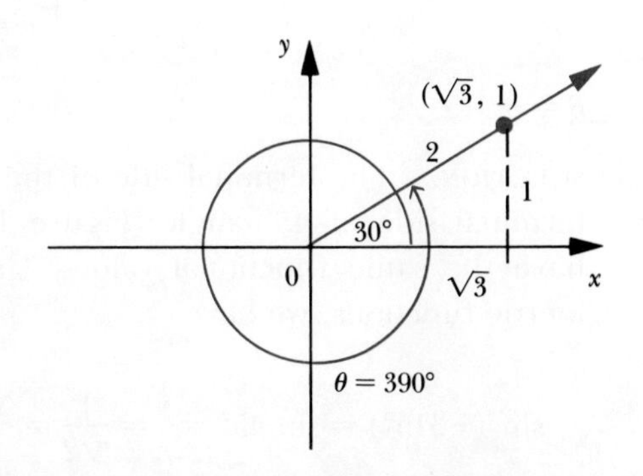

6 $\theta = 39\pi/9$

SOLUTION

$$\frac{39\pi}{9} \text{ has the degree measure } \frac{39\pi}{9}\left(\frac{180^\circ}{\pi}\right) = 780^\circ$$

The terminal side of the 780° angle is the same as the terminal side of the 60° angle (Figure 13). Therefore, the two angles have the same functional values. Using the definition of trigonometric functions, we have

$$\sin\frac{39\pi}{9} = \sin 780^\circ = \sin 60^\circ = \frac{y}{r} = \frac{\sqrt{3}}{2}$$
$$\cos\frac{39\pi}{9} = \cos 780^\circ = \cos 60^\circ = \frac{x}{r} = \frac{1}{2}$$
$$\tan\frac{39\pi}{9} = \tan 780^\circ = \tan 60^\circ = \frac{y}{x} = \frac{\sqrt{3}}{1} = \sqrt{3}$$
$$\cot\frac{39\pi}{9} = \cot 780^\circ = \cot 60^\circ = \frac{x}{y} = \frac{1}{\sqrt{3}} = \frac{\sqrt{3}}{3}$$
$$\sec\frac{39\pi}{9} = \sec 780^\circ = \sec 60^\circ = \frac{r}{x} = \frac{2}{1} = 2$$
$$\csc\frac{39\pi}{9} = \csc 780^\circ = \csc 60^\circ = \frac{r}{y} = \frac{2}{\sqrt{3}} = \frac{2\sqrt{3}}{3}$$

Figure 13

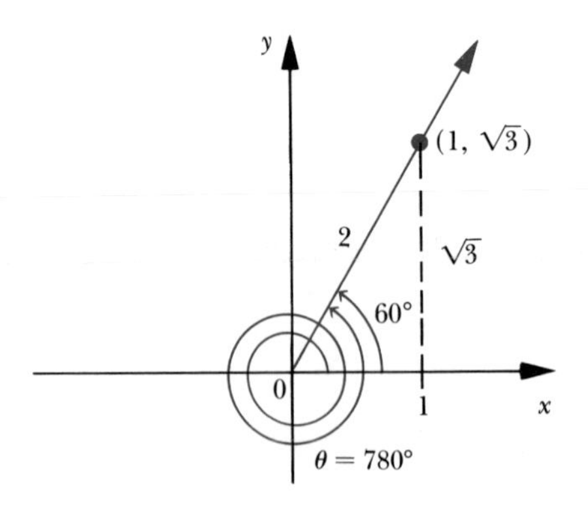

7 $\theta = -315^\circ$

SOLUTION. The terminal side of the -315° angle is the same as the terminal side of 45° angle (Figure 14). Therefore, the two angles have the same functional values. Using the definition of trigonometric functions, we have

$$\sin(-315^\circ) = \sin 45^\circ = \frac{y}{r} = \frac{1}{\sqrt{2}} = \frac{\sqrt{2}}{2}$$

$$\cos(-315°) = \cos 45° = \frac{x}{r} = \frac{1}{\sqrt{2}} = \frac{\sqrt{2}}{2}$$

$$\tan(-315°) = \tan 45° = \frac{y}{x} = \frac{1}{1} = 1$$

$$\cot(-315°) = \cot 45° = \frac{x}{y} = \frac{1}{1} = 1$$

$$\sec(-315°) = \sec 45° = \frac{r}{x} = \frac{\sqrt{2}}{1} = \sqrt{2}$$

$$\csc(-315°) = \csc 45° = \frac{r}{y} = \frac{\sqrt{2}}{1} = \sqrt{2}$$

Figure 14

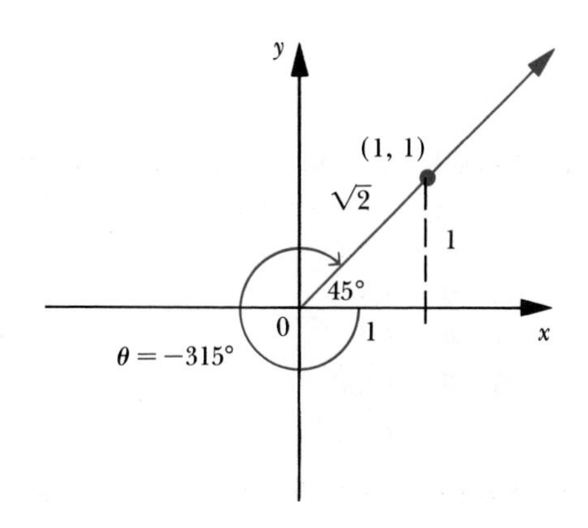

8 $\theta = 5\pi/4$

SOLUTION. Since $5\pi/4$ corresponds to 225°, we select the point $(-1, -1)$ on the terminal side of the 225° angle that is constructed in standard position so that $x = -1$, $y = -1$, and $r = \sqrt{2}$ (Figure 15). Using the definition of the trigonometric functions, we have,

$$\sin\frac{5\pi}{4} = \frac{y}{r} = \frac{-1}{\sqrt{2}} = -\frac{\sqrt{2}}{2}$$

$$\cos\frac{5\pi}{4} = \frac{x}{r} = \frac{-1}{\sqrt{2}} = -\frac{\sqrt{2}}{2}$$

$$\tan\frac{5\pi}{4} = \frac{y}{x} = \frac{-1}{-1} = 1$$

$$\cot\frac{5\pi}{4} = \frac{x}{y} = \frac{-1}{-1} = 1$$

$$\sec\frac{5\pi}{4} = \frac{r}{x} = \frac{\sqrt{2}}{-1} = -\sqrt{2}$$

$$\csc\frac{5\pi}{4} = \frac{r}{y} = \frac{\sqrt{2}}{-1} = -\sqrt{2}$$

Figure 15

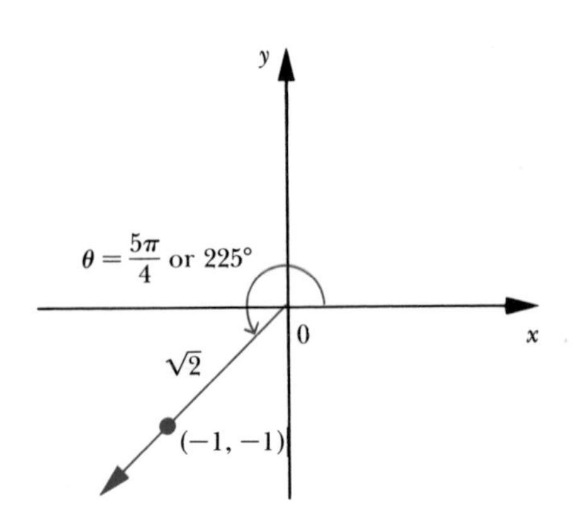

PROBLEM SET 3

1 Sketch the following angles in standard position and determine the values of the six trigonometric functions for these angles.

a) $\theta = -180°$ b) $\theta = -270°$
c) $\theta = 360°$ d) $\theta = 720°$
e) $\theta = 810°$ f) $\theta = -810°$
g) $\theta = 1080°$ h) $\theta = -1080°$
i) $\theta = 11\pi/2$ j) $\theta = 5\pi$

2 Sketch the following angles in standard position and determine the values of the six trigonometric functions for these angles.

a) 135° b) 210° c) −300° d) −390°
e) $11\pi/6$ f) $-5\pi/4$ g) −60° h) $-2\pi/3$
i) −240° j) $7\pi/6$ k) $13\pi/6$ l) 225°
m) 585° n) 315°

3 Complete the following table.

θ	210°	225°	240°	300°	315°	330°
$\sin\theta$						
$\cos\theta$						
$\tan\theta$						
$\cot\theta$						
$\sec\theta$						
$\csc\theta$						

4 Make a table similar to the one in Problem 3 using the radian measure for each angle θ.

5 Answer true or false.

a) $\sin 60° < 2 \sin 30°$
b) $\sin 30° + \sin 60° = \sin 90°$
c) $2 \sin 30° \cos 30° < \sin 60°$
d) $\sin 210° = \sin 180° + \sin 30°$

6 Determine each of the following values.

a) $\sin(-150°)$
b) $\cos 660°$
c) $\tan(-120°)$
d) $\cot 570°$
e) $\sec(-750°)$
f) $\csc 870°$
g) $\tan(-405°)$
h) $\tan(13\pi/4)$
i) $\sin(11\pi/3)$
j) $\cos(61\pi/6)$
k) $\sec 840°$
l) $\csc(59\pi/6)$

5 Evaluation of the Trigonometric Functions of Any Angle

In Section 4 exact values of the trigonometric functions of special angles were calculated. In this section we wish to discuss a method of determining the values of the trigonometric functions for any angle in terms of an acute positive angle. Then we shall evaluate approximately the trigonometric functions of any acute positive angle.

5.1 Reference Angles

In order to evaluate trigonometric functions at any given angle, we can find another angle θ coterminal with the given angle so that the angle θ is between 0° and 360°. For example, since the 420° angle in standard position is coterminal with the 60° angle in standard position, the values of the trigonometric functions of 420° are the same as the values of the trigonometric functions of 60° (Figure 1).

Figure 1

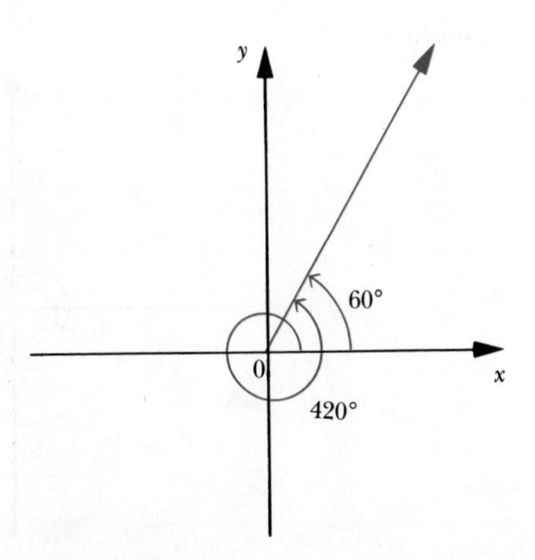

Thus, the value of the trigonometric functions of any angle is the same as the values of the trigonometric functions of a positive angle θ, where θ is coterminal with the given angle and $0 \leq \theta < 360°$.

EXAMPLES

Find an angle θ coterminal with each given angle such that $0 \leq \theta < 360°$.

1 850°

SOLUTION. Since $850° = 2(360°) + 130°$, then an angle of 850° is coterminal with $\theta = 130°$ (Figure 2).

Figure 2

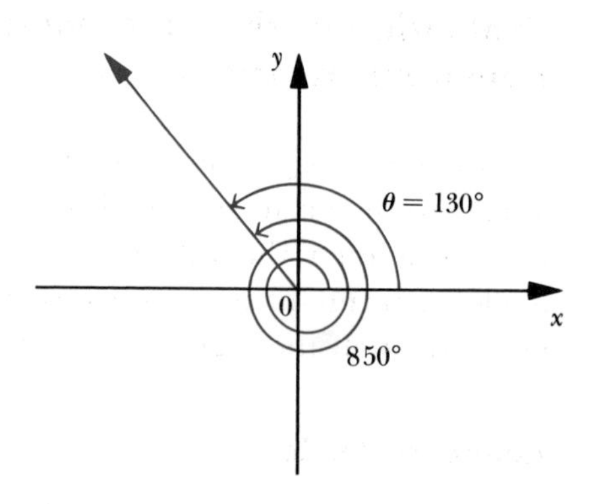

2 −1237°

SOLUTION. Since $-1237° = 3(-360°) + (-157°)$, then an angle of $-1237°$ is coterminal with an angle of $-157°$, which in turn is coterminal with an angle of 203°. Hence, an angle of $-1237°$ is coterminal with $\theta = 203°$ (Figure 3).

Figure 3

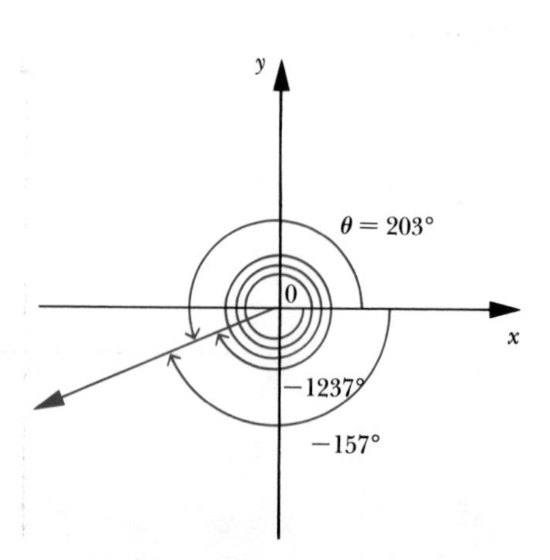

Next we will see how the evaluation of the trigonometric functions for a positive angle θ (different from 0°, 90°, 180°, and 270°) between 0° and 360° can be reduced to an evaluation of the functions at a positive *acute* angle θ_R between 0° and 90°, with the necessary adjustments in signs. θ_R is called the *reference angle* associated with θ and is defined as the smallest positive acute angle determined by the terminal side of θ and the x axis (Figure 4).

Figure 4

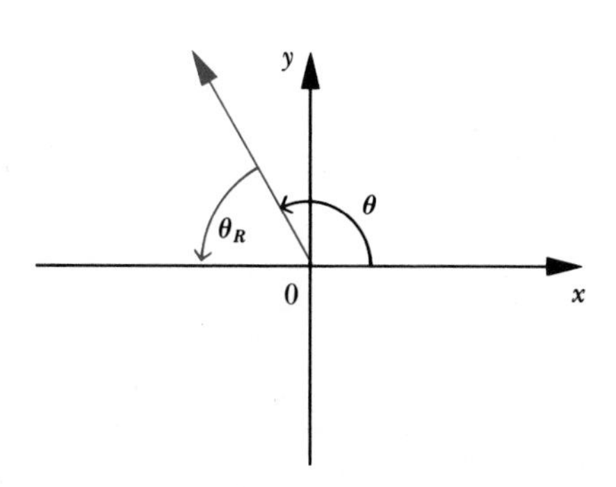

We shall now illustrate how to find the values of the trigonometric functions of θ using the reference angle θ_R by considering the following four cases.

CASE 1. If θ lies in quadrant I, that is, if θ is between 0° and 90°, then the reference angle θ_R is the same as the angle θ (Figure 5).

Figure 5

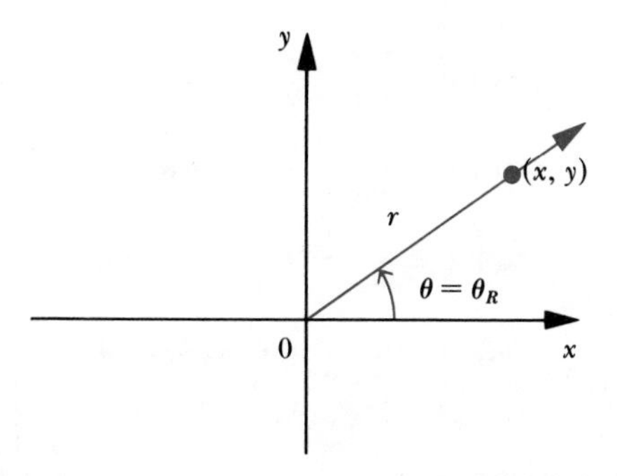

CASE 2. If θ lies in quadrant II, that is, if θ is between 90° and 180°, then $\theta_R = 180° - \theta$ (Figure 6). Choose a point (x, y) on the terminal side of θ_R, redrawn in standard position. Then $\cos \theta_R = x/y$. However, notice that the point $(-x, y)$ now lies on the terminal side of θ itself. Therefore, using the definition of trigonometric functions, we get

$$\sin \theta = \frac{y}{r} = \sin \theta_R$$

$$\cos\theta = \frac{-x}{r} = -\frac{x}{r} = -\cos\theta_R$$

$$\tan\theta = \frac{y}{-x} = -\frac{y}{x} = -\tan\theta_R$$

$$\cot\theta = \frac{-x}{y} = -\frac{x}{y} = -\cot\theta_R$$

$$\sec\theta = \frac{r}{-x} = -\frac{r}{x} = -\sec\theta_R$$

$$\csc\theta = \frac{r}{y} = \csc\theta_R$$

Figure 6

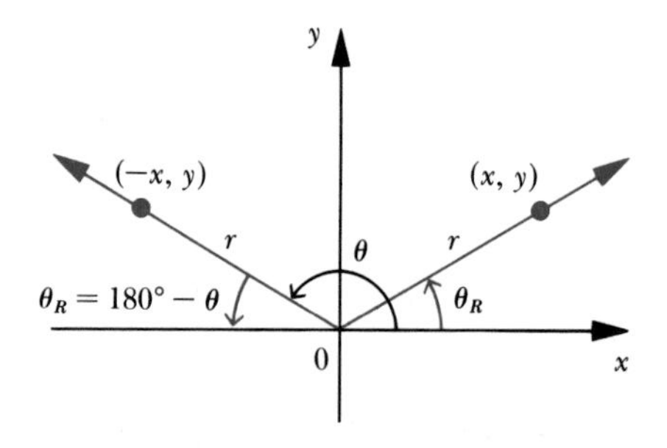

CASE 3. If θ lies in quadrant III, that is, if θ is between 180° and 270°, then $\theta_R = \theta - 180°$ (Figure 7). Here

$$\sin\theta = \frac{-y}{r} = -\frac{y}{r} = -\sin\theta_R$$

$$\cos\theta = \frac{-x}{r} = -\frac{x}{r} = -\cos\theta_R$$

$$\tan\theta = \frac{-y}{-x} = \frac{y}{x} = \tan\theta_R$$

$$\cot\theta = \frac{-x}{-y} = \frac{x}{y} = \cot\theta_R$$

$$\sec\theta = \frac{r}{-x} = -\frac{r}{x} = -\sec\theta_R$$

$$\csc\theta = \frac{r}{-y} = -\frac{r}{y} = -\csc\theta_R$$

Figure 7

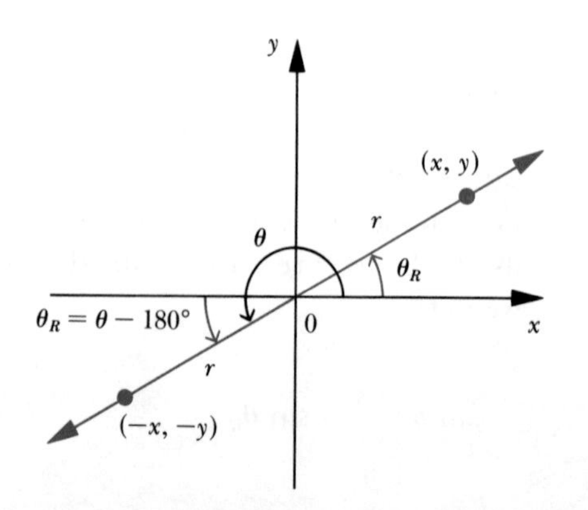

CASE 4. If θ lies in quadrant IV, that is, if θ is between 270° and 360°, then $\theta_R = 360° - \theta$ (Figure 8). Here

$$\sin\theta = \frac{-y}{r} = -\frac{y}{r} = -\sin\theta_R$$

$$\cos\theta = \frac{x}{r} = \cos\theta_R$$

$$\tan\theta = \frac{-y}{x} = -\frac{y}{x} = -\tan\theta_R$$

$$\cot\theta = \frac{x}{-y} = -\frac{x}{y} = -\cot\theta_R$$

$$\sec\theta = \frac{r}{x} = \sec\theta_R$$

$$\csc\theta = \frac{r}{-y} = -\frac{r}{y} = -\csc\theta_R$$

Figure 8

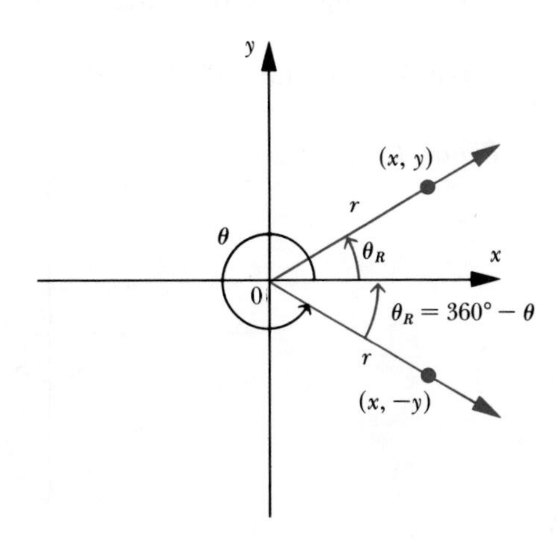

EXAMPLES

Reduce the evaluation of the given trigonometric functions to an evaluation of the same function at its reference angle θ_R.

1 sin 240° and cos 240°

SOLUTION. Since the terminal side of 240° is in quadrant III, the reference angle for 240° is $\theta_R = 240° - 180° = 60°$ (Figure 9). Also, the sine is negative and the cosine is negative in quadrant III. Thus

$$\sin 240° = -\sin 60° \qquad \text{and} \qquad \cos 240° = -\cos 60°$$

Figure 9

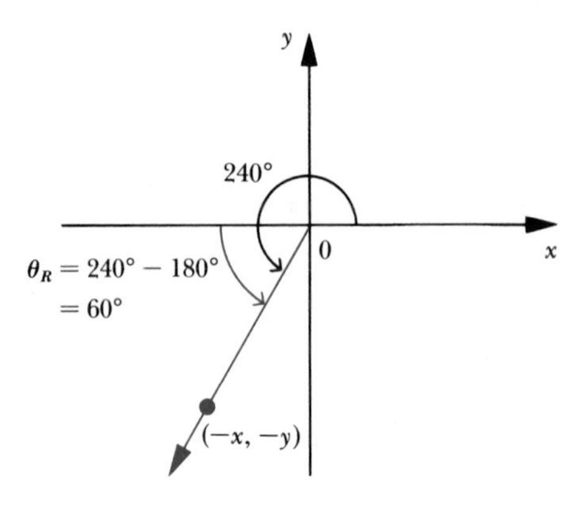

2 tan 320° and cot 320°

SOLUTION. Since the terminal side of 320° lies in quadrant IV, the reference angle for 320° is $\theta_R = 360° - 320° = 40°$ (Figure 10). Also, both the tangent and cotangent are negative in quadrant IV, so that

$$\tan 320° = -\tan 40° \qquad \text{and} \qquad \cot 320° = -\cot 40°$$

Figure 10

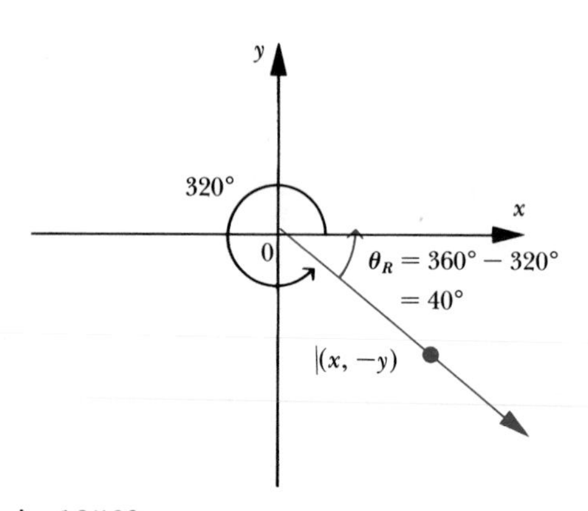

3 cos 1059° and sin 1059°

SOLUTION. Since $1059° = 2(360°) + 339°$, we see that an angle of 1059° is generated by rotating the positive x axis in a counterclockwise direction two complete revolutions and then through an angle of 339°. Therefore, the terminal side of an angle of 1059° lies in quadrant IV and the reference angle $\theta_R = 360° - 339° = 21°$ (Figure 11). Thus, since the cosine is positive and the sine is negative in quadrant IV, we have

$$\cos 1059° = \cos 339° = \cos 21°$$

and

$$\sin 1059° = \sin 339° = -\sin 21°$$

Figure 11

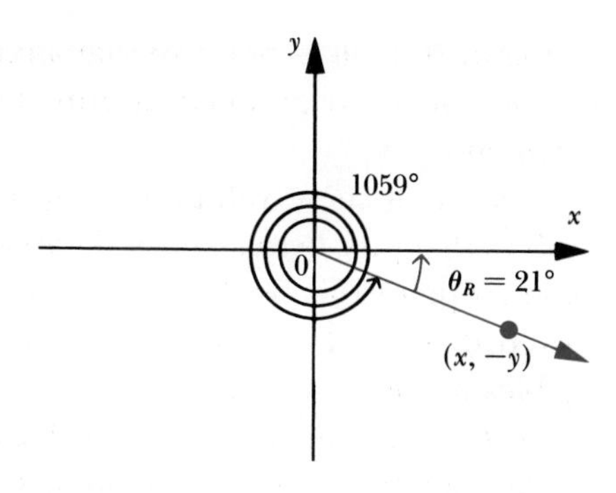

4 sec (−220°) and csc (−220°)

SOLUTION. The terminal side of the −220° angle is the same as the terminal side of the 140° angle and lies in quadrant II. Thus, the reference angle for −220° is the same as the reference angle for 140°. That is, $\theta_R = 180° - 140° = 40°$ (Figure 12). Also the secant is negative and the cosecant is positive in quadrant II, so

$$\sec(-220°) = \sec 140° = -\sec 40°$$

and

$$\csc(-220°) = \csc 140° = \csc 40°$$

Figure 12

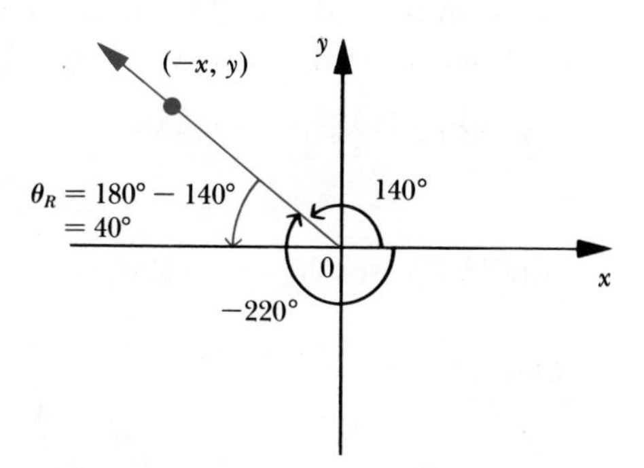

5.2 Tables of Values of Trigonometric Functions

By applying the process illustrated in Section 5.1, we are able to express the trigonometric functions of any nonquadrantal angle θ in terms of the trigonometric functions of a related acute positive angle θ_R $(0 < \theta_R < 90°)$. However, up to this point we have been able to evaluate trigonometric functions only for the special angles 30°, 45°, and 60°. For example, the sin $30° = \frac{1}{2}$, whereas sin 20° is still unknown. If θ_R is not a special angle, we determine the approximate values of the trigonometric functions of θ_R by referring to Table I, Appendix A.

Note that these values are approximations only. There are larger tables that give better approximations, and there are methods beyond the scope of this book that can be used to find the values to any desired degree of accuracy. (Such methods were used to construct the tables in the first place.)

Note: From here on, we shall assume that the table values of the trigonometric function are approximated correct to four decimal places. Hence, we shall omit the word *approximately* when we use the tables.

The following examples illustrate the use of Table I.

EXAMPLES

Use Table I to determine the value of each of the given trigonometric functions.

1 $\cos 380°$ and $\sin 380°$.

SOLUTION Since $380° = 360° + 20°$, we have

$$\cos 380° = \cos 20° \qquad \text{and} \qquad \sin 380° = \sin 20°$$

The terminal side of 20° angle in standard position lies in quadrant I and the reference angle $\theta_R = 20°$ (Figure 13). Thus,

$$\cos 380° = \cos 20° = 0.9397$$

and

$$\sin 380° = \sin 20° = 0.3420$$

Figure 13

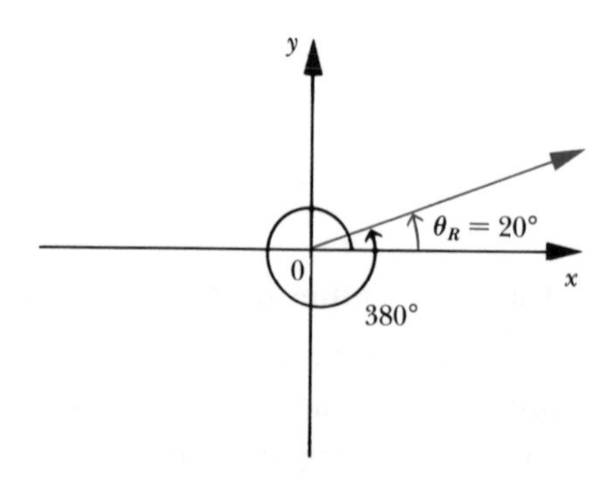

2 $\tan 228°$ and $\cot 228°$

SOLUTION. The terminal side of the 228° angle is in quadrant III where the tangent and cotangent are positive and the reference angle $\theta_R = 228° - 180° = 48°$ (Figure 14). Hence

$$\tan 228° = \tan 48° = 1.111$$

and

$$\cot 228° = \cot 48° = 0.9004$$

Figure 14

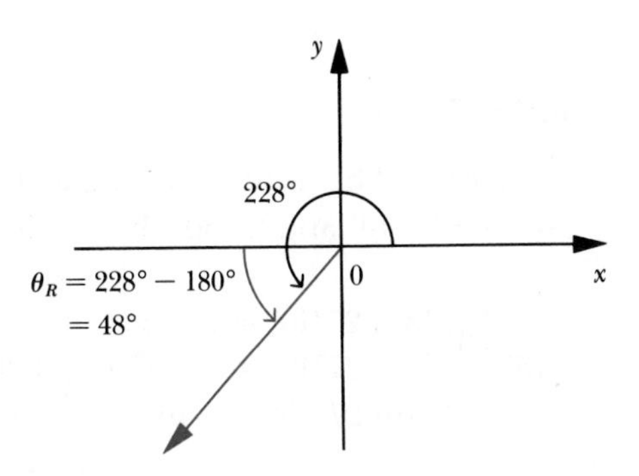

Table I does not contain entries for all values between 0° and 90°. For example, sin 24°15′ and cos 24°15′ cannot be found in Table I. However, as illustrated in the following examples, linear interpolation (see Appendix A) can be used to determine some reasonably accurate approximations.

EXAMPLES

Use Table I and linear interpolation to approximate each of the following values.

1 cos 31°24′

SOLUTION. The reference angle θ_R is 31°24′ (Figure 15); 31°24′ is between 31°20′ and 31°30′; by using Table I, we get

$$10'\left[\begin{array}{l} 4'\left[\begin{array}{l}\cos 31°20' = 0.8542 \\ \cos 31°24' = \quad ?\end{array}\right] d \\ \quad \cos 31°30' = 0.8526\end{array}\right] 0.0016$$

Hence,

$$\frac{d}{0.0016} = \frac{4}{10}$$

so that $d = 0.0006$ (correct to four places) and

$$\cos 31°24' = \cos 31°20' - 0.0006$$
$$= 0.8542 - 0.0006 = 0.8536$$

Figure 15

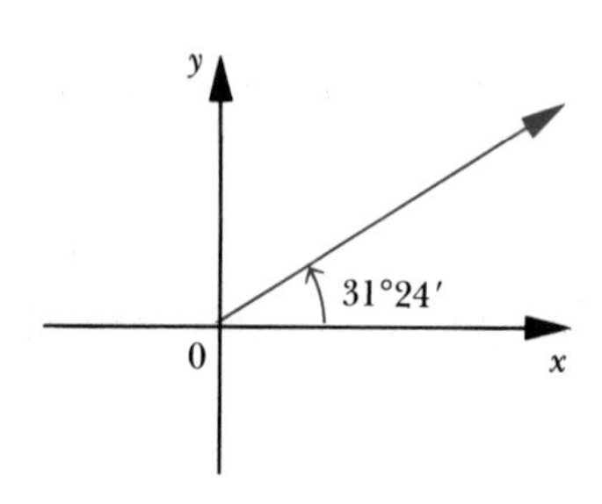

2 sin (−27°43′)

SOLUTION. The reference angle $\theta_R = 27°43'$ (Figure 16). 27°43′ is between 27°40′ and 27°50′; hence, by linear interpolation,

$$10' \left[3' \left[\begin{matrix} \sin 27°40' = 0.4643 \\ \sin 27°43' = \quad ? \end{matrix} \right] d \atop \sin 27°50' = 0.4669 \right] 0.0026$$

Therefore,

$$\frac{d}{0.0026} = \frac{3}{10}$$

so

$$d = 0.3(0.0026) = 0.00078 = 0.0008$$

Hence,

$$\sin 27°43' = 0.4643 + 0.0008 = 0.4651$$

but since −27°43′ is in quadrant IV where the sine is a negative number, we have

$$\sin(-27°43') = -0.4651$$

Figure 16

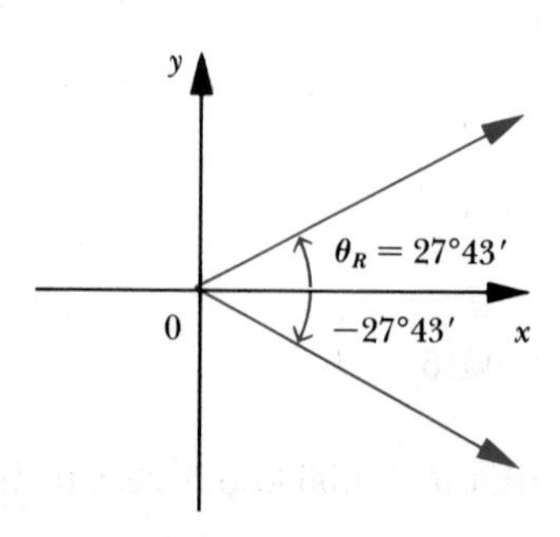

3 tan (200°25′)

SOLUTION. An angle of 200°25′ in standard position has its terminal side in quadrant III. The reference angle is 20°25′ (Figure 17). Since the tangent is positive in quadrant III, we have tan 200°25′ = tan 20°25′. But tan 20°25′ can be approximated by linear interpolation as follows:

$$10' \left[5' \begin{bmatrix} \tan 20°20' = 0.3706 \\ \tan 20°25' = \quad ? \end{bmatrix} d \atop \tan 20°30' = 0.3739 \right] 0.0033$$

Figure 17

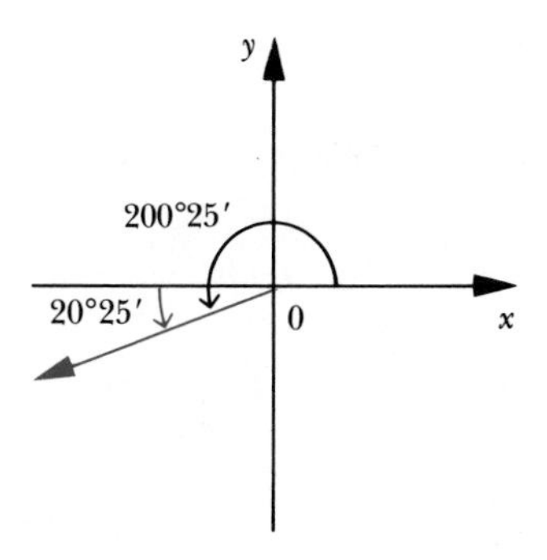

Therefore,

$$\frac{d}{0.0033} = \frac{5}{10}$$

so

$$d = 0.0017$$

Hence,

$$\begin{aligned} \tan 20°25' &= 0.3706 + 0.0017 \\ &= 0.3723 \end{aligned}$$

4 sec (−118°42′)

SOLUTION. Since −118°42′ is coterminal with 241°18′, sec (−118°42′) is negative and the reference angle $\theta_R = 241°18' - 180° = 61°18'$ (Figure 18). Thus sec (−118°42′) = −sec 61°18′. Using Table I and linear interpolation.

$$10' \left[8' \begin{bmatrix} \sec 61°10' = 2.074 \\ \sec 61°18' = \quad ? \end{bmatrix} d \atop \sec 61°20' = 2.085 \right] 0.011$$

Hence,

$$\frac{d}{0.011} = \frac{8}{10}$$

so

$$d = 0.009$$

and

$$\begin{aligned} \sec(-118°42') = -\sec 61°18' &= -(2.074 + 0.009) \\ &= -2.083 \end{aligned}$$

Figure 18

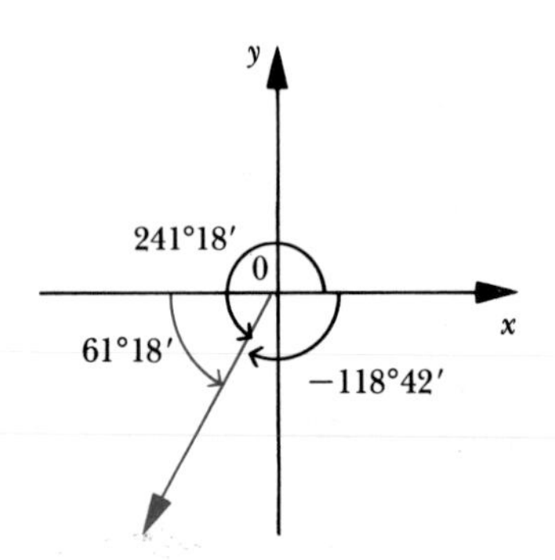

In summary, the steps for finding $T(\theta)$, where θ is any nonquadrantal angle and T is any trigonometric function, are as follows:

1. Determine the quadrant in which the terminal side of θ lies.
2. Find the reference angle θ_R, an acute angle formed by the terminal side of θ and the x axis.
3. Use Table I and linear interpolation if necessary to find $|T(\theta)| = T(\theta_R)$.

4 Adjust the sign of $T(\theta)$ according to the quadrant that contains the terminal side of θ.

PROBLEM SET 4

1 Find an angle θ coterminal with each given angle such that $0 \leq \theta < 360°$.

a) 910° b) 1027°
c) −700° d) 2727°
e) 1433° f) −1111°
g) −3000° h) −875°

2 Find the reference angle θ_R for each angle in Problem 1.

3 Express each of the following trigonometric functions in terms of the same function of the reference angle.

a) sin (−70°) b) cos 320° c) tan 290°
d) sec 200° e) cos (−160°) f) cot (−195°)
g) csc 315° h) cos (−135°) i) cot 170°
j) sin 1045° k) cot 3205° l) csc 623°
m) cos 798° n) tan 1375° o) sec (−495°)

4 Use Table I to find each of the values of the trigonometric functions in Problem 3.

5 Use Table I, and linear interpolation if necessary, to find each of the following values.

a) sin 135°20′ b) cos 125°34′ c) tan 185°47′
d) sec 127°53′ e) csc 241°20′ f) tan 160°38′
g) cot 201°27′ h) sin 333°28′ i) csc 241°52′
j) sin 237°15′ k) cos 175°55′ l) cot 326°15′

6 What conclusion can you obtain when you compare the value of θ and sin θ if θ is in radian measure and θ is close to zero?

REVIEW PROBLEM SET

1 Find the radian measure for each of the following degree measures.

a) −30° b) 810° c) 80° d) −355°
e) 27° f) 765° g) 65° h) 570°
i) −310° j) −50°

2 Find the degree measure for each of the following radian measures.

a) $-11\pi/6$ b) $17\pi/3$ c) $35\pi/3$ d) $-11\pi/3$
e) $25\pi/2$ f) $17\pi/4$ g) $-13\pi/4$ h) $-7\pi/18$
i) $18\pi/5$ j) $51\pi/4$

3 Find the area and arc length of the sector of the circle with the given radius r and given subtended central angle t.

a) $r = 10$ feet and $t = \pi/5$ b) $r = 7$ inches and $t = 11\pi/14$
c) $r = 20$ inches and $t = 60°$ d) $r = 5$ feet and $t = 135°$

4 Find the measure (in radians) of the angles that subtend the given arc.

a) $r = 5$ inches and $s = 7.3$ inches
b) $r = 11.5$ feet and $s = 4.3$ feet
c) $r = 12$ inches and $s = 3\pi$ inches
d) $r = 15$ feet and $s = 20\pi$ feet

5 Sketch the following angles in standard position and give the quadrant that contains the terminal side of each angle.

a) $\pi/6$ b) $23\pi/24$ c) $210°$ d) $13\pi/6$
e) $-5\pi/3$ f) $-300°$ g) $-17\pi/3$ h) $13\pi/6$
i) $-13\pi/8$ j) $500°$

6 Evaluate the trigonometric functions of θ in standard position if the terminal side contains the following points.

a) $(5, 12)$ b) $(-3, 5)$ c) $(\frac{1}{5}, 2\sqrt{6}/5)$
d) $(-7, -6)$ e) $(2, -3)$ f) $(\sqrt{3}/3, \frac{2}{3})$

7 Determine the quadrant in which the terminal side of θ lies if θ is in standard position and if

a) $\sin \theta > 0$ and $\cos \theta < 0$ b) $\tan \theta < 0$ and $\sec \theta > 0$
c) $\sin \theta < 0$ and $\cot \theta > 0$ d) $\csc \theta < 0$ and $\tan \theta < 0$
e) $\cos \theta < 0$ and $\tan \theta > 0$ f) $\sin \theta < 0$ and $\sec \theta > 0$

8 Evaluate the remaining trigonometric functions of θ, where θ is in standard position and the terminal side of θ is in the given quadrant.

a) $\sin \theta = -\frac{5}{13}$, quadrant IV b) $\csc \theta = -\sqrt{2}$, quadrant IV
c) $\tan \theta = \frac{3}{2}$, quadrant III d) $\sec \theta = 2$, quadrant I
e) $\sin \theta = -\frac{12}{13}$, quadrant III f) $\cos \theta = \frac{12}{13}$, quadrant IV

9 Express each of the following trigonometric functions in terms of the same function of the reference angle.

a) $\sin 310°$ b) $\cot 700°$ c) $\sec(-250°)$
d) $\cos 1000°$ e) $\csc(-400°)$ f) $\tan 800°$

10 Use Table I to find the values of the functions in Problem 9.

11 Use Table I and linear interpolation to find the values of each of the following trigonometric functions.

a) $\sin 63°37'$	b) $\cos 223°43'$	c) $\tan 312°18'$
d) $\csc 117°25'$	e) $\cot 52°56'$	f) $\sec 85°12'$
g) $\cos(-117°47')$	h) $\tan 265°15'$	i) $\cot 71°12'$
j) $\sec 228°13'$	k) $\csc 16°45'$	l) $\cos 389°25'$

CHAPTER 3

Trigonometric Functions Defined on Real Numbers

CHAPTER 3

Trigonometric Functions Defined on Real Numbers

1 Introduction

In Chapter 2 we discussed trigonometric functions whose domains are sets of angles. In this chapter we shall investigate trigonometric functions whose domains are sets of real numbers instead of angles. These functions, called *circular functions*, are used to study periodic phenomena such as the phases of the moon, alternating current, sound waves, and business cycles. We shall begin the discussion by defining circular functions.

2 Circular Functions

Let us consider the position of a point P on a *unit circle*—a circle of radius 1 with center (0, 0) (Figure 1). As we know, its equation is

Figure 1

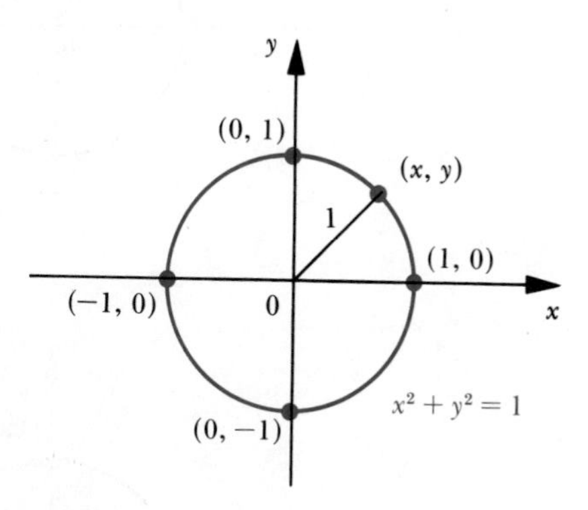

$x^2 + y^2 = 1$ (see Section 5, Chapter 1) and its circumference 2π. Let t be any real number. Locate the point P on the unit circle by measuring t units along the circumference of the unit circle starting from (1, 0). (Let us agree to measure counterclockwise if t is positive and clockwise by the amount $|t|$ if t is negative.)

Notice that if $t = 0$, then $P = (1, 0)$, and as t increases, the position of P moves counterclockwise around the circle, so that by the time $t = 2\pi$, $P = (1, 0)$ again, having moved once around the circle (Figure 2).

Figure 2

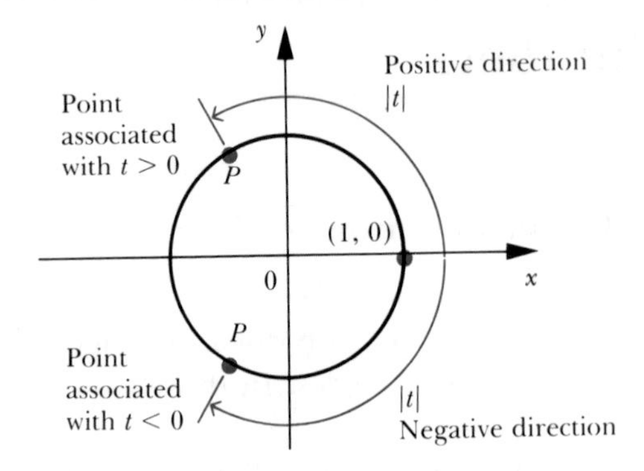

Since every point on the circle $x^2 + y^2 = 1$ has associated with it an ordered pair (x, y) of real numbers as coordinates, we may say that the position of the point P defines a function W. With each real number t, we associate an ordered pair of real numbers (x, y), the coordinates of the point P (Figure 3). Abbreviating, we may write $W(t) = (x, y)$.

Figure 3

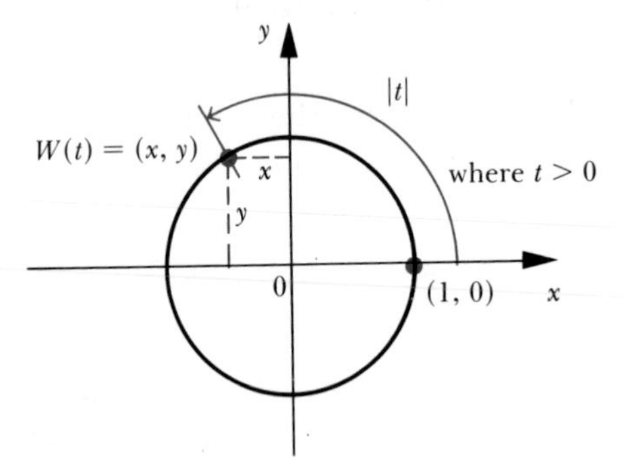

Figure 4

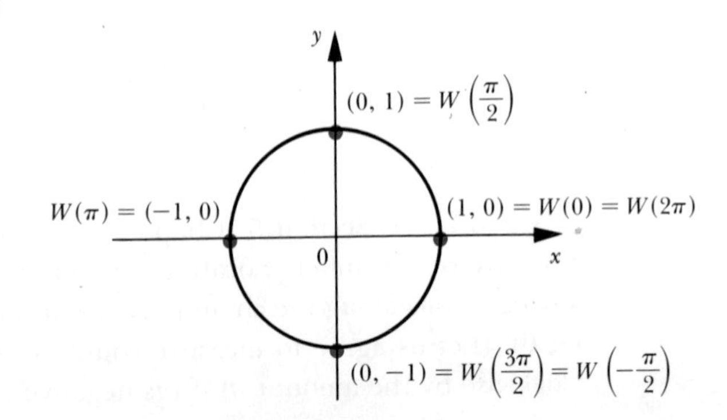

For example, we see in Figure 4 that $W(0) = (1, 0)$, $W(\pi/2) = (0, 1)$, $W(\pi) = (-1, 0)$, $W(3\pi/2) = (0, -1)$, $W(-\pi/2) = (0, -1)$, and $W(2\pi) = (1, 0)$.

Since it is difficult to develop the properties of a function whose range is a set of ordered pairs rather than a set of single numbers, we shall use the function W to construct other functions. These functions are defined as follows.

Definition—The Circular Functions

Assume that $W(t) = (x, y)$. Then

$\cos t = x$	$\cot t = \dfrac{x}{y}$ (if $y \neq 0$)
$\sin t = y$	$\sec t = \dfrac{1}{x}$ (if $x \neq 0$)
$\tan t = \dfrac{y}{x}$ (if $x \neq 0$)	$\csc t = \dfrac{1}{y}$ (if $y \neq 0$)

Notice that the cosine and sine functions are constructed by "separating" the ordered pairs in the range of the function W (Figure 5).

Figure 5

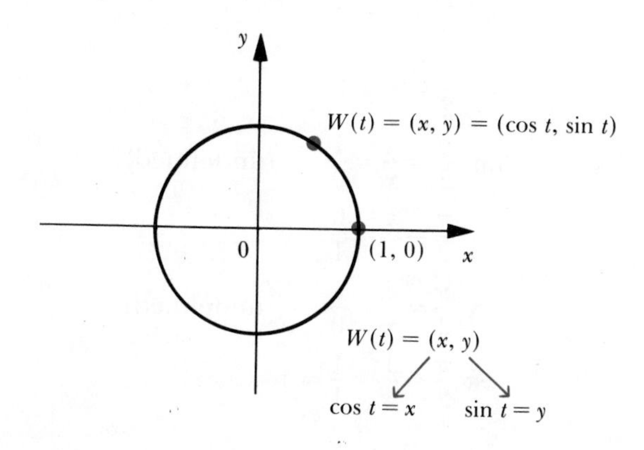

EXAMPLES

Determine the values of the six circular functions for the given value of t.

1 $t = 0$

SOLUTION. Since $W(0) = (1, 0)$ (Figure 6), we have

$\cos 0 = x = 1$

$\sin 0 = y = 0$

$$\tan 0 = \frac{y}{x} = \frac{0}{1} = 0$$

$$\cot 0 = \frac{x}{y} = \frac{1}{0} \qquad \text{(undefined)}$$

$$\sec 0 = \frac{1}{x} = \frac{1}{1} = 1$$

$$\csc 0 = \frac{1}{y} = \frac{1}{0} \qquad \text{(undefined)}$$

Figure 6

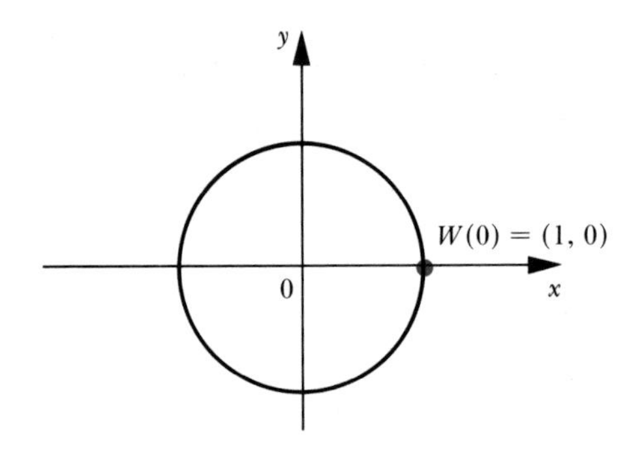

2 $t = \pi/2$

SOLUTION. Since $W(\pi/2) = (0, 1)$ (Figure 7), we have

$$\cos \frac{\pi}{2} = x = 0$$

$$\sin \frac{\pi}{2} = y = 1$$

$$\tan \frac{\pi}{2} = \frac{y}{x} = \frac{1}{0} \qquad \text{(undefined)}$$

$$\cot \frac{\pi}{2} = \frac{x}{y} = \frac{0}{1} = 0$$

$$\sec \frac{\pi}{2} = \frac{1}{x} = \frac{1}{0} \qquad \text{(undefined)}$$

$$\csc \frac{\pi}{2} = \frac{1}{y} = \frac{1}{1} = 1$$

Figure 7

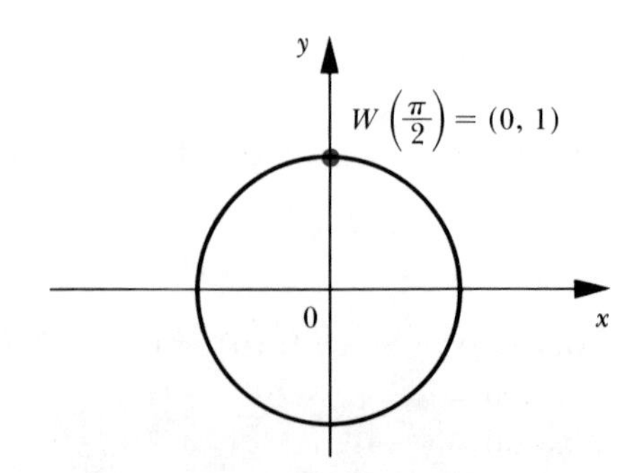

3 Find the values of the six circular functions of t if the coordinates of $W(t)$ are given by $(1/\sqrt{10},\ 3/\sqrt{10})$. (The reader should check that this point is on the unit circle.)

SOLUTION. Since $W(t) = (x,\ y) = (1/\sqrt{10},\ 3/\sqrt{10})$ (Figure 8), we have

$$\cos t = x = \frac{1}{\sqrt{10}} = \frac{\sqrt{10}}{10}$$

$$\sin t = y = \frac{3}{\sqrt{10}} = \frac{3\sqrt{10}}{10}$$

$$\tan t = \frac{y}{x} = \frac{\frac{3}{\sqrt{10}}}{\frac{1}{\sqrt{10}}} = 3$$

$$\cot t = \frac{x}{y} = \frac{\frac{1}{\sqrt{10}}}{\frac{3}{\sqrt{10}}} = \frac{1}{3}$$

$$\sec t = \frac{1}{x} = \frac{1}{\frac{1}{\sqrt{10}}} = \sqrt{10}$$

$$\csc t = \frac{1}{y} = \frac{1}{\frac{3}{\sqrt{10}}} = \frac{\sqrt{10}}{3}$$

Figure 8

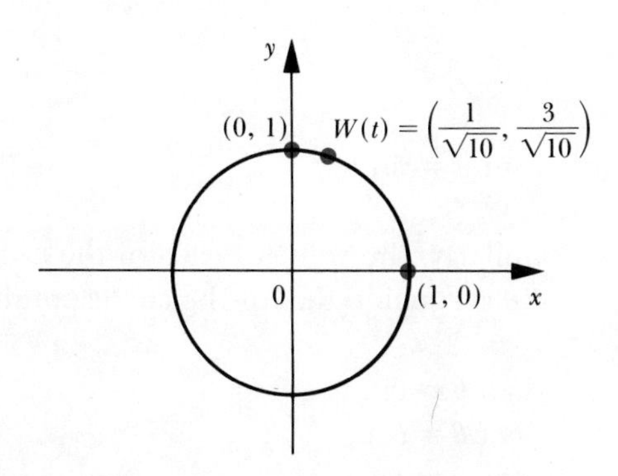

2.1 Relationships Between Trigonometric and Circular Functions

Earlier in this section we defined the six circular functions of real numbers, whose names are the same as those of the six trigonometric functions defined in Chapter 2. We now investigate the relationship between these two sets of functions.

Suppose that t is a real number and $W(t) = (x, y)$; then as we have just seen, the circular functions yield $\cos t = x$ and $\sin t = y$ (Figure 9). Now let θ be the angle between the positive x axis and the ray through the point $W(t)$ (Figure 9).

Figure 9

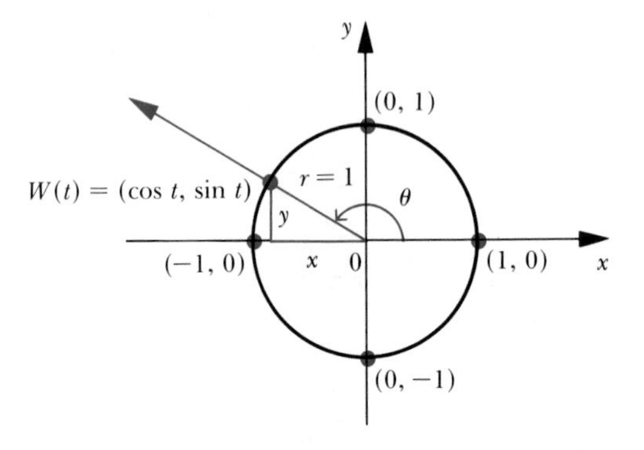

The definition of the *trigonometric* functions on θ yield $\cos \theta = x/r$ and $\sin \theta = y/r$, where $(x, y) \neq (0, 0)$ is a point on the terminal side of θ and $r = \sqrt{x^2 + y^2}$. But with (x, y) selected as above, we have $r = 1$, so that $\cos \theta = x/1 = x$ and $\sin \theta = y/1 = y$ (Figure 9). Thus $\cos \theta = \cos t$ and $\sin \theta = \sin t$. Furthermore, t equals the arc length subtended by θ. (Why?) Hence, using the definition of radian measure, t equals the measure of θ *in radians.*

Summing up: If θ is any angle, and t is its measure in radians, then

$$\cos \theta = \cos t$$

and

$$\sin \theta = \sin t$$

Similarly, the value of each of the remaining trigonometric functions of θ is equal to that of the corresponding circular function of t, that is,

$$\tan \theta = \tan t$$
$$\cot \theta = \cot t$$
$$\sec \theta = \sec t$$
$$\csc \theta = \csc t$$

In general, if T is any one of the six trigonometric functions and C is the corresponding circular function, then $T(\theta) = C(t)$, where θ is an angle with radian measure t.

EXAMPLES

Find the values of the six circular functions for the given value of t.

1 $t = \frac{\pi}{4}$

SOLUTION. Since $C(t) = T(\theta)$ and $\pi/4$ is the radian measure of angle 45° (Figure 10), we have

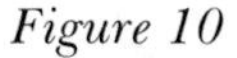
Figure 10

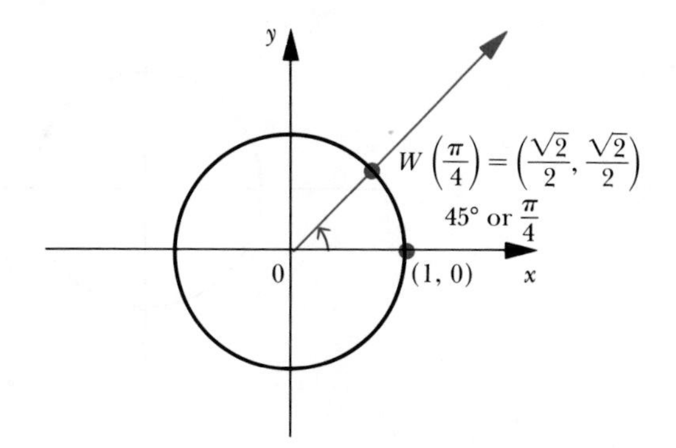

$$\sin \frac{\pi}{4} = \sin 45° = \frac{\sqrt{2}}{2}$$

$$\cos \frac{\pi}{4} = \cos 45° = \frac{\sqrt{2}}{2}$$

$$\tan \frac{\pi}{4} = \tan 45° = 1$$

$$\cot \frac{\pi}{4} = \cot 45° = 1$$

$$\sec \frac{\pi}{4} = \sec 45° = \sqrt{2}$$

$$\csc \frac{\pi}{4} = \csc 45° = \sqrt{2}$$

2 $t = \frac{\pi}{6}$

SOLUTION. $C(t) = T(\theta)$ and $\pi/6$ is the radian measure of a 30° angle (Figure 11) so that

$$\sin \frac{\pi}{6} = \sin 30° = \frac{1}{2}$$

$$\cos \frac{\pi}{6} = \cos 30° = \frac{\sqrt{3}}{2}$$

$$\tan \frac{\pi}{6} = \tan 30° = \frac{\sqrt{3}}{3}$$

$$\cot\frac{\pi}{6} = \cot 30° = \sqrt{3}$$

$$\sec\frac{\pi}{6} = \sec 30° = \frac{2}{\sqrt{3}} = \frac{2\sqrt{3}}{3}$$

$$\csc\frac{\pi}{6} = \csc 30° = 2$$

Figure 11

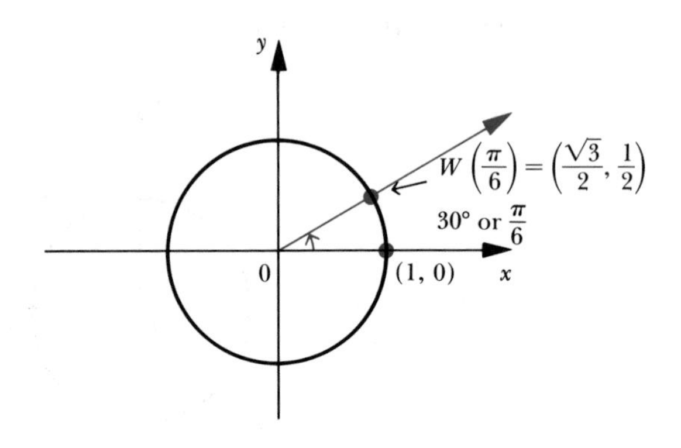

3 $t = \dfrac{\pi}{3}$

SOLUTION. $C(t) = T(\theta)$ and $\pi/3$ is the radian measure of a 60° angle (Figure 12) so that

$$\sin\frac{\pi}{3} = \sin 60° = \frac{\sqrt{3}}{2}$$

$$\cos\frac{\pi}{3} = \cos 60° = \frac{1}{2}$$

$$\tan\frac{\pi}{3} = \tan 60° = \sqrt{3}$$

$$\cot\frac{\pi}{3} = \cot 60° = \frac{\sqrt{3}}{3}$$

$$\sec\frac{\pi}{3} = \sec 60° = 2$$

$$\csc\frac{\pi}{3} = \csc 60° = \frac{2}{\sqrt{3}} = \frac{2\sqrt{3}}{3}$$

Figure 12

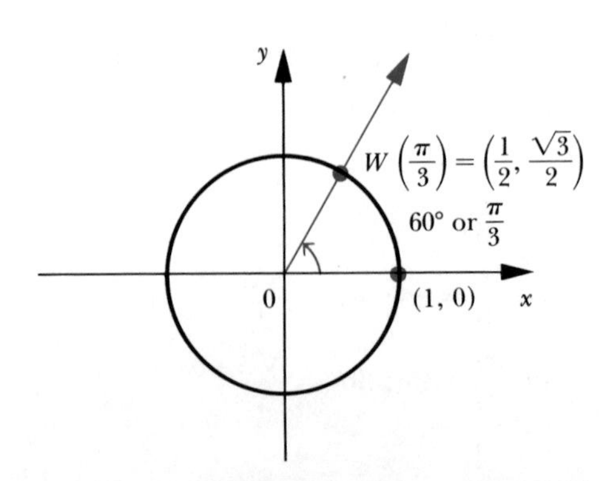

4 $t = \dfrac{3\pi}{4}$

SOLUTION. $3\pi/4$ is the radian measure of a 135° angle (Figure 13) so that

$$\sin \frac{3\pi}{4} = \sin 135° = \frac{\sqrt{2}}{2}$$
$$\cos \frac{3\pi}{4} = \cos 135° = -\frac{\sqrt{2}}{2}$$
$$\tan \frac{3\pi}{4} = \tan 135° = -1$$
$$\cot \frac{3\pi}{4} = \cot 135° = -1$$
$$\sec \frac{3\pi}{4} = \sec 135° = -\sqrt{2}$$
$$\csc \frac{3\pi}{4} = \csc 135° = \sqrt{2}$$

Figure 13

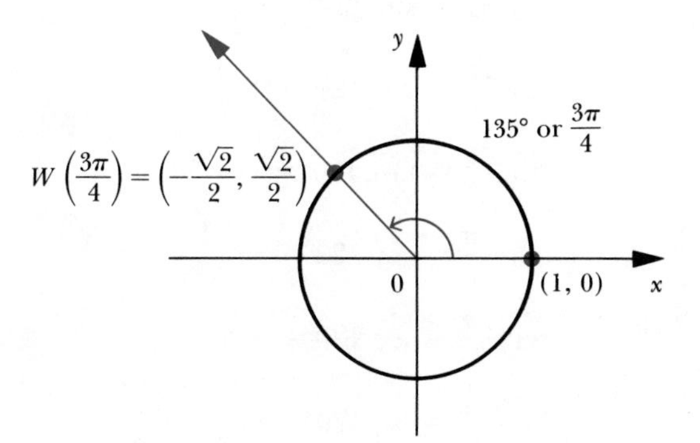

5 $t = \dfrac{5\pi}{6}$

SOLUTION. $5\pi/6$ is the radian measure of a 150° angle (Figure 14) so that

$$\sin \frac{5\pi}{6} = \sin 150° = \frac{1}{2}$$
$$\cos \frac{5\pi}{6} = \cos 150° = -\frac{\sqrt{3}}{2}$$
$$\tan \frac{5\pi}{6} = \tan 150° = -\frac{\sqrt{3}}{3}$$
$$\cot \frac{5\pi}{6} = \cot 150° = -\sqrt{3}$$
$$\sec \frac{5\pi}{6} = \sec 150° = -\frac{2}{\sqrt{3}} = -\frac{2\sqrt{3}}{3}$$
$$\csc \frac{5\pi}{6} = \csc 150° = 2$$

Figure 14

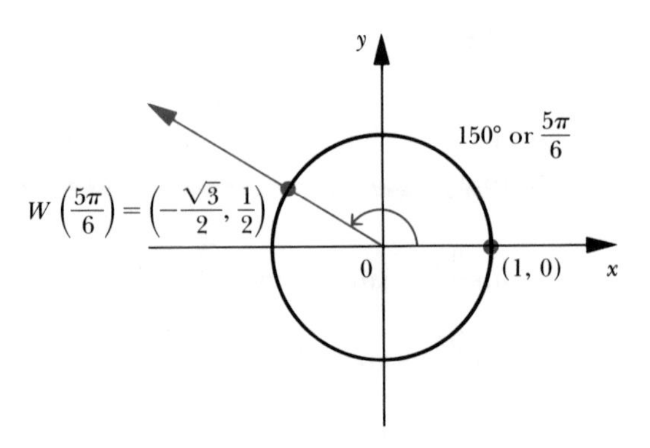

6 $t = \dfrac{2\pi}{3}$

SOLUTION. $2\pi/3$ is the radian measure of a 120° angle (Figure 15) implies that

$$\sin \frac{2\pi}{3} = \sin 120° = \frac{\sqrt{3}}{2}$$

$$\cos \frac{2\pi}{3} = \cos 120° = -\frac{1}{2}$$

$$\tan \frac{2\pi}{3} = \tan 120° = -\sqrt{3}$$

$$\cot \frac{2\pi}{3} = \cot 120° = -\frac{1}{\sqrt{3}} = -\frac{\sqrt{3}}{3}$$

$$\sec \frac{2\pi}{3} = \sec 120° = -2$$

$$\csc \frac{2\pi}{3} = \csc 120° = \frac{2}{\sqrt{3}} = \frac{2\sqrt{3}}{3}$$

Figure 15

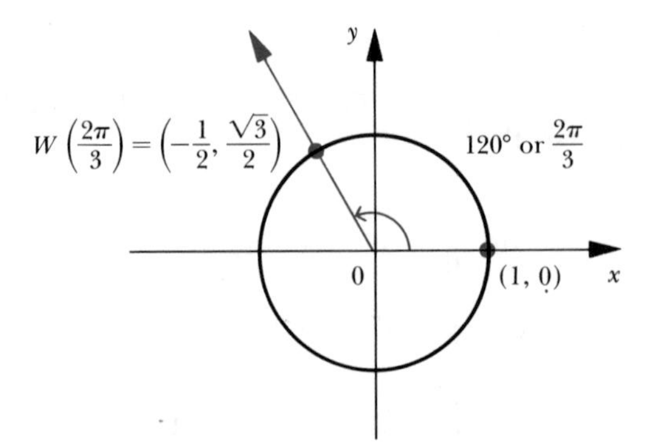

2.3 Domains of the Circular Functions

We have seen that if $W(t) = (x, y)$, then $\cos t = x$ and $\sin t = y$. Since $W(t)$ is defined for any real number t, $\cos t$ and $\sin t$ are defined for any real number; that is, the domain for both the cosine and sine functions is the set of real numbers R.

The ranges of the cosine and sine can be found by using the fact that any point (x, y) on the unit circle satisfies $-1 \leq x \leq 1$ and $-1 \leq y \leq 1$, so $\cos t = x$ has range $[-1, 1]$ and $\sin t = y$ also has range $[-1, 1]$ (Figure 16).

Figure 16

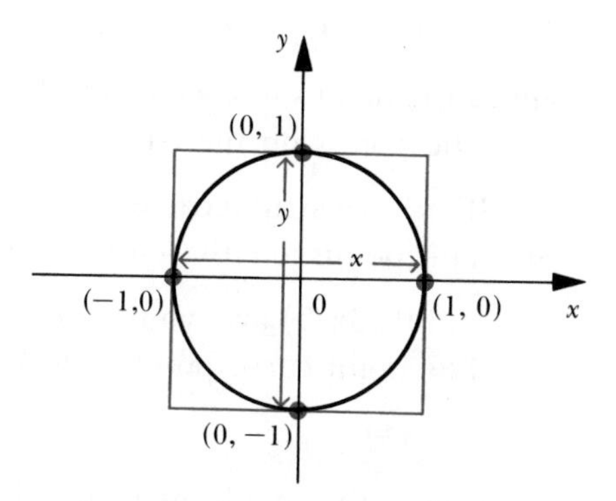

The domains of the other circular functions can be determined as follows. Let t be a real number for which $W(t) = (0, y)$; then $\tan t = y/x$ is not defined for t at that point but is defined for other possible values of $W(t)$. Hence, the domain of the tangent function is the set of real numbers t such that $x \neq 0$. The only points on the unit circle such that $x = 0$ are the points $(0, 1)$ and $(0, -1)$. These points correspond to the set of real numbers $\pi/2 + n\pi$, where n is an integer. Thus, the domain of the tangent function is

$$\{t \mid t \neq (\pi/2) + n\pi,\ n \in I\}$$

Since $\sec t = 1/x$, the domain of the secant is the same as the domain of the tangent; that is, the domain of the secant is given by

$$\{t \mid t \neq (\pi/2) + n\pi,\ n \in I\}.$$

As above, the domain of the cotangent function is the set of real numbers t such that $y \neq 0$. The only points on the unit circle such that $y = 0$ are the points $(1, 0)$ and $(-1, 0)$. These points correspond to the set of real numbers $n\pi$, where n is an integer. Thus, the domain of the cotangent function is

$$\{t \mid t \neq n\pi,\ n \in I\}$$

Since $\csc t = 1/y$, the domain of the cosecant is the same as the domain of the cotangent; namely,

$$\{t \mid t \neq n\pi,\ n \in I\}.$$

The domains of the trigonometric functions can also be expressed in terms of angles (see Problem 12, Problem Set 1).

PROBLEM SET 1

1 Fill in the blanks in each of the following cases.

a) The points on the circle $x^2 + y^2 = 1$ corresponding to $W(0)$ and $W(\pi)$ are symmetric with respect to the ________ axis.

b) The points on the circle $x^2 + y^2 = 1$ corresponding to $W(\pi/2)$ and $W(3\pi/2)$ are symmetric with respect to the ________ axis.

c) The point corresponding to $W(\pi/4)$ on the unit circle lies on the line $y =$ ________.

2 Determine the quadrant in which each of the following points lies. Use the approximation $\pi = 3.14$.

a) $W(2)$ b) $W(5)$ c) $W(7.5)$
d) $W(7.2)$ e) $W(4.7)$ f) $W(0.62)$
g) $W(-11.2)$ h) $W(-\frac{11}{3})$ i) $W(-8.9)$

3 Determine (x, y) if $W(t) = (x, y)$ for each of the following.

a) $W(\pi)$ b) $W(3\pi/2)$ c) $W(2\pi)$
d) $W(-\pi)$ e) $W(-3\pi/2)$ f) $W(-2\pi)$

4 Determine the six circular functions at the given value of t if

a) $t = \pi$ b. $t = 3\pi/2$ c) $t = -\pi/2$
d) $t = -3\pi/2$ e) $t = 2\pi$ f) $t = -2\pi$

5 Find the values of the six circular functions if the coordinates of $W(t)$ are

a) $(1/\sqrt{5}, 2/\sqrt{5})$ b) $(-2/\sqrt{13}, 3/\sqrt{13})$ c) $(-1/2, \sqrt{3}/2)$
d) $(3/\sqrt{10}, -1/\sqrt{10})$ e) $(-12/13, -5/13)$ f) $(\sqrt{2}/2, -\sqrt{2}/2)$

6 a) In each case, with t the radian measure of θ, find the values of the six trigonometric functions.

i) $t = 3\pi/4$ ii) $t = 7\pi/6$ iii) $t = -\pi/4$
iv) $t = 5\pi/4$ v) $t = 4\pi/3$ vi) $t = 5\pi/3$
vii) $t = 7\pi/4$ viii) $t = 11\pi/6$

b) In each case above, with the real number t as given, find the values of the six circular functions.

c) What is the connection between parts (a) and (b)?

7 Explain why the domain of cotangent and cosecant cannot contain even multiples of $\pi/2$.

8 Explain why the domain of tangent and secant cannot contain odd multiples of $\pi/2$.

9 Make a table showing the domain of each of the circular functions.

10 Can we solve the equation $\sin t = 5$ for t? Explain. How about the equation $\cos t = 3$?

11 In the following table, two of the trigonometric functions of an angle have the signs as indicated. Indicate the quadrant in which the terminal side of the angle θ lies and the signs of the other trigonometric functions.

	(a)	(b)	(c)	(d)	(e)	(f)	(g)	(h)	(i)	(j)
Quadrant	II									
$\sin\theta$	+	+				−				
$\cos\theta$	−		+						+	−
$\tan\theta$	−	+	−	+	+		+			+
$\cot\theta$	−								+	
$\sec\theta$	−				−	+		+		
$\csc\theta$	+			−			+	−		

12 Use the relation $C(t) = T(\theta)$ to express the domains of the six trigonometric functions in terms of angles in degree measurement.

3 Some Properties of the Circular Functions

Using $C(t) = T(\theta)$, the following properties derived through the use of the definition of circular functions will also apply to the corresponding trigonometric functions. Although a real number t is used to represent the elements of the domains of the circular functions, the results are also true if t is replaced by an angle θ. We should remember that when we write $C(t) = T(\theta)$, we are assuming that t equals the measure of θ in radians.

3.1 Fundamental Properties of Circular Functions

In this section we shall derive some properties using the definitions of the circular functions.

THEOREM 1

a) $\tan t = \dfrac{\sin t}{\cos t}$ if $\cos t \neq 0$

b) $\cot t = \dfrac{\cos t}{\sin t}$ if $\sin t \neq 0$

c) $\csc t = \dfrac{1}{\sin t}$ if $\sin t \neq 0$

d) $\sec t = \dfrac{1}{\cos t}$ if $\cos t \neq 0$

e) $\cot t = \dfrac{1}{\tan t}$ if $\tan t \neq 0$

PROOF. Using the definition of circular functions we have

a) $\tan t = \dfrac{y}{x} = \dfrac{\sin t}{\cos t}$

b) $\cot t = \dfrac{x}{y} = \dfrac{\cos t}{\sin t}$

c) $\csc t = \dfrac{1}{y} = \dfrac{1}{\sin t}$

d) $\sec t = \dfrac{1}{x} = \dfrac{1}{\cos t}$

e) $\cot t = \dfrac{x}{y} = \dfrac{1}{y/x} = \dfrac{1}{\tan t}$

THEOREM 2

For any real number t,

a) $\cos^2 t + \sin^2 t = 1$

b) $\cos(-t) = \cos t$ (the cosine is an even function)

c) $\sin(-t) = -\sin t$ (the sine is an odd function)

PROOF OF (a). Assume that $W(t) = (x, y)$, so that $\cos t = x$ and $\sin t = y$. Since (x, y) is on the unit circle, we have $x^2 + y^2 = 1$. By substituting $\cos t$ for x and $\sin t$ for y, we have $(\cos t)^2 + (\sin t)^2 = 1$ [the power notation for $(\cos t)^n$ is $\cos^n t$ and for $(\sin t)^n$ t], so that

$$\cos^2 t + \sin^2 t = 1$$

PROOF OF (b) AND (c). Although we will illustrate this result for a value of t that displays $W(t)$ in quadrant I, it is important to remember that the argument can be used for any real number t. Assume that $W(t) = (x, y)$. Then, by definition, $W(t) = (x, y) = (\cos t, \sin t)$

and $W(-t) = (x, -y) = (\cos(-t), \sin(-t))$ (Figure 1). Hence, by symmetry of the circle, $\cos t = x = \cos(-t)$ and $\sin(-t) = -y = -\sin t$.

Figure 1

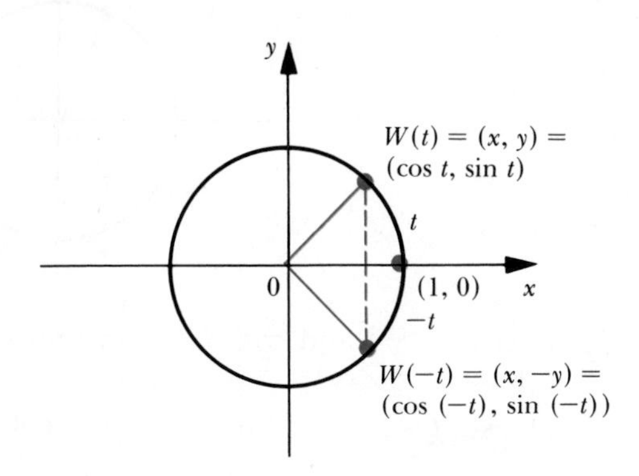

(NOTE: Hereafter we do not emphasize the distinction between trigonometric functions of θ and circular functions of t. This follows because of the relationship established in Section 2.1.)

EXAMPLES

1 If $\sin\theta = \frac{3}{10}$ and θ is between 0° and 90°, find $\cos\theta$.

SOLUTION. Using $\cos^2\theta + \sin^2\theta = 1$, we have $\cos^2\theta = 1 - \sin^2\theta$, so $\cos\theta = \pm\sqrt{1-\sin^2\theta}$. Since angle θ lies in the first quadrant (Figure 2), $\cos\theta > 0$, so that

$$\cos\theta = \sqrt{1-\sin^2\theta} = \sqrt{1-\frac{9}{100}} = \sqrt{\frac{91}{100}} = \frac{\sqrt{91}}{10}$$

Figure 2

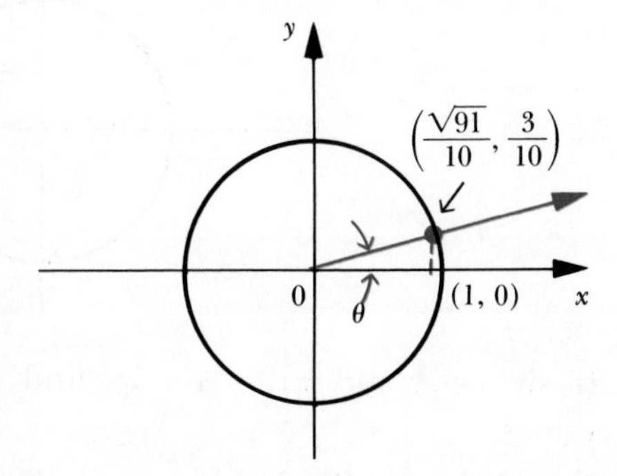

2 If $\cos\theta = -\frac{5}{13}$ and θ is between 90° and 180°, find $\sin\theta$.

SOLUTION. Using $\cos^2\theta + \sin^2\theta = 1$, we have $\sin\theta = \pm\sqrt{1-\cos^2\theta}$. Since angle θ lies in the second quadrant (Figure 3), $\sin\theta > 0$, so that

$$\sin\theta = \sqrt{1-\cos^2\theta} = \sqrt{1-(-\tfrac{5}{13})^2} = \sqrt{\tfrac{144}{169}} = \tfrac{12}{13}$$

Figure 3

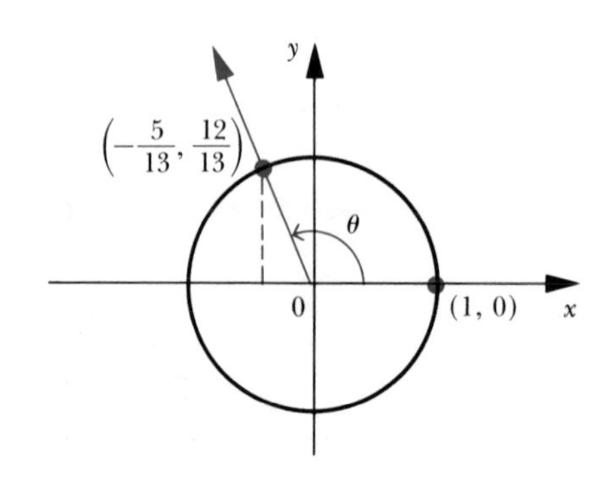

3 If $\cos t = -\frac{4}{5}$ and $\pi/2 < t < \pi$, find each of the following functions values.

a) $\sin t$ b) $\tan t$ c) $\cot t$ d) $\sec t$ e) $\csc t$

SOLUTION. If $\cos t = -\frac{4}{5}$ and $\pi/2 < t < \pi$ (Figure 4), then

a) $\sin t = \sqrt{1 - \cos^2 t} = \sqrt{1 - \frac{16}{25}} = \frac{3}{5}$

b) $\tan t = \dfrac{\sin t}{\cos t} = \dfrac{\frac{3}{5}}{-\frac{4}{5}} = -\dfrac{3}{4}$

c) $\cot t = \dfrac{1}{\tan t} = \dfrac{1}{-\frac{3}{4}} = -\dfrac{4}{3}$

d) $\sec t = \dfrac{1}{\cos t} = \dfrac{1}{-\frac{4}{5}} = -\dfrac{5}{4}$

e) $\csc t = \dfrac{1}{\sin t} = \dfrac{1}{\frac{3}{5}} = \dfrac{5}{3}$

Figure 4

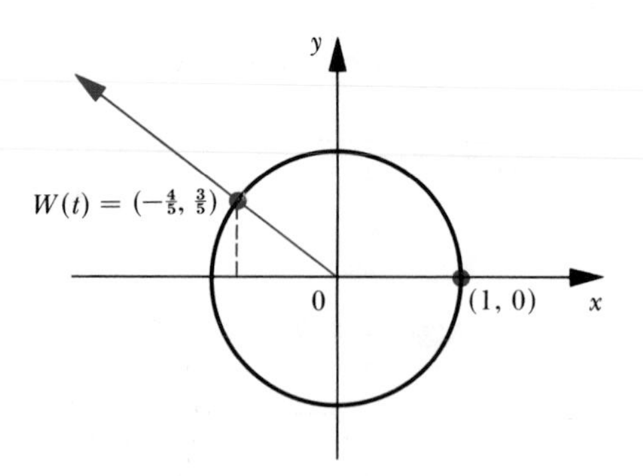

4 If $\sin t = \frac{3}{5}$ and $\pi/2 < t < \pi$, find each of the following function values.

a) $\cos t$ b) $\tan t$ c) $\cot t$ d) $\sec t$ e) $\csc t$

SOLUTION. Since $\cos^2 t + \sin^2 t = 1$, $\cos t = \pm\sqrt{1 - \sin^2 t}$. But, since $W(t)$ is in quadrant II (Figure 5), we have

a) $\cos t = -\sqrt{1 - \sin^2 t} = -\sqrt{1 - \frac{9}{25}} = -\frac{4}{5}$

b) $\tan t = \dfrac{\sin t}{\cos t} = \dfrac{\frac{3}{5}}{-\frac{4}{5}} = -\dfrac{3}{4}$

c) $\cot t = \dfrac{\cos t}{\sin t} = \dfrac{-\frac{4}{5}}{\frac{3}{5}} = -\dfrac{4}{3}$

d) $\sec t = \dfrac{1}{\cos t} = \dfrac{1}{-\frac{4}{5}} = -\dfrac{5}{4}$

e) $\csc t = \dfrac{1}{\sin t} = \dfrac{1}{\frac{3}{5}} = \dfrac{5}{3}$

Figure 5

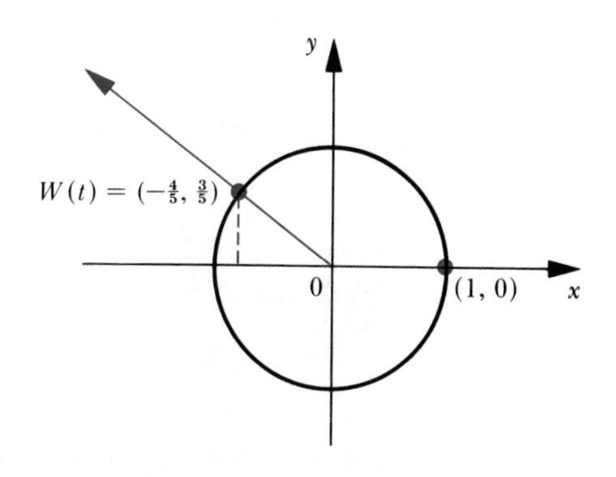

5 If $\cos \theta = -\frac{4}{5}$ and the terminal side of θ lies in the third quadrant, find
a) $\sin \theta$ b) $\tan \theta$ c) $\cot \theta$ d) $\sec \theta$ e) $\csc \theta$

SOLUTION. Since θ lies in quadrant III (Figure 6), we have

a) $\sin \theta = -\sqrt{1 - \cos^2 \theta} = -\sqrt{1 - (\frac{4}{5})^2} = -\frac{3}{5}$

b) $\tan \theta = \dfrac{\sin \theta}{\cos \theta} = \dfrac{-\frac{3}{5}}{-\frac{4}{5}} = \dfrac{3}{4}$

c) $\cot \theta = \dfrac{\cos \theta}{\sin \theta} = \dfrac{-\frac{4}{5}}{-\frac{3}{5}} = \dfrac{4}{3}$

d) $\sec \theta = \dfrac{1}{\cos \theta} = \dfrac{1}{-\frac{4}{5}} = -\dfrac{5}{4}$

e) $\csc \theta = \dfrac{1}{\sin \theta} = \dfrac{1}{-\frac{3}{5}} = -\dfrac{5}{3}$

Figure 6

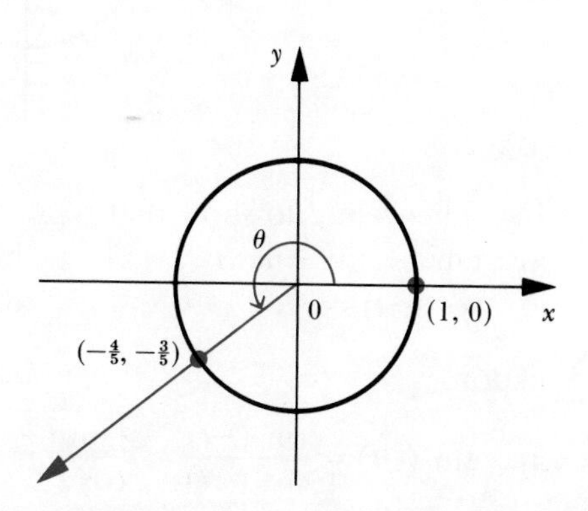

6 If $\cos\theta = 1/\sqrt{2}$ and the terminal side of θ is in quadrant IV, find each of the following.

a) $\sin\theta$ b) $\tan\theta$ c) $\cot\theta$ d) $\sec\theta$ e) $\csc\theta$

SOLUTION

a) Using $\sin^2\theta + \cos^2\theta = 1$ and the fact that $\sin\theta < 0$ in quadrant IV (Figure 7), we have

$$\sin\theta = -\sqrt{1-\cos^2\theta} = -\sqrt{1-\left(\frac{1}{\sqrt{2}}\right)^2} = -\sqrt{1-\frac{1}{2}} = -\frac{1}{\sqrt{2}}$$

b) $$\tan\theta = \frac{\sin\theta}{\cos\theta} = \frac{-\frac{1}{\sqrt{2}}}{\frac{1}{\sqrt{2}}} = -1$$

c) $$\cot\theta = \frac{\cos\theta}{\sin\theta} = \frac{\frac{1}{\sqrt{2}}}{-\frac{1}{\sqrt{2}}} = -1$$

d) $$\sec\theta = \frac{1}{\cos\theta} = \frac{1}{\frac{1}{\sqrt{2}}} = \sqrt{2}$$

e) $$\csc\theta = \frac{1}{\sec\theta} = \frac{1}{-\frac{1}{\sqrt{2}}} = -\sqrt{2}$$

Figure 7

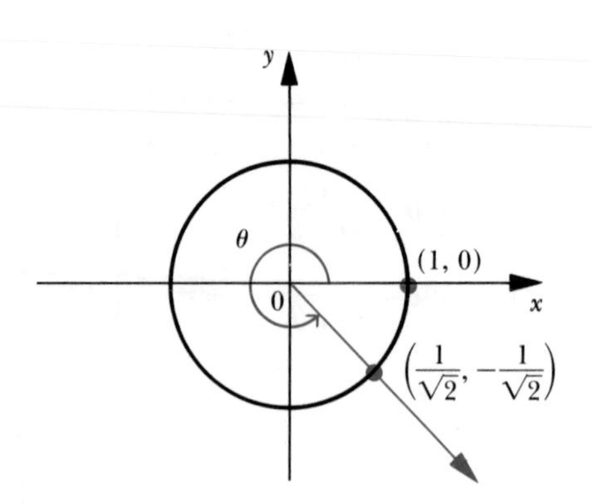

7 Use Theorem 2 to show that

a) $\tan(-t) = -\tan t$ b) $\cot(-t) = -\cot t$

c) $\sec(-t) = \sec t$ d) $\csc(-t) = -\csc t$

PROOF

a) $$\tan(-t) = \frac{\sin(-t)}{\cos(-t)} = \frac{-\sin t}{\cos t} = -\tan t$$

b) $\cot(-t) = \dfrac{\cos(-t)}{\sin(-t)} = \dfrac{\cos t}{-\sin t} = -\cot t$

c) $\sec(-t) = \dfrac{1}{\cos(-t)} = \dfrac{1}{\cos t} = \sec t$

d) $\csc(-t) = \dfrac{1}{\sin(-t)} = \dfrac{1}{-\sin t} = -\csc t$

3.2 Periodicity

Since the circumference of the unit circle is 2π, we have $W(t) = W(t+2\pi)$ for any real number t; hence, the function W associates more than one real number with the same point on the unit circle (Figure 8).

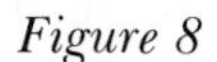
Figure 8

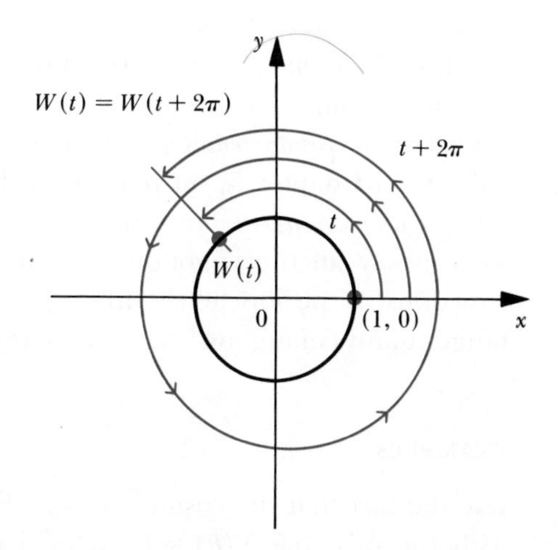

For example, $W(0) = W(2\pi) = W(4\pi) = W(6\pi) = (1, 0)$. In any case, the function values of W repeat every 2π units; that is, $W(t) = W(t+2\pi) = W(t+4\pi) = W(t+6\pi)$. Functions with this repetitive property are called *periodic functions*. In general, a function f is said to be a *periodic function* with period a, where a is a nonzero constant, if, for all x in the domain of f, $x + a$ is also in the domain of f, and $f(x+a) = f(x)$.

The smallest *positive* period a is called the *fundamental period* of f. Consequently, the function W is considered to be a periodic function with period $2\pi, 4\pi, 6\pi, \ldots$, in fact any multiple of 2π. However, since 2π is the smallest positive period of W, then 2π is considered to be the fundamental period of W. Using the fact that $W(t) = (x, y) = W(t+2\pi)$, it follows that $\cos t = x = \cos(t+2\pi)$ and $\sin t = y = \sin(t+2\pi)$. Hence, the cosine and sine functions are periodic functions of period 2π. For example, $\cos(37\pi/6) = \cos(\pi/6 + 3 \cdot 2\pi) = \cos(\pi/6)$ and $\sin(37\pi/6) = \sin(\pi/6 + 3 \cdot 2\pi) = \sin(\pi/6)$. Using the periodicity of the sine and cosine functions, we can determine the periodicity of the tangent, cotangent, secant, and cosecant as follows.

$$\tan(t+2\pi)=\frac{\sin(t+2\pi)}{\cos(t+2\pi)}=\frac{\sin t}{\cos t}=\tan t$$
$$\cot(t+2\pi)=\frac{\cos(t+2\pi)}{\sin(t+2\pi)}=\frac{\cos t}{\sin t}=\cot t$$
$$\sec(t+2\pi)=\frac{1}{\cos(t+2\pi)}=\frac{1}{\cos t}=\sec t$$
$$\csc(t+2\pi)=\frac{1}{\sin(t+2\pi)}=\frac{1}{\sin t}=\csc t$$

These derivations suggest that all the trigonometric functions are periodic functions of period 2π. But is 2π the smallest positive number for which each of these functions repeat? The answer is yes, for the sine, cosine, secant, and cosecant functions, and in this case we say that the *fundamental period of these functions is* 2π. However, we will see in Chapter 5, on page 196, that the *fundamental period of the tangent and cotangent functions is* π. Using the fact that $C(t)=T(\theta)$, we can say all the trigonometric functions of angle θ are periodic functions of period 360°, although the fundamental period of the tangent and cotangent functions is 180°.

EXAMPLES

Use the fact that the cosine and sine functions are periodic, together with the fact that $T(\theta)=C(t)$, to determine each of the following values.

1 $\cos\left(\frac{43\pi}{4}\right)$ and $\sin\left(\frac{43\pi}{4}\right)$

SOLUTION. Since the cosine and the sine are periodic functions of period 2π, we have

$$\cos\left(\frac{43\pi}{4}\right)=\cos\left(\frac{3\pi}{4}+5\cdot 2\pi\right)=\cos\left(\frac{3\pi}{4}\right)=\cos 135°=-\frac{\sqrt{2}}{2}$$
$$\sin\left(\frac{43\pi}{4}\right)=\sin\left(\frac{3\pi}{4}+5\cdot 2\pi\right)=\sin\left(\frac{3\pi}{4}\right)=\sin 135°=\frac{\sqrt{2}}{2}$$

2 $\cos\left(\frac{141\pi}{2}\right)$ and $\sin\left(\frac{141\pi}{2}\right)$

SOLUTION. Since the cosine and sine are periodic functions of period 2π, we have

$$\cos\left(\frac{141\pi}{2}\right) = \cos\left(\frac{\pi}{2} + 35 \cdot 2\pi\right) = \cos\frac{\pi}{2} = \cos 90° = 0$$

$$\sin\left(\frac{141\pi}{2}\right) = \sin\left(\frac{\pi}{2} + 35 \cdot 2\pi\right) = \sin\frac{\pi}{2} = \sin 90° = 1$$

3 $\cot\left(-\frac{35\pi}{6}\right)$ and $\tan\left(-\frac{35\pi}{6}\right)$

SOLUTION. Since the cotangent and tangent are periodic functions of period 2π, we have

$$\cot\left(-\frac{35\pi}{6}\right) = \cot\left(-\frac{35\pi}{6} + 3 \cdot 2\pi\right) = \cot\left(\frac{\pi}{6}\right) = \cot 30° = \frac{\sqrt{3}}{3}$$

$$\tan\left(-\frac{35\pi}{6}\right) = \tan\left(-\frac{35\pi}{6} + 3 \cdot 2\pi\right) = \tan\left(\frac{\pi}{6}\right) = \tan 30° = \sqrt{3}$$

4 $\csc\left(\frac{49\pi}{3}\right)$ and $\sec\left(\frac{49\pi}{3}\right)$

SOLUTION. Since the cosecant and secant are periodic functions of period 2π, we have

$$\csc\left(\frac{49\pi}{3}\right) = \csc\left(\frac{\pi}{3} + 8 \cdot 2\pi\right) = \csc\frac{\pi}{3} = \csc 60° = \frac{2\sqrt{3}}{3}$$

$$\sec\left(\frac{49\pi}{3}\right) = \sec\left(\frac{\pi}{3} + 8 \cdot 2\pi\right) = \sec\frac{\pi}{3} = \sec 60° = 2$$

5 Use the fact that the cosine and sine functions are periodic functions of period 2π to find the period of each of the following functions.
a) $f(x) = \sin 2x$ b) $f(x) = \cos 3x$ c) $f(x) = \sin x/8$

SOLUTION. Note that we shall, hereafter, often use x instead of t to denote a real number. In this usage, of course, x can be thought of as representing an arc length along the unit circle, just as t did.

a) $f(x) = \sin 2x = \sin(2x + 2\pi) = \sin 2(x + \pi) = f(x + \pi)$; consequently, $\sin 2x$ repeats every π units because the function evaluation on x has the same value as the function evaluation on $x + \pi$. Hence, $\sin 2x$ is a periodic function of period π.
b) $f(x) = \cos 3x = \cos(3x + 2\pi) = \cos 3(x + 2\pi/3) = f(x + 2\pi/3)$, so that f is a periodic function of period $2\pi/3$.
c) $f(x) = \sin(x/8) = \sin(x/8 + 2\pi) = \sin \frac{1}{8}(x + 16\pi) = f(x + 16\pi)$, so that f is a periodic function of period 16π.

PROBLEM SET 2

1 Fill in the blanks in each of the following cases.

a) If $x^2 + y^2 = 1$ and $y = x$, then $x = \cos(\pi/4) =$ ____________ and $y = \sin(\pi/4) =$ ____________.

b) Since $(\cos t, \sin t)$ is on the unit circle, for every number t we have $\cos^2 t + \sin^2 t =$ ________.

c) Since $\cos (t + 2\pi) = \cos t$ and $\sin (t + 2\pi) =$ ________ for all real numbers t, cosine and sine are called ________ functions of ________.

d) The circular functions, $\sec t$, $\csc t$, and $\cot t$ are, respectively, the reciprocals of ________, ________, and ________.

e) The fact that $\sin (-t) = -\sin t$ implies that $\csc (-t) =$ ________ and the fact that $\cos (-t) = \cos t$ implies that $\sec (-t) =$ ________.

2 a) Find the values of $\cos \theta$, $\tan \theta$, $\cot \theta$, and $\csc \theta$ if $\sin \theta = \frac{5}{13}$ and θ is an angle with terminal side in quadrant II.
b) Find the values of $\sin \theta$, $\tan \theta$, $\cot \theta$, $\sec \theta$, and $\csc \theta$ if $\cos \theta = \frac{5}{7}$ and θ is an angle with terminal side in quadrant IV.
c) Find the values of $\sin \theta$, $\csc \theta$, and $\tan \theta$ if $\cos \theta = -\frac{24}{25}$ and θ is an angle with terminal side in quadrant III.
d) Find the values of $\cos \theta$, $\tan \theta$, and $\csc \theta$ if $\sin \theta = -\frac{3}{5}$ and θ is an angle with terminal side in quadrant III.

3 Find $\sin t$ and $\csc t$ in each of the following cases.
a) If $\cos t = -\sqrt{3}/2$ and $W(t)$ is in quadrant II.
b) If $\cos t = \frac{1}{2}$ and $W(t)$ is in quadrant IV.
c) If $\cos t = -\frac{5}{13}$ and $W(t)$ is in quadrant III.

4 Find $\cos t$ and $\sec t$ in each of the following cases.
a) If $\sin t = \frac{12}{13}$ and $W(t)$ is in quadrant II.
b) If $\sin t = -\frac{2}{5}$ and $W(t)$ is in quadrant IV.
c) If $\sin t = -\frac{5}{24}$ and $W(t)$ is in quadrant III.

5 a) Find cos 1.41 given that sin 1.41 = 0.9871.
b) Find sin 1.09 given that cos 1.09 = 0.4625.

6 Find $\tan t$, $\cot t$, $\sec t$, and $\csc t$ if
a) $\sin t = \frac{4}{5}$ and $\cos t > 0$
b) $\cos t = \frac{3}{5}$ and $\sin t > 0$
c) $\sin t = -\frac{5}{13}$ and $\cos t > 0$
d) $\cos t = \frac{12}{13}$ and $\sin t < 0$
e) $\cos t = -\frac{24}{25}$ and $\sin t > 0$
f) $\cos t = -\frac{2}{5}$, $\pi/2 < t < \pi$

7 Use the fact that $C(t) = T(\theta)$ and the cosine and sine are periodic functions of period 2π, together with the fact that the cosine is an even function and the sine is an odd function to find $\cos t$ and $\sin t$ for each of the following values of t.

a) $13\pi/6$ b) $-5\pi/4$ c) -5π d) $-4\pi/3$
e) $41\pi/2$ f) $59\pi/6$ g) $-31\pi/6$ h) $67\pi/4$

8 Use the fact that $C(t) = T(\theta)$ to determine each of the following values.
a) $\tan(-4\pi/3)$ b) $\cot(-13\pi/6)$ c) $\sec(-5\pi/6)$
d) $\csc(8\pi/3)$ e) $\tan(25\pi/6)$ f) $\cot(-41\pi/4)$
g) $\sec(76\pi/3)$ h) $\csc(-91\pi/6)$

9 Use the fact that the cosine and sine are periodic functions of period 2π to find the period of each of the following functions.
a) $f(x) = \sin 5x$ b) $f(x) = \cos(x/4)$
c) $f(x) = \sin(2x/3)$ d) $f(x) = \cos(2x - 1)$
e) $f(x) = \sin(3x/4)$

10 If $W(t)$ is in quadrant IV, which of the following equations are true if $0 < t_1 < \pi/2$ and $W(t_1)$ is symmetric to $W(t)$ about the x axis? $W(t)$ about the x axis?
a) $\cos t = -\cos t_1$ b) $\cos t = \cos t_1$
c) $\sin t = -\sin t_1$ d) $\sin t = \sin t_1$

11 Which of the circular functions are even functions and which are odd functions? Explain.

12 Prove Theorem 2 for $W(t)$ in
a) quadrant II b) quadrant III c) quadrant IV

4 Evaluation of the Circular Functions

The circular functions are periodic functions of period 2π. Hence, the evaluation of any circular function can always be reduced to an evaluation of the function at a number in $[0, 2\pi)$.

For example, the evaluation of the circular functions at $41\pi/5$ can be reduced to an evaluation of the same functions at $\pi/5$ in the interval $[0, 2\pi)$, since $41\pi/5 = (\pi/5) + 8\pi = (\pi/5) + 4 \cdot 2\pi$. Thus,

$$\sin \frac{41\pi}{5} = \sin\left(\frac{\pi}{5} + 4 \cdot 2\pi\right) = \sin \frac{\pi}{5}$$

$$\cos \frac{41\pi}{5} = \cos\left(\frac{\pi}{5} + 4 \cdot 2\pi\right) = \cos \frac{\pi}{5}$$

$$\tan \frac{41\pi}{5} = \tan\left(\frac{\pi}{5} + 4 \cdot 2\pi\right) = \tan \frac{\pi}{5}$$

$$\cot \frac{41\pi}{5} = \cot\left(\frac{\pi}{5} + 4 \cdot 2\pi\right) = \cot \frac{\pi}{5}$$

$$\sec \frac{41\pi}{5} = \sec \left(\frac{\pi}{5} + 4 \cdot 2\pi\right) = \sec \frac{\pi}{5}$$
$$\csc \frac{41\pi}{5} = \csc \left(\frac{\pi}{5} + 4 \cdot 2\pi\right) = \csc \frac{\pi}{5}$$

EXAMPLES

In each of the following examples use the fact that the circular functions are even or odd functions, together with the fact that the circular functions are periodic functions of period 2π, to reduce the evaluation of the given function at the given number t to an evaluation of the same function at a number in $[0, 2\pi)$. Use 3.14 as the approximate value of π. Also sketch $W(t)$.

1 a) $\cos(-7)$ b) $\cos 81$ c) $\sin(-15)$

SOLUTION

a) $$\begin{aligned} \cos(-7) &= \cos 7 \\ &= \cos(0.72 + 6.28) \\ &= \cos 0.72 \qquad \text{(Figure 1)} \end{aligned}$$

Figure 1

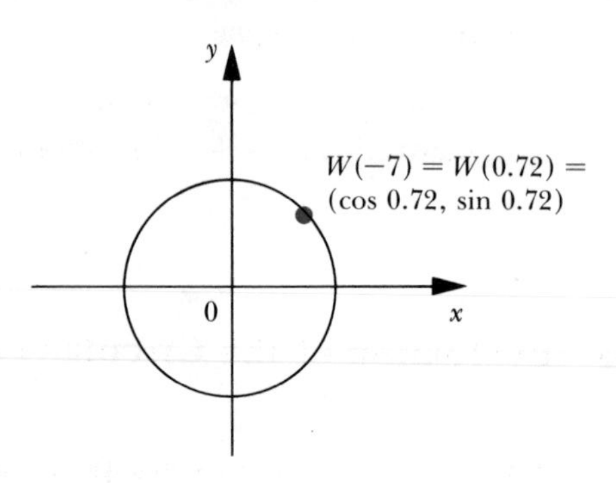

b) $$\begin{aligned} \cos 81 &= \cos [5.64 + 12(6.28)] \\ &= \cos 5.64 \qquad \text{(Figure 2)} \end{aligned}$$

Figure 2

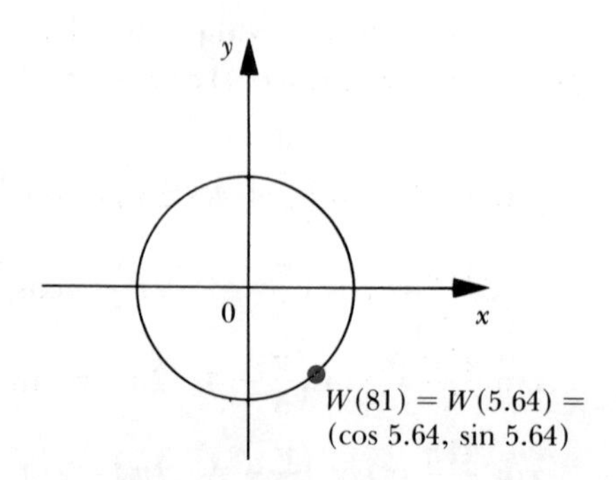

c) $\sin(-15) = -\sin(15)$
$= -\sin[2.44 + 2(6.28)]$
$= -\sin 2.44$ (Figure 3)

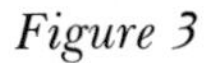

Figure 3

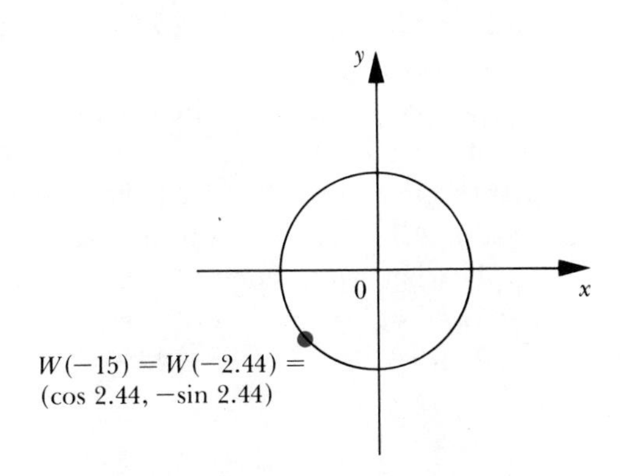

An alternate method proceeds as follows.

$$\sin(-15) = \sin[-15 + 3(6.28)]$$
$$= \sin 3.84$$

2 a) $\tan 26$ b) $\cot 64$ c) $\sec 19$ d) $\csc(-13)$

SOLUTION

a) $\tan 26 = \tan(0.88 + 4 \cdot 6.28) = \tan 0.88$
b) $\cot 64 = \cot(1.20 + 10 \cdot 6.28) = \cot 1.20$
c) $\sec 19 = \sec(0.16 + 3 \cdot 6.28) = \sec 0.16$
d) $\csc(-13) = -\csc 13 = -\csc(0.44 + 2 \cdot 6.28) = -\csc 0.44$

3 a) $\tan 7$ b) $\sin(-10)$ c) $\sec 60$
d) $\cot(-45)$ e) $\csc(-30)$ f) $\cos 21$

SOLUTION

a) $\tan 7 = \tan(0.72 + 6.28) = \tan 0.72$
b) $\sin(-10) = -\sin 10 = -\sin(3.72 + 6.28) = -\sin 3.72$
c) $\sec 60 = \sec(3.48 + 9 \cdot 6.28) = \sec 3.48$
d) $\cot(-45) = -\cot 45 = -\cot(1.04 + 7 \cdot 6.28) = -\cot 1.04$
e) $\csc(-30) = -\csc 30 = -\csc(4.88 + 4 \cdot 6.28) = -\csc 4.88$
f) $\cos 21 = \cos(2.16 + 3 \cdot 6.28) = \cos 2.16$

4.1 Reference Numbers

Since $T(\theta) = C(t)$, where t is the radian measure of θ, then the evaluation of the circular functions for any number t parallels exactly the evaluation of the trigonometric functions for any angle θ. Given any real number t, then the evaluation of the circular functions at t can be reduced to an evaluation of the functions at 0, $\pi/2$, π, $3\pi/2$, or at a *reference number* t_R in the interval $(0, \pi/2)$, with the necessary adjustment in sign. The reference number t_R can be found just as we found a reference angle for the trigonometric functions in Chapter 2, Section 5. All we do is think of t as an angle measured in radians and apply what we learned there. For example, the reference number t_R for $5\pi/3$ is $t_R = 2\pi - (5\pi/3) = \pi/3$ (Figure 4*a*) and the reference number t_R for 2.93 is $t_R = \pi - 2.93 = 3.14 - 2.93 = 0.21$ (Figure 4*b*).

Figure 4

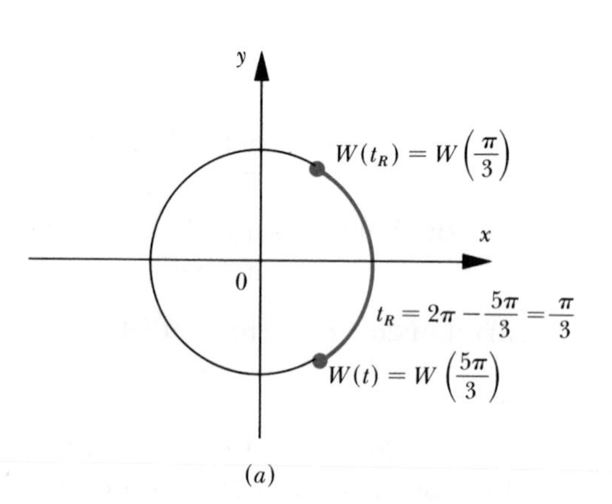

(*a*)

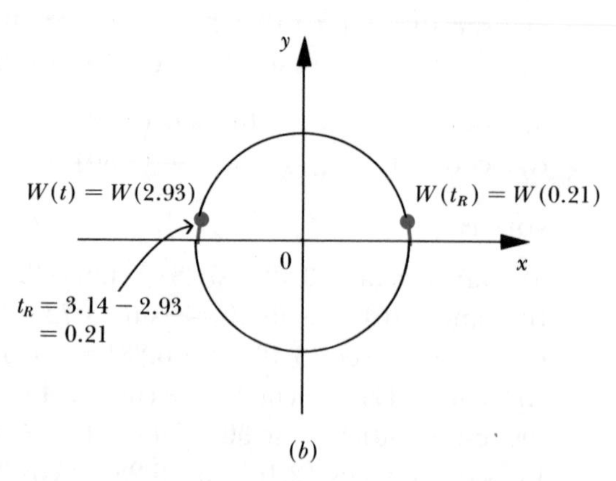

(*b*)

As with the trigonometric functions, the value of the given circular function at the given real number t is found from the value of the function at the reference number, with the necessary adjustment in sign.

EXAMPLES

In each of the following parts, reduce the evaluation of the given function to an evaluation of the same function at a reference number, that is, a number t_R such that $t_R \in (0, \pi/2)$ (use $\pi = 3.14$).

1 sin 2.5 and cos 2.5

SOLUTION. The reference number is $t_R = 3.14 - 2.50 = 0.64$ (Figure 5). Since $W(2.5)$ is in quadrant II, the sine is positive and the cosine is negative, so that

$$\sin 2.5 = \sin 0.64 \qquad \text{and} \qquad \cos 2.5 = -\cos 0.64$$

Figure 5

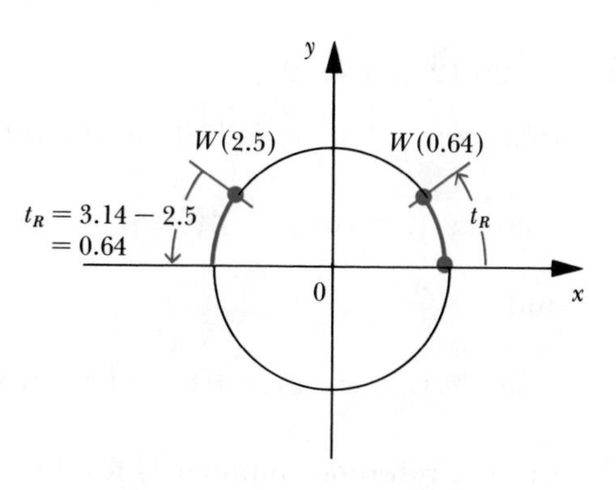

2 tan 4 and cot 4

SOLUTION. The reference number is $t_R = 4 - 3.14 = 0.86$ (Figure 6). Since $W(4)$ is in quadrant III, where the tangent is positive and the cotangent is positive, $\tan 4 = \tan 0.86$ and $\cot 4 = \cot 0.86$.

Figure 6

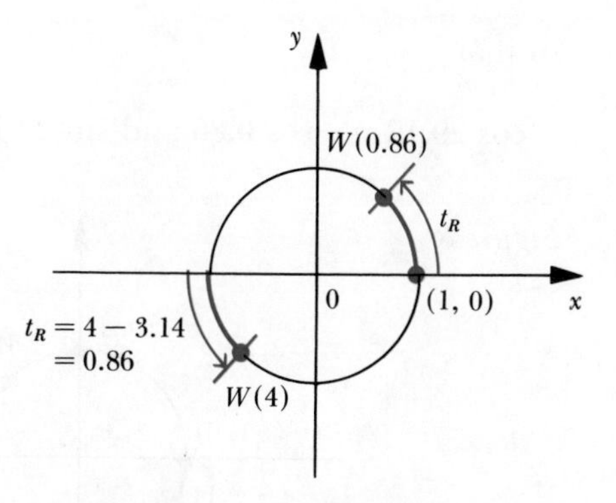

3 sec 6 and csc 6

SOLUTION. Since $t_R = 6.28 - 6 = 0.28$, $W(6)$ is in quadrant IV

(Figure 7), where the secant is positive and the cosecant is negative, so that $\sec 6 = \sec 0.28$ and $\csc 6 = -\csc 0.28$.

Figure 7

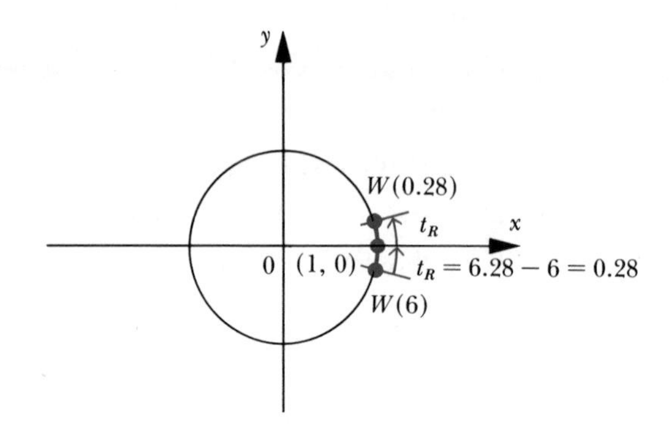

4 cos 29.12 and sin 29.12

SOLUTION. The periodicity of the cosine and sine can be used to get

$$\cos 29.12 = \cos [4 + 4(6.28)] = \cos 4$$

and

$$\sin 29.12 = \sin [4 + 4(6.28)] = \sin 4$$

But the reference number t_R for 4 is

$$t_R = 4 - 3.14 = 0.86 \qquad \text{(Figure 8)}$$

Since the cosine and sine are both negative in quadrant III,

$$\cos 4 = -\cos 0.86 \qquad \text{and} \qquad \sin 4 = -\sin 0.86$$

so that

$$\cos 29.12 = -\cos 0.86 \text{ and } \sin 29.12 = -\sin 0.86$$

Figure 8

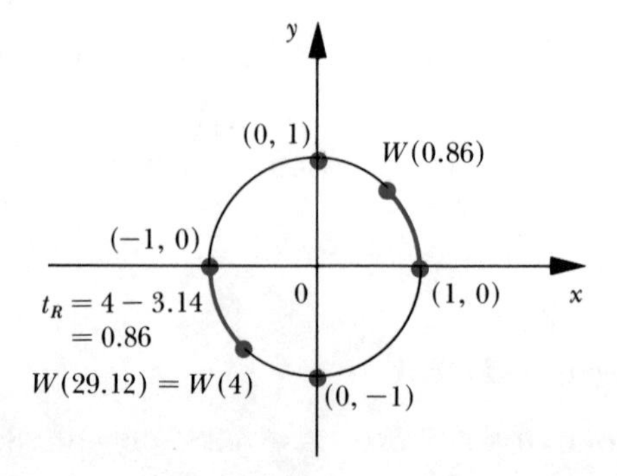

4.2 Tables

The evaluations of the circular functions parallel the evaluations of the trigonometric functions. That is, the evaluation of the circular functions for any number t can always be reduced to an evaluation at a reference number t_R, where $t_R \in (0, \pi/2)$, with the proper adjustment in sign, or at 0, $\pi/2$, π, or $3\pi/2$. The only question that remains is how to determine the values of the circular functions for $0 < t_R < \pi/2$. As we often commented, this is done by thinking of t as an angle measure in radians. Thus, approximations to these values can be found in Table II of Appendix A.

EXAMPLE

Use Table II to determine each of the following values.

a) sin 0.65 b) cos 0.65 c) tan 0.89
d) cot 0.94 e) sec 0.76 f) csc 0.44

SOLUTION

a) $\sin 0.65 = 0.6052$ b) $\cos 0.65 = 0.7961$
c) $\tan 0.89 = 1.235$ d) $\cot 0.94 = 0.7303$
e) $\sec 0.76 = 1.380$ f) $\csc 0.44 = 2.348$

It is obvious that Table II does not contain entries for all values between 0 and $\pi/2$. (For example, sin 1.514 and sin 0.005 cannot be found in Table II.) However, exactly as before, *linear interpolation* can be used to give reasonably accurate approximations.

Let us now summarize the steps for finding $C(t)$, where $C(t)$ is any of the six circular functions.

1 Locate the quadrant in which $W(t)$ lies.

2 Determine the reference number t_R associated with t.

3 Use Table II and linear interpolation if necessary to find $|C(t)| = C(t_R)$.

4 Adjust the sign of $C(t)$ according to the quadrant in which $W(t)$ lies.

EXAMPLES

Use Table II and linear interpolation to approximate each of the following values (use $\pi = 3.14$).

1 sin 1.514

SOLUTION. $W(1.514)$ is in quadrant I, so the reference number is

1.514. Table II, together with linear interpolation, can be used to approximate sin 1.514 as follows.

$$0.01\left[\begin{array}{ll} & \sin 1.51 \ \ = 0.9982 \\ 0.006 & \left[\begin{array}{l}\sin 1.514 = \quad ? \\ \sin 1.52 \ \ = 0.9987\end{array}\right] d \end{array}\right] 0.0005$$

$$\frac{0.006}{0.01} = \frac{d}{0.0005}$$

so

$$d = 0.0003$$

Hence,

$$\begin{aligned}\sin 1.514 &= 0.9987 - 0.0003 \\ &= 0.9984\end{aligned}$$

2 sin (−2.955)

SOLUTION. Since the sine is an odd function, sin (−2.955) = −sin 2.955. But sin 2.955 = sin 0.185 (Figure 9) since the sine is positive in quadrant II and the reference number for 2.955 is 0.185, so that sin (−2.955) = −sin 0.185. Using linear interpolation, sin 0.185 can be approximated as follows.

$$0.01\left[\begin{array}{ll} & \sin 0.18 \ \ = 0.1790 \\ 0.005 & \left[\begin{array}{l}\sin 0.185 = \quad ? \\ \sin 0.19 \ \ = 0.1889\end{array}\right] d \end{array}\right] 0.0099$$

$$\frac{0.005}{0.01} = \frac{d}{0.0099}$$

Figure 9

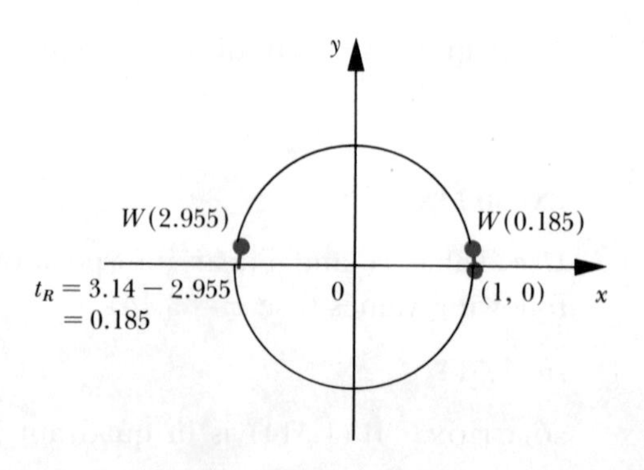

Hence,

$$d = 0.0050$$

so that

$$\sin 0.185 = 0.1889 - 0.005 = 0.1839$$

Hence,

$$\sin(-2.955) = -0.1839$$

3 cos 14.073

SOLUTION. $W(14.073)$ is in quadrant I, and the reference number t_R for 14.073 is 1.513 (Figure 10), since $14.073 = 2(6.28) + 1.513$. Therefore,

$$\begin{aligned} \cos 14.073 &= \cos\,[1.513 + 2(6.28)] \\ &= \cos 1.513 \end{aligned}$$

Figure 10

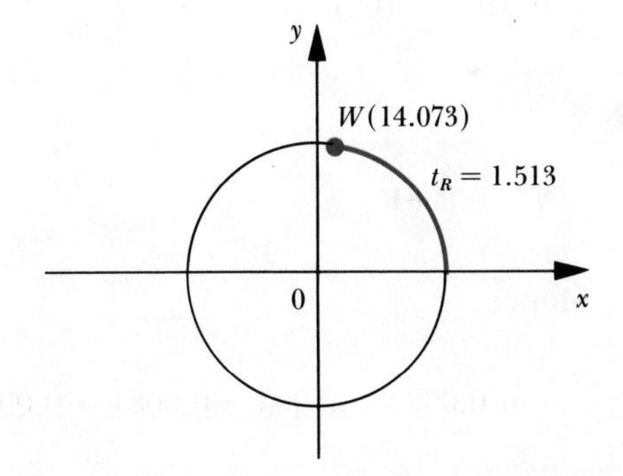

Now, linear interpolation can be used to determine cos 1.513 as follows.

$$0.01\left[\begin{array}{l} \quad\;\; \cos 1.51 \;\; = 0.0608 \\ 0.007 \left[\begin{array}{l} \cos 1.513 = \quad ? \\ \cos 1.52 \;\; = 0.0508 \end{array}\right] d \end{array}\right] 0.01$$

$$\frac{0.007}{0.01} = \frac{d}{0.01}$$

so that

$$d = 0.007$$

Hence,

$$\cos 1.513 = 0.0508 + 0.0070 = 0.0578$$

(Notice here that d is added to 0.0508. Compare this to Examples 1 and 2 above.) Since $W(14.073)$ is in quadrant I, it follows that

$$\cos 14.073 = 0.0578$$

4 tan 3.048

SOLUTION. $W(3.048)$ is in quadrant II, and the reference number t_R is $3.14 - 3.048 = 0.092$ (Figure 11), so that $\tan 3.048 = -\tan 0.092$. Next, linear interpolation can be used to evaluate tan 0.092 as follows.

$$0.01\left[\begin{array}{l} \qquad \tan 0.09 \;\; = 0.0902 \\ 0.008\left[\begin{array}{l}\tan 0.092 = \\ \tan 0.10 \;\; = 0.1003\end{array}\right] d \end{array}\right] 0.0101$$

$$\frac{d}{0.0101} = \frac{0.008}{0.01}$$

so that

$$d = 0.0081$$

Hence,

$$\tan 0.092 = 0.1003 - 0.0081 = 0.0922$$

so that

$$\tan 3.048 = -0.0922$$

Figure 11

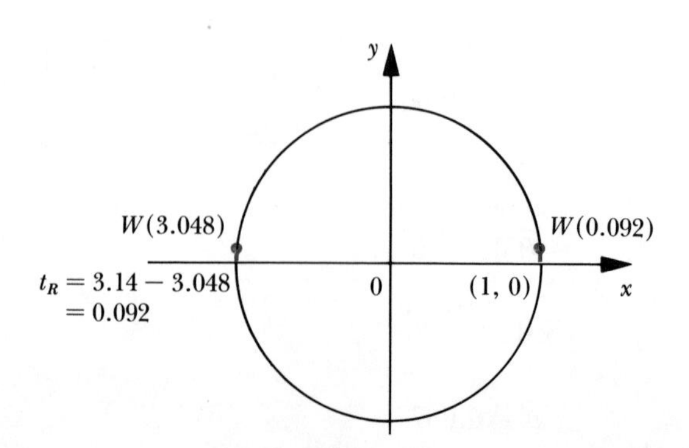

5 cot (4.045)

SOLUTION

$$\cot(4.045) = \cot(4.045 - 3.14) = \cot(0.905) \qquad \text{(Figure 12)}$$

Figure 12

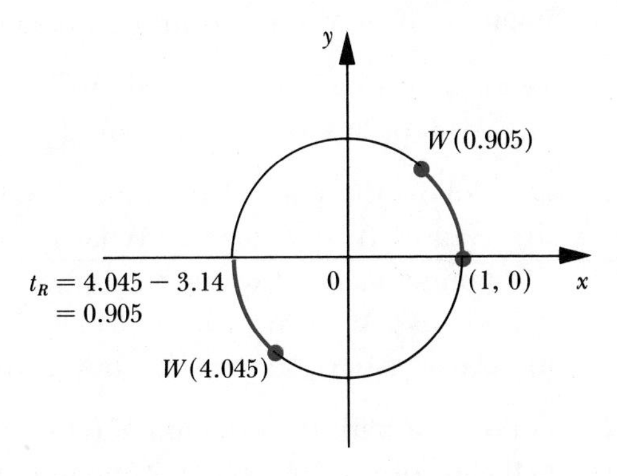

Using linear interpolation, we have

$$0.01\left[\begin{array}{l} \qquad\quad \cot 0.900 = 0.7936 \\ 0.005\left[\begin{array}{l} \cot 0.905 = \quad ? \\ \cot 0.910 = 0.7774 \end{array}\right] d \end{array}\right] 0.0162$$

$$\frac{0.005}{0.01} = \frac{d}{0.0162}$$

Hence,

$$d = 0.0081$$

so that

$$\cot(4.045) = \cot(0.91) + 0.0081 = 0.7774 + 0.0081 = 0.7855$$

PROBLEM SET 3

1 Use the fact that the cosine and sine functions are periodic functions of period 2π together with the fact that the cosine is an even function and the sine is an odd function to express each of the following as a function of a number in the interval $[0, 2\pi)$. (Use $\pi = 3.14$.)

a) sin 3.82 b) cos 4.76 c) sin 6.195
d) tan (−5.41) e) cot 45 f) cos (−47)
g) sec (−14.125) h) csc (−10.535) i) cot 60.155
j) tan (−5.32) k) sin 11.58 l) csc (−12.3)
m) cot 3.745 n) csc (−5.145) o) sec (−85.125)

2 Find the reference number for each of the parts of Problem 1, then express the given evaluation in terms of an evaluation at the reference number.

3 Use the results of Problem 2 together with Table II and linear interpolation, if necessary, to find each value in Problem 1.

4 Determine which is greater than the other:

a) $\cos 6$ or $\cos 7$ b) $\sin 6$ or $\sin 7$ c) $\cos 9$ or $\cos 10$

5 a) Follow the path of the point $W(t)$ around the unit circle as t increases from 0 to $\pi/2$. What is the behavior of $\cos t$? Next, examine Table II to see the behavior of $\cos t$ as t increases from 0 to 1.57. What do you notice?

b) Answer part (a) for $\sin t$, $\tan t$, $\cot t$, $\sec t$, and $\csc t$.

6 Follow the path of the point $W(t)$ on the unit circle to determine the behavior of the six circular functions as

a) t increases from $\pi/2$ to π

b) t increases from π to $3\pi/2$

c) t increases from $3\pi/2$ to 2π

Do the table values show these behaviors? Explain.

REVIEW PROBLEM SET

1 Find t in each of the following cases, where $t \in [0, 2\pi)$ and

a) $W(t) = (0, 1)$ b) $W(t) = (-1, 0)$

c) $W(t) = (0, -1)$ d) $W(t) = (1, 0)$

2 Find the values of the six circular functions if

a) $W(t) = \left(\frac{1}{2}, -\frac{\sqrt{3}}{2}\right)$ b) $W(t) = \left(-\frac{1}{\sqrt{2}}, \frac{1}{\sqrt{2}}\right)$

c) $W(t) = \left(\frac{5}{13}, \frac{12}{13}\right)$ d) $W(t) = \left(-\frac{4}{5}, \frac{3}{5}\right)$

3 If $W(t) = \left(-\frac{1}{3}, y\right)$ is a point on a unit circle, find the values of y.

4 Find each of the following values without the use of tables.

a) $\sec(-\pi/3)$ b) $\cot(-5\pi/6)$ c) $\cos(47\pi/6)$

d) $\sin(-89\pi/2)$ e) $\tan(4\pi/3)$ f) $\csc(25\pi/6)$

g) $\sin(125\pi/6)$ h) $\cos(-325\pi/3)$ i) $\sec(-29\pi/4)$

j) $\csc(47\pi/6)$ k) $\cot(-31\pi/3)$ l) $\tan(79\pi/6)$

5 Find the remaining five circular functions if

a) $\sin t = \frac{1}{3}$ and $\cos t < 0$

b) $\cos t = -\frac{\sqrt{3}}{3}$ and $\sin t > 0$

c) $\cos t = \frac{1}{4}$ and $\sin t < 0$

d) $\sin t = -\frac{24}{25}$ and $\cos t > 0$

6 Find $\cos t$, $\tan t$, $\cot t$, $\sec t$, and $\csc t$ in each of the following cases.

a) If $\sin t = \frac{2}{3}$ and $W(t)$ is in quadrant II.
b) If $\sin t = -\frac{3}{8}$ and $W(t)$ is in quadrant IV.
c) If $\sin t = -\frac{1}{3}$ and $W(t)$ is in quadrant III.
d) If $\sin t = \dfrac{2}{\sqrt{5}}$ and $W(t)$ is in quadrant I.

7 Find the period of each of the following functions.

a) $f(x) = \cos 7x$
b) $f(x) = \sin (4x/5)$
c) $f(x) = \sin (3x + 4)$
d) $f(x) = \cos (3 - 2x)$

8 Express each of the following as a function of a positive number in the interval $[0, 2\pi)$. Use $\pi = 3.14$.

a) $\sin 13.4$
b) $\cos (-20)$
c) $\tan 12.45$
d) $\cos 15.725$
e) $\sin (-7.513)$
f) $\cot 16.1951$
g) $\sec (-11.495)$
h) $\csc 65.545$
i) $\tan 55.135$

9 Find the reference number for each of the parts of Problem 8, then express the given evaluation in terms of an evaluation at the reference number.

10 Use Table II and linear interpolation, if necessary, to find each of the following values. Use $\pi = 3.14$.

a) $\sin 0.937$
b) $\cos 1.234$
c) $\cos (-0.378)$
d) $\sin (-0.115)$
e) $\tan 1.34$
f) $\cot 0.826$
g) $\csc (-0.324)$
h) $\sec 1.414$
i) $\tan 26.4$
j) $\cot 63.3$
k) $\sec (-13)$
l) $\csc 19.1$

11 Which quadrant will $W(t)$ be in if

a) $\sin t < 0$ and $\cos t < 0$
b) $\sin t < 0$ and $\cos t > 0$
c) $\tan t > 0$ and $\cos t < 0$
d) $\sec t > 0$ and $\cot t < 0$

CHAPTER 4

Graphs of the Trigonometric Functions and Inverse Functions

CHAPTER 4

Graphs of the Trigonometric Functions and Inverse Functions

1 Introduction

The purpose of this chapter is to investigate the graphs of the trigonometric functions. It was proved in Chapter 3, Section 3, that

$$\cos(-t) = \cos t$$
$$\sin(-t) = -\sin t$$
$$\tan(-t) = -\tan t$$
$$\sec(-t) = \sec t$$
$$\csc(-t) = -\csc t$$
$$\cot(-t) = -\cot t$$

Consequently, we can conclude that the cosine and secant functions are even functions and therefore that their graphs are symmetric with respect to the y axis. The other four trigonometric functions—the sine, tangent, cosecant, and cotangent—are odd functions and their graphs are symmetric with respect to the origin.

Besides symmetry, we shall also use some general properties of these functions which we discussed in Chapters 2 and 3 and plot some specific points in order to construct the graphs. Finally, we will investigate inverses of the trigonometric functions.

2 Graphs of the Sine and Cosine Functions

In this section we consider the graphs of $f(x) = \sin x$ and $g(x) = \cos x$. We already know that the sine and cosine functions are periodic functions of period 2π. Consequently, since the function values repeat every 2π units, we can restrict our attention to values of x, where $x \in [0, 2\pi]$. Then the graphs of $f(x) = \sin x$ and $g(x) = \cos x$ for $x \in R$ can be extended as far as we like by repeating the graph every 2π units (Figure 1). That part of the graph which occurs in the interval $[0, 2\pi]$ is called a *cycle* or *wavelength* of the curve.

Figure 1

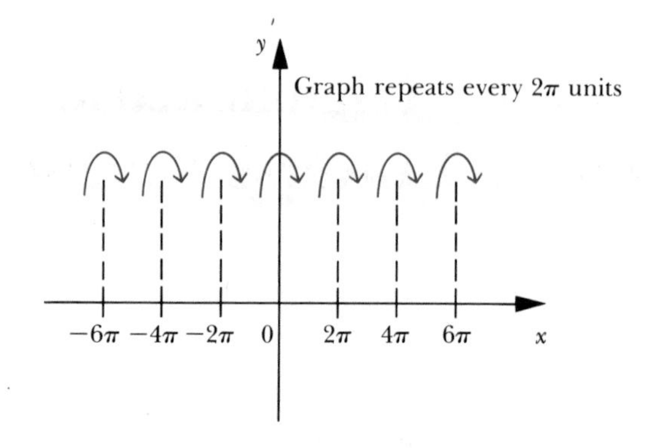

Also, from Chapter 3, we know that the domain of each of the cosine and sine functions is the set of real numbers, R, whereas the range of each function is the interval $[-1, 1]$. Thus the graphs of one cycle of both the sine and cosine functions are restricted in the plane between $y = -1$ and $y = 1$ (Figure 2).

Figure 2

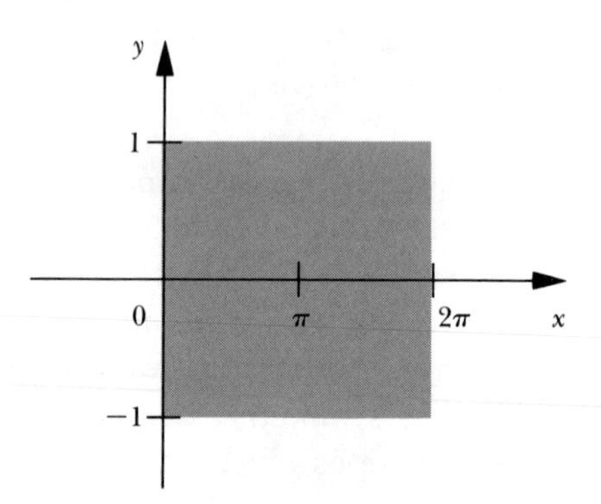

2.1 Graph of the Sine Function

In Chapter 2, Section 3.1 on page 70, it was noted that the sine of x is a positive number if $0 < x < \pi$ and the sine of x is a negative number if $\pi < x < 2\pi$. This means that the graph of $f(x) = \sin x$ is above the x axis for $x \in (0, \pi)$ and below the x axis for $x \in (\pi, 2\pi)$. Hence, the graph of $f(x) = \sin x$ is further restricted to the shaded region shown in Figure 3. Locating the ordered pairs $(x, \sin x)$ for specific values of x and sketching the curve suggested by these points, we obtain the graph of $f(x) = \sin x$ for $x \in [0, 2\pi]$ (Figure 4). Note that the graph of $f(x) = \sin x$ lies in the shaded region previously illus-

Figure 3

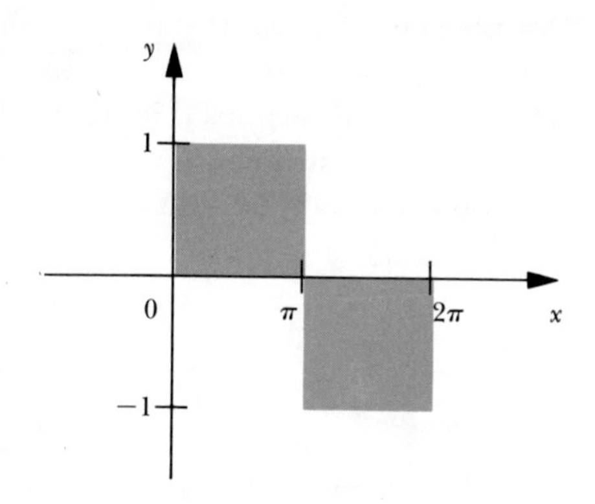

Figure 4

x	$\sin x$
0	0
$\frac{\pi}{6}$	$\frac{1}{2}$
$\frac{\pi}{4}$	$\frac{\sqrt{2}}{2}$
$\frac{\pi}{2}$	1
$\frac{2\pi}{3}$	$\frac{\sqrt{3}}{2}$
π	0
$\frac{4\pi}{3}$	$-\frac{\sqrt{3}}{2}$
$\frac{3\pi}{2}$	-1
$\frac{5\pi}{3}$	$-\frac{\sqrt{3}}{2}$
$\frac{11\pi}{6}$	$-\frac{1}{2}$
2π	0

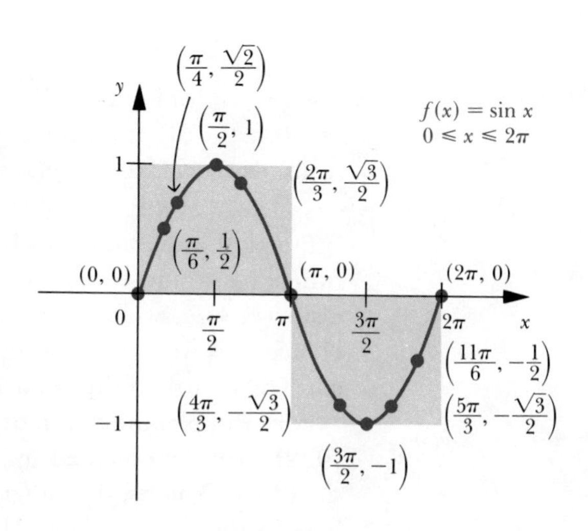

trated by Figure 3. This basic graph is the cycle of the sine function. Notice that the function is increasing in the interval $(0, \pi/2)$, decreasing in the interval $(\pi/2, 3\pi/2)$, and again is increasing in the interval $(3\pi/2, 2\pi)$. This increasing behavior of the function values for the interval $(0, \pi/2)$ can also be seen by examining the function values for the sine in Table II. Since the sine function is periodic of period 2π, the graph of the cycle repeats itself over intervals of length 2π units in both directions (Figure 5). Also, we notice that the graph of $f(x) = \sin x$ is symmetric with respect to the origin; that is, the sine function is an odd function.

Figure 5

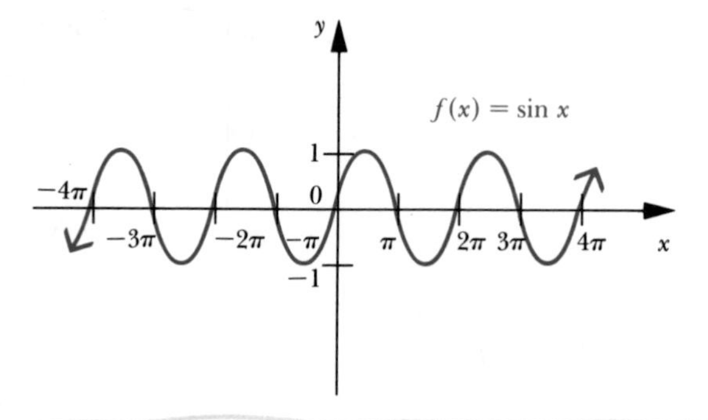

The graph of $f(x) = \sin x$ in Figure 5 also reveals the zeros of the sine function, that is, those values of x for which $\sin x = 0$. The zeros occur wherever the sine function intersects the x axis, namely, at $x = n\pi$, where n is any integer.

Practical applications of the trigonometric functions often involve functions of the forms $f(x) = a \sin x$, $f(x) = a \sin kx$, and $f(x) = a \sin (kx + b)$, where a, k, and b are constants and $a \neq 0$ and $k \neq 0$. Hence, it is worthwhile to investigate the geometric effects that these constants have on the behavior of the graph of $y = \sin x$.

For example, the graph of one cycle of $f(x) = 4 \sin x$ in the interval $[0, 2\pi)$ can be obtained by multiplying each of the ordinates of $y = \sin x$ by 4. That is, the range of $f(x) = 4 \sin x$ is the set of real numbers in the interval $[-4, 4]$ (Figure 6). 4 is called the *amplitude* of $f(x) = 4 \sin x$.

Figure 6

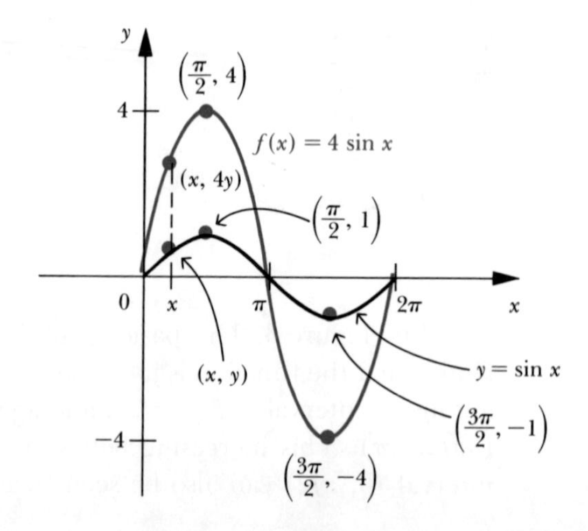

In general, since the graph of $y = \sin x$ contains all points of the form $(x, \sin x)$, the graph of $f(x) = a \sin x$ contains all points of the form $(x, a \sin x)$. That is, for each value of x, each ordinate of the graph of $f(x) = a \sin x$ is a times the corresponding ordinate of the graph of $y = \sin x$. The *amplitude* of $f(x) = a \sin x$ is $|a|$ and the range of f is the interval $[-|a|, |a|]$.

EXAMPLES

Use the graph of $y = \sin x$ to graph one cycle of each of the following functions. Indicate the amplitude in each case.

1 $f(x) = \frac{1}{2} \sin x$

SOLUTION. The amplitude of this function is $\frac{1}{2}$. Consequently, the graph of one cycle of $f(x) = \frac{1}{2} \sin x$ can be thought of as a "vertical contraction" of the graph of $y = \sin x$ by $\frac{1}{2}$. In other words, each ordinate of the graph of $y = \sin x$ is multiplied by $\frac{1}{2}$ to get the graph of f (Figure 7).

Figure 7

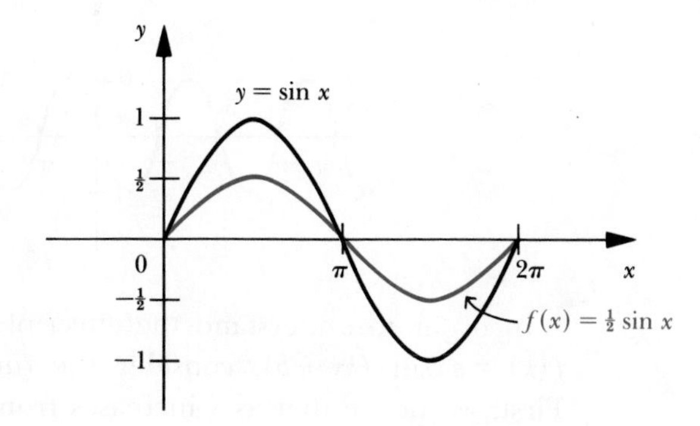

2 $f(x) = -3 \sin x$

Figure 8

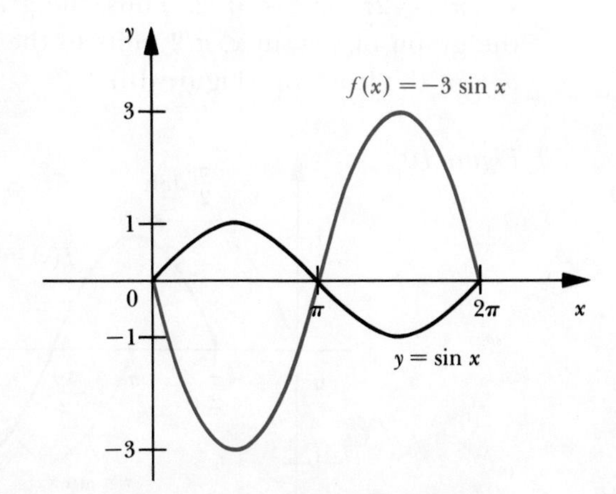

SOLUTION. In this case each ordinate of the graph of $y = \sin x$ is multiplied by -3 in order to obtain the graph of $f(x) = -3 \sin x$. This multiplication has the geometric effect of reflecting the graph of $y = \sin x$ across the x axis (because of the negative value) and also of "stretching vertically" the graph by a multiple of 3 so that the amplitude of f is 3 (Figure 8).

Next consider the graph of $f(x) = \sin 2x$. Since the sine function is periodic of period 2π, we know that the function f generates one cycle whenever $2x$, called the *argument* of the function, covers an interval of 2π units. In other words, $f(x) = \sin 2x$ generates one sine cycle if $0 \leqslant 2x \leqslant 2\pi$, that is, if $0 \leqslant x \leqslant \pi$. This means that $f(x) = \sin 2x$ generates one cycle over an interval of length $2\pi/2 = \pi$. In this case, π is the period of f.

In general, if $f(x) = \sin kx$, one cycle is generated over the interval $2\pi/|k|$. In other words, $2\pi/|k|$ is the period of $f(x) = \sin kx$.

For example, the period of the function $f(x) = \sin 3x$ is $2\pi/3$, that is, the graph repeats every $2\pi/3$ units (Figure 9).

Figure 9

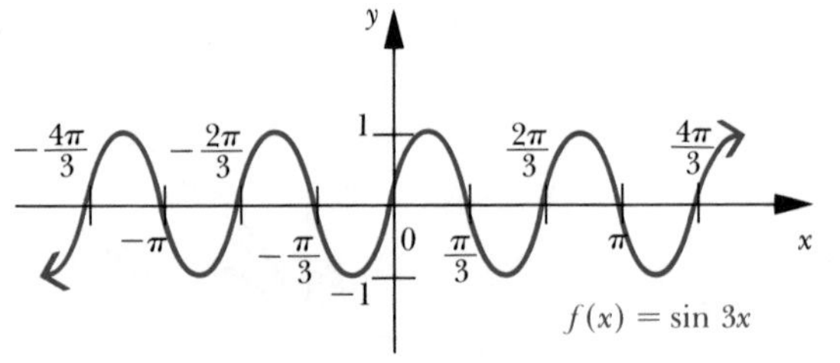

In order to understand the effect of the constant b in the form $f(x) = a \sin (kx + b)$, consider the function $f(x) = \sin (x - \pi/2)$. First, we notice that as x increases from $\pi/2$ to $(2\pi + \pi/2)$, $x - \pi/2$ increases from 0 to 2π. In this case, the graph of $f(x) = \sin (x - \pi/2)$ generates a cycle that begins with $x - \pi/2 = 0$ or $x = \pi/2$ and ends with $x - \pi/2 = 2\pi$ or $x = 5\pi/2$. Thus, the graph is obtained by "shifting" the graph of $y = \sin x$, $\pi/2$ units to the right. $\pi/2$ is called the *phase shift* of the function (Figure 10).

Figure 10

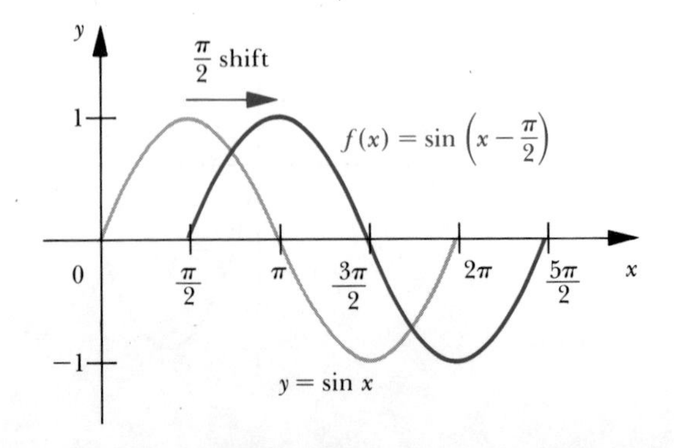

In general, if one cycle of a function of the form $f(x) = a \sin(kx + b)$ is to be graphed, we have the following properties.

1 The *amplitude* is $|a|$.

2 The *period* is $2\pi/|k|$.

3 The *phase shift* is $-b/k$.

EXAMPLES

Use the graph of $y = \sin x$ to graph a cycle of each of the following functions of the form $f(x) = a \sin(kx + b)$. Indicate the amplitude, the period, the phase shift, and the zeros of f for one cycle in each case.

1 $f(x) = 2 \sin x$

SOLUTION. Since $a = 2$, $k = 1$, and $b = 0$, the amplitude is 2, the period is $2\pi/1 = 2\pi$, and the phase shift is 0. We can obtain the graph of $f(x) = 2 \sin x$ from the graph of $y = \sin x$ by doubling the ordinates. The zeros of f are 0, π, and 2π (Figure 11).

Figure 11

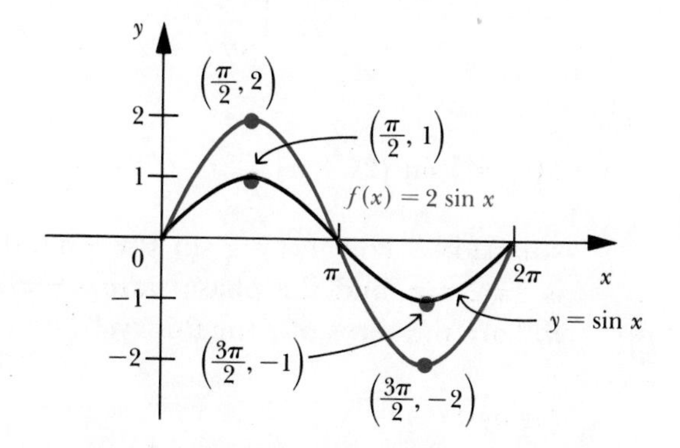

2 $f(x) = \sin \frac{1}{2}x$

SOLUTION. Here $a = 1$, $k = \frac{1}{2}$, and $b = 0$, so that the amplitude is 1, the period of f is $2\pi/\frac{1}{2} = 4\pi$, and the phase shift is 0 (Figure 12). The graph displays the zeros of f to be 0, 2π, and 4π in the cycle.

Figure 12

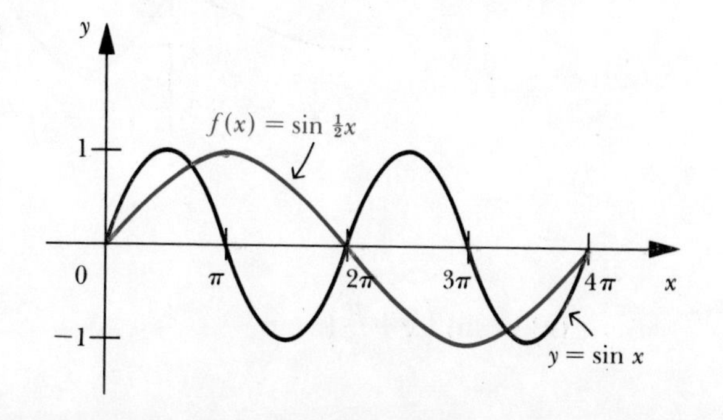

3 $f(x) = 3 \sin (x - \pi)$

SOLUTION. For $f(x) = 3 \sin (x - \pi)$, $a = 3$, $k = 1$, and $b = -\pi$, so that the amplitude is 3, the period is 2π, and the phase shift is given by $-(-\pi)/1 = \pi$ (Figure 13). Here π, 2π, and 3π are the zeros of f for this cycle.

Figure 13

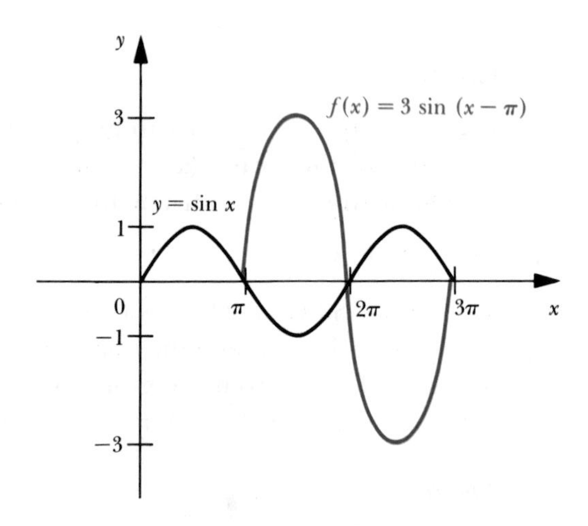

4 $f(x) = \frac{3}{2} \sin (2x + \pi)$

SOLUTION. For $f(x) = \frac{3}{2} \sin (2x + \pi)$, the amplitude is $\frac{3}{2}$, the period is $2\pi/2 = \pi$, and the phase shift is $-\pi/2$ (Figure 14). $-\pi/2$, 0, and $\pi/2$ are the zeros of f for this cycle.

Figure 14

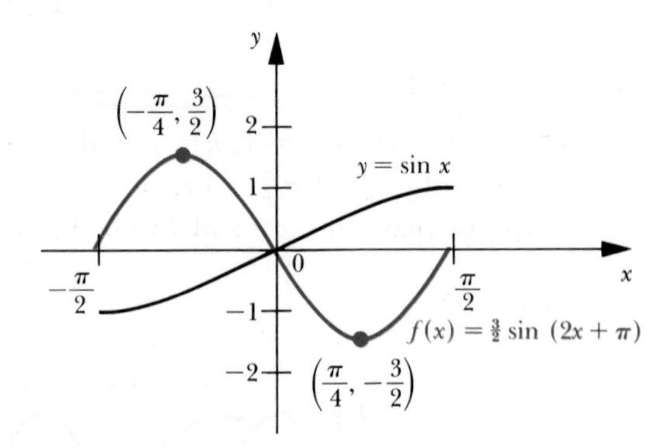

5 $f(x) = \sin \left(x + \frac{\pi}{2}\right)$

SOLUTION. Here the amplitude is 1, the period is 2π, and the phase shift is $-\pi/2$ (Figure 15). The zeros of f for one cycle are $-\pi/2$, $\pi/2$ and $3\pi/2$.

Figure 15

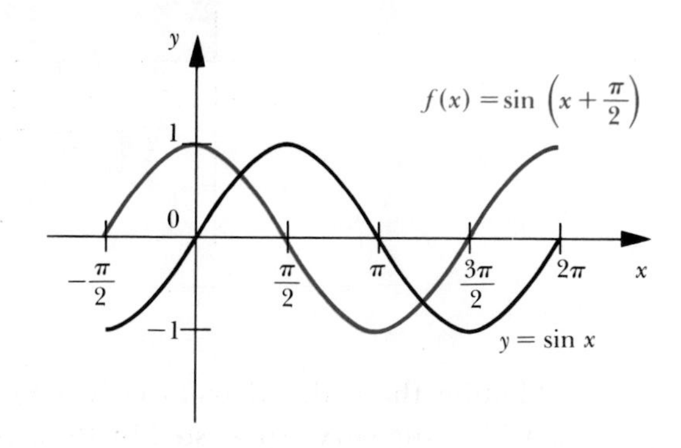

2.2 Graph of the Cosine Function

The cosine function can be graphed in the same way as the sine function. Since the cosine function is periodic of period 2π, it is enough to graph it in $[0, 2\pi]$. Also, the range of the cosine function is the set of real numbers in the interval $[-1, 1]$. Hence, a cycle of $g(x) = \cos x$ occurs in the shaded region of Figure 16.

Figure 16

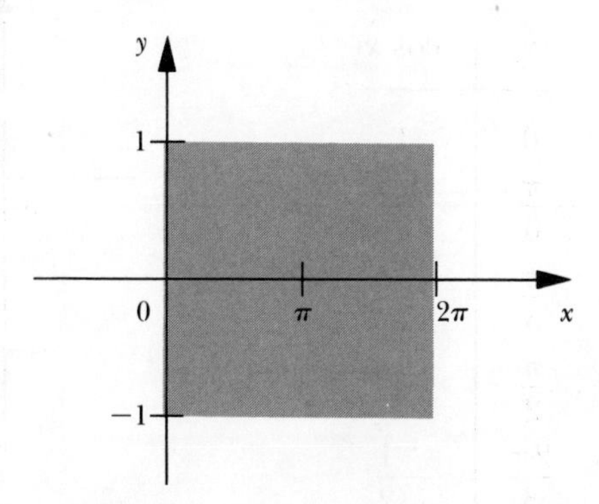

In Chapter 2, Section 3.1 on page 70, it was shown that the cosine of x is positive for $0 < x < \pi/2$; the cosine of x is negative for $\pi/2 < x < 3\pi/2$; and the cosine of x is again positive for $3\pi/2 < x < 2\pi$. Consequently, the graph of $g(x) = \cos x$ is above the x axis for $x \in [0, \pi/2)$, below the x axis for $x \in (\pi/2, 3\pi/2)$, and above the x axis for $x \in (3\pi/2, 2\pi)$. Hence, the graph is further restricted to the shaded region in Figure 17.

Figure 17

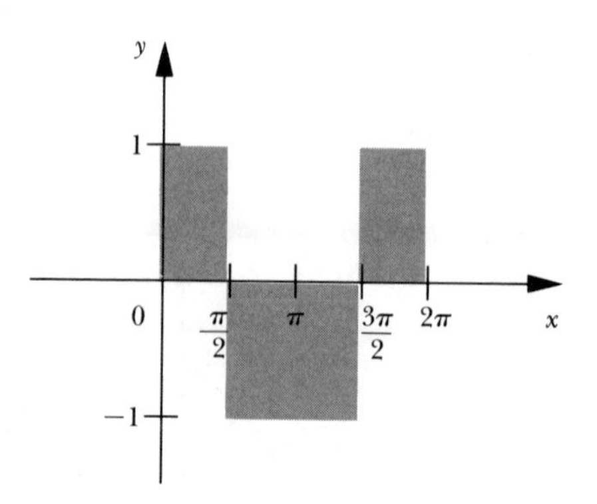

Plotting the ordered pairs $(x, \cos x)$ for specific values of x and sketching the curve suggested by these points we obtain the graph of $g(x) = \cos x$ for $x \in [0, 2\pi]$ (Figure 18). Note that the graph of $g(x) = \cos x$ lies in the shaded region illustrated in Figure 17. The graph of g shows that the cosine function decreases in the interval $[0, \pi)$ and increases in the interval $(\pi, 2\pi]$. Table II also displays this behavior in function values for interval $(0, \pi/2)$.

Figure 18

x	$\cos x$
0	1
$\frac{\pi}{6}$	$\frac{\sqrt{3}}{2}$
$\frac{\pi}{3}$	$\frac{1}{2}$
$\frac{\pi}{2}$	0
$\frac{2\pi}{3}$	$-\frac{1}{2}$
π	-1
$\frac{4\pi}{3}$	$-\frac{1}{2}$
$\frac{3\pi}{2}$	0
$\frac{5\pi}{3}$	$\frac{1}{2}$
2π	1

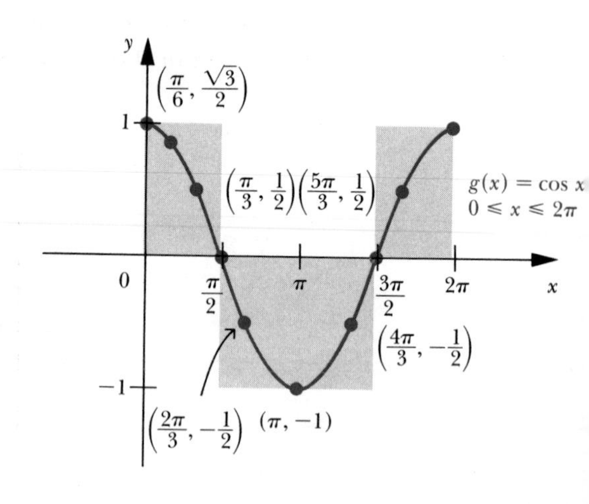

Because the cosine function is periodic of period 2π, this graph repeats itself over intervals of length 2π units in both directions. Also, since $\cos x = \cos(-x)$, the graph of $g(x) = \cos x$ is symmetric with respect to the y axis (Figure 19).

Figure 19

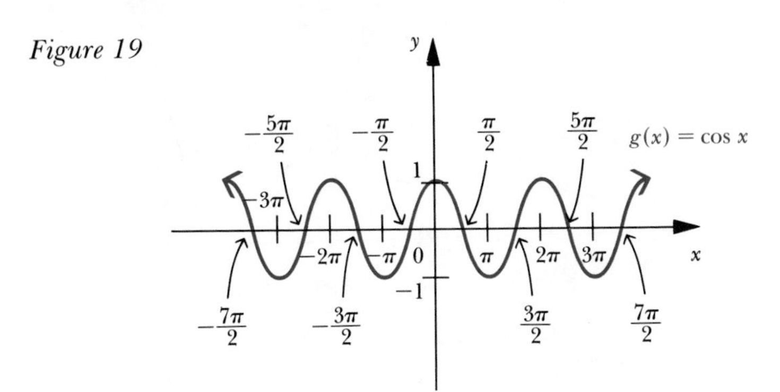

The graph of $g(x) = \cos x$ in Figure 19 reveals that the zeros of $g(x) = \cos x$ are $x = (\pi/2) + n\pi$, where n is any integer. Note that the graph of $f(x) = \sin [x + (\pi/2)]$ (Example 5, on page 151) is the same as that of $g(x) = \cos x$. In other words, we have shown that for all x, $\sin [x + (\pi/2)] = \cos x$. In Chapter 5 we shall discuss this "identity" and others like it.

As with the sine function, we can use the graph of $g(x) = \cos x$ to obtain the graph of functions of the form $g(x) = a \cos x$, $g(x) = a \cos kx$, or $g(x) = a \cos (kx + b)$, where a, k, and b are constant real numbers and $a \neq 0$ and $k \neq 0$. The geometric effect that these constants have on the behavior of the graph of g are similar to those of the sine function. That is, $|a|$ is the amplitude, $2\pi/|k|$ is the period, and $-b/k$ is the phase shift.

EXAMPLES

Use the graph of $y = \cos x$ to graph one cycle of each of the following functions. Indicate the amplitude, the period, the phase shift, and the zeros of g for each graph in the following cases.

1 $g(x) = 3 \cos x$

SOLUTION. Here the technique is the same as that used for the sine graphs. The amplitude is 3, the period is $2\pi/1 = 2\pi$, and the phase shift is 0. The graph of g can be obtained from the graph of $y = \cos x$ by multiplying the ordinates by 3 (Figure 20). Since the amplitude is 3, the range of g is $[-3, 3]$. The zeros of g for one cycle are $\pi/2$ and $3\pi/2$.

Figure 20

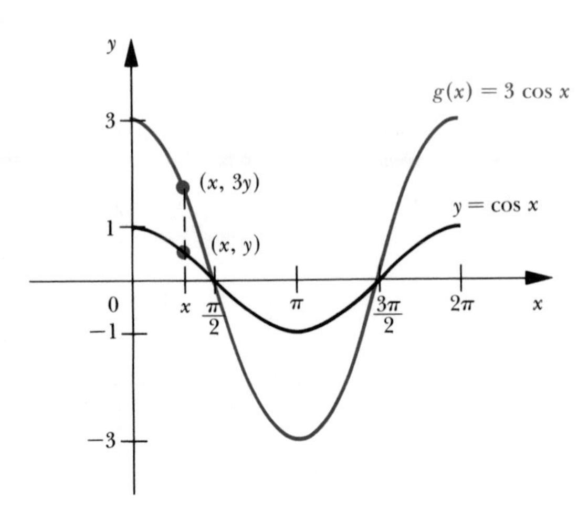

2 $g(x) = \cos 3x$

SOLUTION. The amplitude is 1, the period is $2\pi/3$, and the phase shift is 0. The range is $[-1, 1]$ since the amplitude is 1. The period of $2\pi/3$ indicates that three cycles of g would occur in an interval of length 2π. The zeros of g are $\pi/6$, $\pi/2$, $5\pi/6$, $7\pi/6$, and $3\pi/2$ for the three cycles graphed in Figure 21.

Figure 21

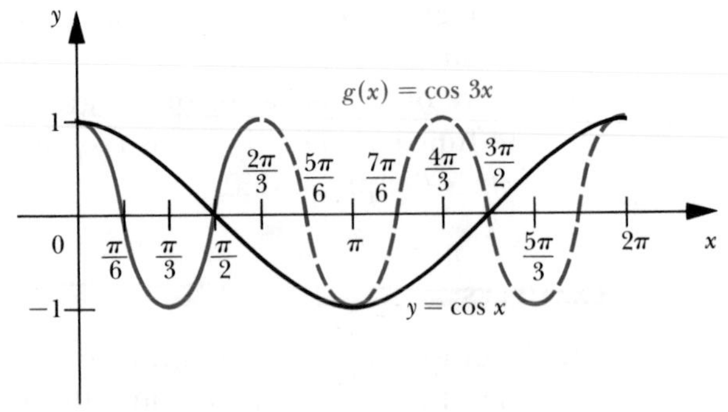

3 $g(x) = \frac{1}{2} \cos \left(x + \frac{\pi}{4}\right)$

SOLUTION. The amplitude is $\frac{1}{2}$, the period is 2π, and the phase shift is $-\pi/4$ (Figure 22). The range is $[-\frac{1}{2}, \frac{1}{2}]$. For this cycle, the zeros are $x = \pi/4$ and $5\pi/4$.

Figure 22

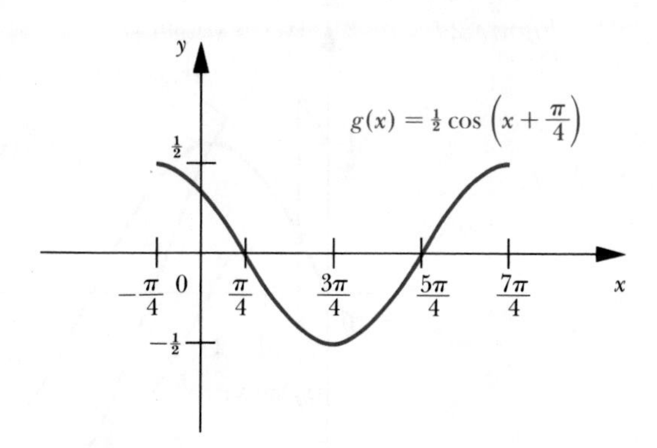

4 $g(x) = 2\cos(3x - \pi)$

SOLUTION. Since

$$\begin{aligned} g(x) &= 2\cos(3x - \pi) \\ &= 2\cos[3x + (-\pi)] \end{aligned}$$

the amplitude is 2, the period is $2\pi/3$, and the phase shift is $-(-\pi)/3 = \pi/3$ (Figure 23). The range is $[-2, 2]$. $x = \pi/2$ and $5\pi/6$ are the zeros for this cycle.

Figure 23

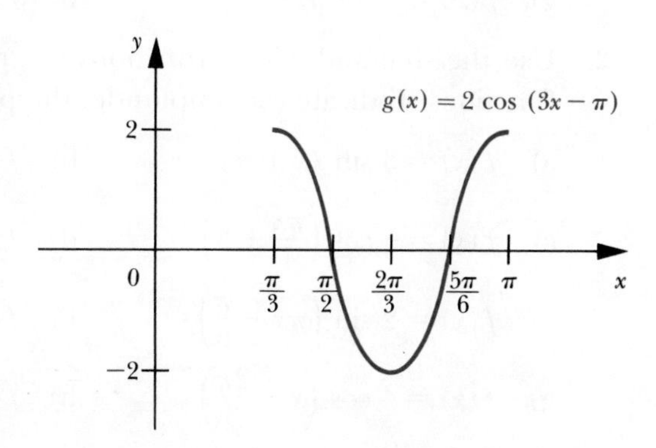

5 Use the graphs of $g(x) = \cos x$ and $h(x) = \sin x$ to graph $f(x) = \cos x + \sin x$.

SOLUTION. First, we graph $g(x) = \cos x$ and $h(x) = \sin x$ on the same coordinate system (Figure 24). Next, we can select values on the x axis as the abscissas and add graphically the ordinates $g(x)$ and $h(x)$ to obtain a sketch of the graph of $f(x) = g(x) + h(x) = \cos x + \sin x$ (Figure 24).

Figure 24

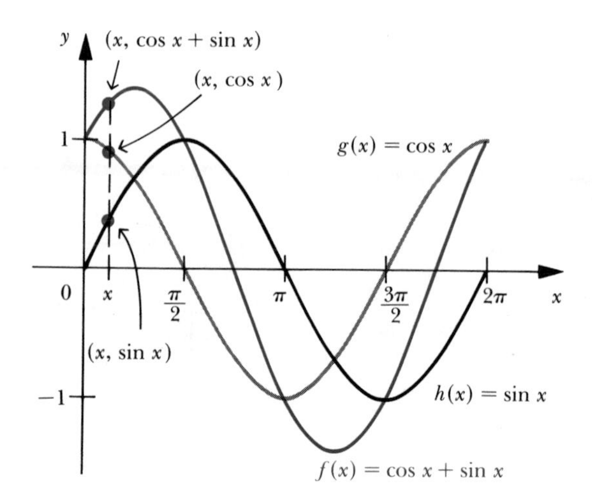

PROBLEM SET 1

1 Use the sine and cosine functions to graph a cycle of the following functions. Indicate the amplitude and the period.

a) $f(x) = 6 \cos x$
b) $f(x) = \frac{1}{2} \sin 4x$
c) $f(x) = \frac{3}{4} \sin 2x$
d) $f(x) = 5 \cos (-2x)$
e) $f(x) = 3 \cos (\pi x/3)$
f) $f(x) = 3 \sin (-x/5)$
g) $f(x) = \frac{1}{3} \sin \pi x$
h) $f(x) = -2 \cos 2\pi x$

2 Use the sine and cosine functions to graph a cycle of the following functions. Indicate the amplitude, the period, and the phase shift.

a) $f(x) = 3 \sin (x + \pi)$
b) $f(x) = 4 \cos \left(\frac{x}{3} - 2\right)$
c) $f(x) = -\cos \left(\frac{\pi x}{4} + 3\right)$
d) $f(x) = -2 \sin \left(\frac{2x}{3} - \frac{\pi}{8}\right)$
e) $f(x) = 2 \sin \left(\pi x - \frac{\pi}{6}\right)$
f) $f(x) = -\frac{1}{3} \cos \left(x - \frac{\pi}{4}\right)$
g) $f(x) = 4 \cos \left(x - \frac{2\pi}{3}\right)$
h) $f(x) = 2 \sin (\pi - x)$

3 Show that the graph of $f(x) = \cos x$ is symmetric with respect to the y axis (see Figure 19 on page 153).

4 Show that the graph of $f(x) = \sin x$ is symmetric with respect to the origin (see Figure 5 on page 146).

5 Consider $f(x) = \cos x - \cos 2x$.

a) Sketch the graph of $g(x) = \cos x$ and $h(x) = \cos 2x$ on the same set of axes.

b) Select appropriate points on the x axis and subtract graphically the ordinates $g(x)$ and $h(x)$ of these points to obtain a sketch of the graph of $f(x) = g(x) - h(x)$.

6 Use the graphs of the sine and cosine functions to find the zeros of each of the following functions.

a) $f(x) = 3 \sin 2x$ b) $f(x) = 4 \sin \frac{x}{2}$

c) $g(x) = 2 \cos 4x$ d) $g(x) = 3 \cos \frac{x}{3}$

3 Graphs of the Tangent, Cotangent, Secant, and Cosecant Functions

The graphs of the functions defined by $f(x) = \tan x$, $f(x) = \cot x$, $f(x) = \sec x$, and $f(x) = \csc x$ display the properties of being periodic, increasing or decreasing on restricted intervals, and even or odd, that is, symmetric with respect to either the y axis or origin.

3.1 Graph of the Tangent Function

Let us begin by graphing the tangent function $f(x) = \tan x$. It was shown in Chapter 3, Section 2.3 on page 115, that the domain of the tangent function is $\{x \mid x \neq (\pi/2) + n\pi,\ n \in I\}$. Consequently, the graph of the tangent function cannot intersect any line of the form $x = (\pi/2) + n\pi$, $n \in I$ (Figure 1).

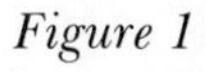
Figure 1

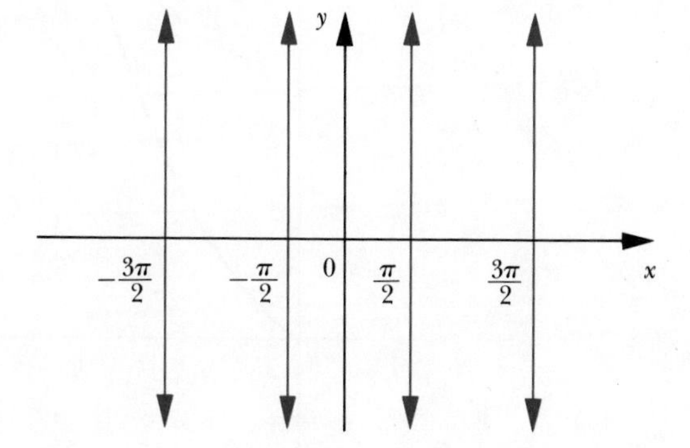

Since $\tan x = \sin x/\cos x$, $f(x) = \tan x$ is increasing in the interval $[0, \pi/2)$ because of the behavior of the sine and cosine functions in this interval. This behavior of f can be observed by examining the values in the following Table.

Table

x	$\sin x$	$\cos x$	$\tan x = \dfrac{\sin x}{\cos x}$
0	0	1	0
0.14	0.1395	0.9902	0.1409
0.28	0.2764	0.9611	0.2876
0.42	0.4078	0.9131	0.4466
0.88	0.7707	0.6372	1.2100
1.05	0.8674	0.4976	1.7430
1.44	0.9915	0.1304	7.6020
1.55	0.9998	0.0208	48.0780
1.56	0.9999	0.0108	92.6200
1.57	1.0000	0.0008	1,255.8000
$\frac{\pi}{2}$	1.0000	0.0000	(undefined)

The values in Table 1 suggest that $f(x) = \tan x$ increases indefinitely as x gets "closer" to $\pi/2$ since the numerator (sine) increases to 1 as the denominator (cosine) is simultaneously decreasing to 0. Also, there is no point on the graph corresponding to $x = \pi/2$ because $\tan \pi/2$ is not defined. The line $x = \pi/2$ is called a *vertical asymptote* of the graph of $f(x) = \tan x$ (Figure 2).

Figure 2

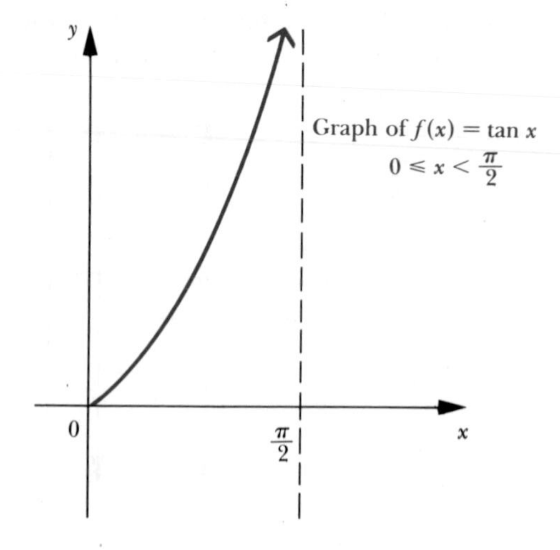

Since $f(x) = \tan x$ is an odd function, we can use the fact that f is symmetric with respect to the origin to obtain the graph of f for the interval $-\pi/2 < x \leqslant 0$ from the graph of f for $0 \leqslant x < \pi/2$ (Figure 3).

Figure 3

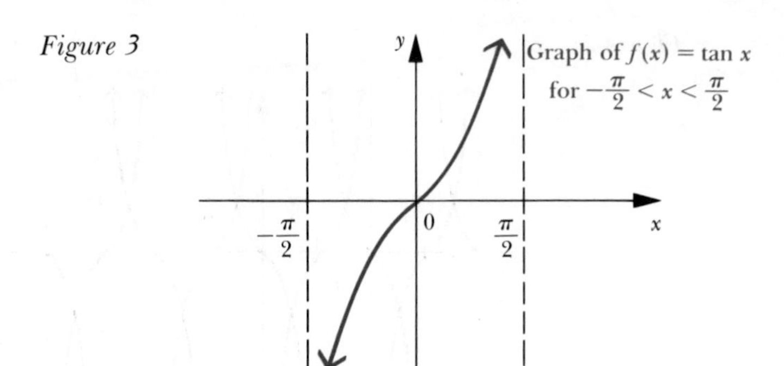

It will be proved in Chapter 5, on page 196, that the fundamental period of $f(x) = \tan x$ is π. Consequently, the graph in Figure 3 represents one cycle of the graph of $f(x) = \tan x$ which repeats every π units in either direction (Figure 4).

Figure 4

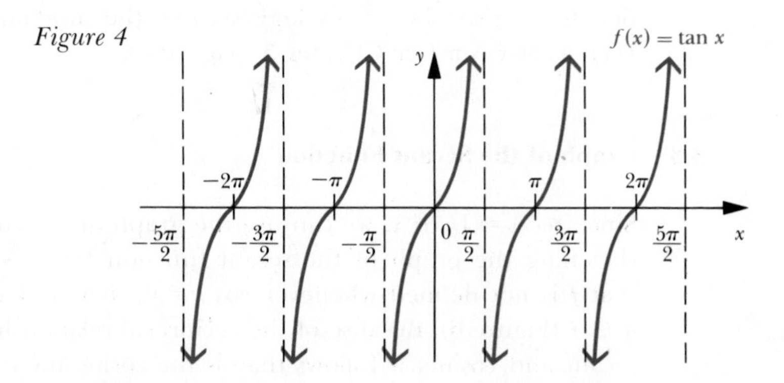

The graph of $f(x) = \tan x$ shows that the range of the tangent is the set of real numbers R. Also, the graph suggests that the zeros of $f(x) = \tan x$ are given by $x = n\pi$, where n is any integer. Also, the function increases between any two consecutive asymptotes.

3.2 Graph of the Cotangent Function

It is possible to use the graph of the tangent function, together with the fact that $\cot x = 1/\tan x$, to obtain the graph of the cotangent function $f(x) = \cot x$.

First, the values of x for which $f(x) = \cot x = 1/\tan x$ is not defined occur if $\tan x = 0$, that is, if $x = n\pi$, where n is any integer. Second, since the cotangent is the reciprocal of the tangent, the function values of $f(x) = \cot x$ decrease as the tangent values increase. Using these two results, together with the fact that the zeros of $f(x) = \cot x$ occur if $x = (\pi/2) + n\pi$, where n is any integer, we can sketch the graph of $f(x) = \cot x$ (Figure 5).

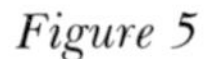

Figure 5

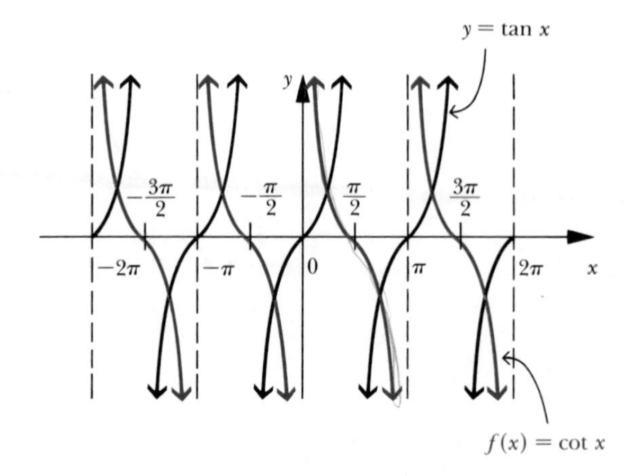

The graph of $f(x) = \cot x$ displays the fact that the cotangent is symmetric with respect to the origin and that the range of the function is R. Also, the graph suggests that the fundamental period of $f(x) = \cot x$ is π (see Chapter 5, page 196).

3.3 Graph of the Secant Function

Since $\sec x = 1/\cos x$, we can use the graph of $y = \cos x$ as an aid in sketching the graph of the secant function $f(x) = \sec x$. We know that f is not defined whenever $\cos x = 0$, that is, if $x = (\pi/2) + n\pi$, $n \in I$ (Figure 6). Because of the reciprocal relationship between the secant and cosine, it follows that as the cosine increases, the secant decreases, and as the cosine decreases, the secant increases (Figure 6). Notice that the graph of $f(x) = \sec x$ displays the fact that the range of the secant function is $(-\infty, -1] \cup [1, \infty)$ and also that the function is an even function or symmetric with respect to the y axis. It is apparent from the graph of $f(x) = \sec x$ that f does not have any zeros nor do its values ever fall between -1 and 1.

Figure 6

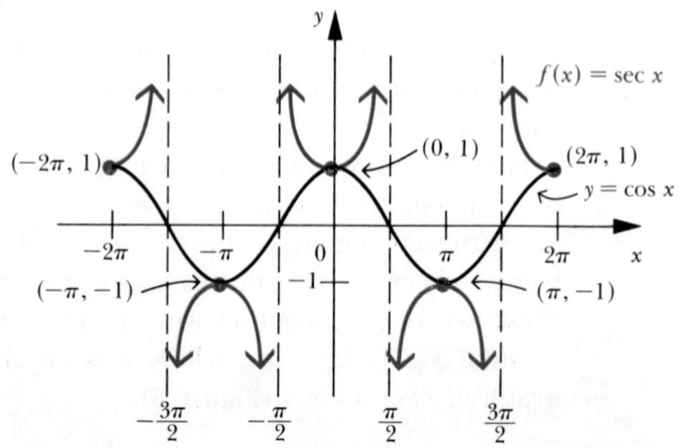

A derivation of the graph of the cosecant function $f(x) = \csc x$ is outlined in Problem 1.

EXAMPLES

Sketch the graph of each of the following functions in the interval $[0, 2\pi)$.

1 $f(x) = \tan\left(x - \frac{\pi}{6}\right)$

SOLUTION. As with the sine and cosine functions, $x - \pi/6$ indicates a phase shift of $\pi/6$ so that the graph of $f(x) = \tan(x - \pi/6)$ can be obtained by shifting the graph of $y = \tan x$, $\pi/6$ units to the right (Figure 7).

Figure 7

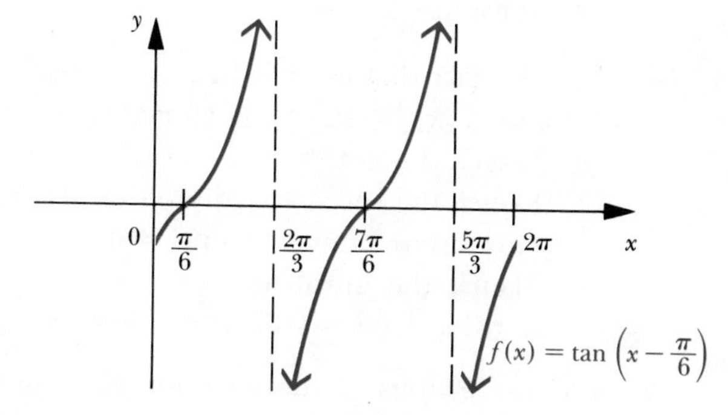

2 $f(x) = \cot 3x$

SOLUTION. Since the period of $y = \cot x$ is π, the period of $f(x) = \cot 3x$ is $\pi/3$ (Figure 8).

Figure 8

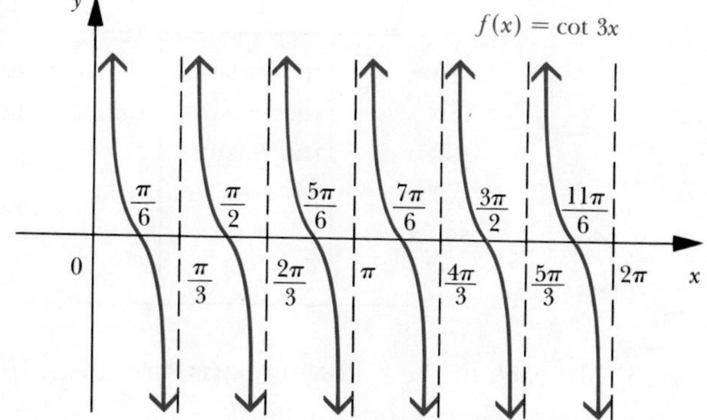

3 $f(x) = 3 \sec x$

SOLUTION. The graph of $f(x) = 3 \sec x$ can be obtained from the graph of $y = \sec x$ by multiplying each ordinate in the graph of $y = \sec x$ by 3 (Figure 9).

Figure 9

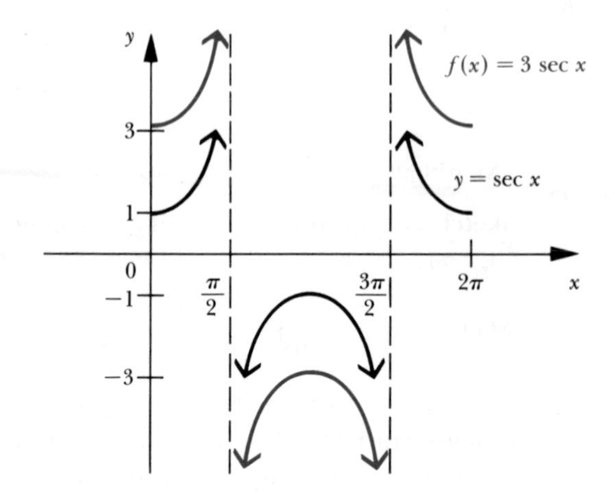

PROBLEM SET 2

1 Use the fact that $\csc x = 1/\sin x$ to sketch the graph of the cosecant function $f(x) = \csc x$ by performing the following steps.

a) Sketch the graph of $y = \sin x$.

b) Sketch the vertical asymptotes of $f(x) = \csc x$ by noting that these lines occur whenever $\sin x = 0$.

c) Sketch the graph of $f(x) = \csc x$ by using the reciprocal relationship $\csc x = 1/\sin x$ together with the sine graph.

2 Use the graphs of the trigonometric functions to complete the following table.

x increasing from ______ to ______	$0 \to \frac{\pi}{2}$	$\frac{\pi}{2} \to \pi$	$\pi \to \frac{3\pi}{2}$	$\frac{3\pi}{2} \to 2\pi$
$\sin x$	increasing	decreasing	decreasing	
$\cos x$	decreasing	decreasing		increasing
$\tan x$	increasing			
$\cot x$			decreasing	
$\sec x$				decreasing
$\csc x$				

3 In each of the following parts, use the graph to answer the questions.

i) What is the domain f?

ii) What is the range of f?

iii) Is f even or odd?

iv) What kind of symmetry does the graph of f display?

v) What is the fundamental period of f?

a) $f(x) = \tan x$ b) $f(x) = \cot x$
c) $f(x) = \sec x$ d) $f(x) = \csc x$

4 Sketch the graphs of each of the following functions in the interval $[0, 2\pi]$.
a) $f(x) = 3 \tan 2x$ b) $f(x) = \frac{1}{2} \sec x$
c) $f(x) = \frac{5}{2} \cot x$ d) $f(x) = 4 \sec (-2x)$
e) $f(x) = \tan (x/3)$ f) $f(x) = \frac{1}{4} \csc 3x$
g) $f(x) = 2 \tan (x - \pi/3)$ h) $f(x) = 2 \sec (x - \frac{1}{2})$

5 Use the graphs of the trigonometric functions to answer true or false for each of the following statements.
a) If $\pi/2 < t_1 < t_2 < \pi$, then $\sin t_1 < \sin t_2$.
b) If $0 < t_1 < t_2 < \pi/2$, then $\tan t_1 < \tan t_2$.
c) If $\pi < t_1 < t_2 < 3\pi/2$, then $\cot t_1 > \cot t_2$.
d) If $3\pi/2 < t_1 < t_2 < 2\pi$, then $\sec t_1 > \sec t_2$.
e) If $-\pi/2 < t_1 < t_2 < \pi/2$, then $\cos t_1 > \cos t_2$.
f) If $30\pi < t_1 < t_2 < 61\pi/2$, then $\csc t_1 > \csc t_2$.

6 Use the graph of $f(x) = \sin x$ to approximate each of the following values.
a) $\sin 2$ b) $\sin 1.5$ c) $\sin 2.1$ d) $\sin 1.57$

4 Inverse Functions

Our next goal is to investigate the notion of the inverses of the trigonometric functions. Before investigating this important topic, we will first study the definition of an inverse function, the formation of inverse functions, and conditions for determining whether or not a function has an inverse.

4.1 Invertible Functions

Suppose that f is a function defined by $f(x) = x + 2$ and g is the function defined by $g(x) = x - 2$. Now observe what happens when we "apply" these two functions in succession.

First, we apply f, then g, to the number 2; that is, $2 \xrightarrow{f} f(2) \xrightarrow{g} g[f(2)]$; and since $f(2) = 4$, we have $g[f(2)] = g(4) = 4 - 2 = 2$.

In general, $g[f(x)]$ is the result obtained when we first "apply" f to an element x and then "apply" g to the result. The function defined by $y = g[f(x)]$ is called a *composition* of f by g.
$x + 2$ and $g(x) = x - 2$, we have $f[g(x)] = f(x - 2) = (x - 2) + 2 = x$ and $g[f(x)] = g(x + 2) = x + 2 - 2 = x$. Here $f[g(x)] = g[f(x)]$.

Let us consider another example where $f(x) = 2x - 5$ and $g(x) = (x + 5)/2$; then

$$f[g(x)] = f\left(\frac{x+5}{2}\right) = 2\left(\frac{x+5}{2}\right) - 5 = x + 5 - 5 = x$$

and

$$g[f(x)] = g(2x-5) = \frac{(2x-5)+5}{2} = \frac{2x}{2} = x$$

Once again we have a situation in which $f[g(x)] = g[f(x)]$.

It is *not true* in general that $f[g(x)] = g[f(x)]$. For example, if $f(x) = x^2$ and $g(x) = x + 1$, we have

$$f[g(x)] = f(x+1) = (x+1)^2 = x^2 + 2x + 1$$

and

$$g[f(x)] = g(x^2) = x^2 + 1$$

Since $x^2 + 2x + 1 \neq x^2 + 1$ for all x, $f[g(x)] \neq g[f(x)]$ for these two functions f and g.

The functions f and g defined in the first two examples are called "invertible functions." These examples can be generalized as follows.

Definition

Let f and g be two functions so related that $f[g(x)] = x$ for every element x in the domain of g, and $g[f(x)] = x$ for every element x in the domain of f, then f and g are said to be *invertible,* and each is said to be the *inverse* of the other. In symbols, we write $g = f^{-1}$ or $f = g^{-1}$. In general, $f[f^{-1}(x)] = x$ and $f^{-1}[f(x)] = x$.

Thus, from the first example, we can write $f^{-1}(x) = x - 2$ for $f(x) = x + 2$, or we can write $g^{-1}(x) = x + 2$ for $g(x) = x - 2$. Similarly, in the second example, we can write $f^{-1}(x) = (x+5)/2$ for $f(x) = 2x - 5$ or $g^{-1}(x) = 2x - 5$ for $g(x) = (x+5)/2$.

EXAMPLES

1 Suppose that $f(x) = x^3$ and $g(x) = \sqrt[3]{x}$; show that f and g are invertible. (Recall that $\sqrt[3]{x}$ is defined for all x in R.)

SOLUTION. Let us examine $f[g(x)]$ and $g[f(x)]$.

$$f[g(x)] = f(\sqrt[3]{x}) = (\sqrt[3]{x})^3 = x$$

and

$$g[f(x)] = g(x^3) = \sqrt[3]{x^3} = x$$

Since $f[g(x)] = g[f(x)] = x$, it follows that $f = g^{-1}$ and $g = f^{-1}$.

2 Suppose that $f(x) = 5x + 9$ and $g(x) = (x - 9)/5$; show that f and g are invertible.

SOLUTION.

$$f[g(x)] = f\left(\frac{x-9}{5}\right) = 5\left(\frac{x-9}{5}\right) + 9 = x - 9 + 9 = x$$

and

$$g[f(x)] = g(5x+9) = \frac{(5x+9)-9}{5} = \frac{5x}{5} = x$$

Hence, $f = g^{-1}$ and $g = f^{-1}$.

4.2 Existence of Inverse Functions

Consider the two functions $f = \{(1, 2), (3, 1), (4, 3)\}$ and $g = \{(1, 2), (3, 2), (4, 5)\}$. Notice that for each member of the domain of f, there is one and only one corresponding member of the range. Similarly, for each member of the range of f there is one and only one corresponding member of the domain of f, that is,

$$f \quad \begin{array}{l} 1 \longleftrightarrow 2 \\ 3 \longleftrightarrow 1 \\ 4 \longleftrightarrow 3 \end{array}$$

whereas, for g, 2 in the range corresponds to more than one member of the domain, that is,

$$g \quad \begin{array}{l} 1 \longrightarrow 2 \\ 3 \nearrow \\ 4 \longleftrightarrow 5 \end{array}$$

We say that f is a "one-to-one" function, whereas g is not one-to-one.

In general, a function f is *one-to-one* if each member of the range of the function corresponds to one and only one member of the domain; that is, if $x_1 \neq x_2$, then $f(x_1) \neq f(x_2)$.

We can use the graph of f to decide whether or not a function f is one-to-one. For example, the function f defined by $f(x) = x^3$ is one-to-one since for each y there is one and only one x. This means that any possible horizontal lines intersect the graph no more than once (Figure 1).

Figure 1

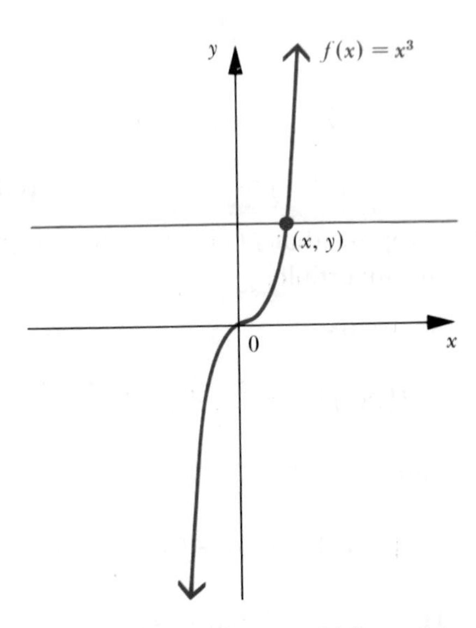

However, the function f defined by $f(x) = x^2$ is not one-to-one, since any horizontal line above the x axis intersects the graph twice (Figure 2). Consequently, f is not invertible.

Figure 2

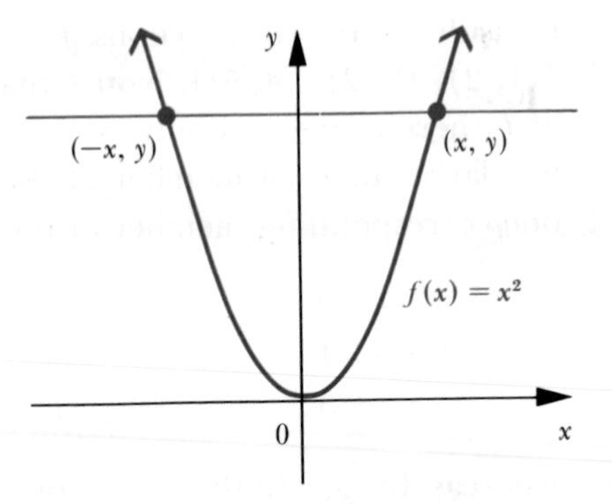

The function $f(x) = \sin x$ is not one-to-one, since it is possible to draw a horizontal line that intersects the graph in more than one point (Figure 3).

Figure 3

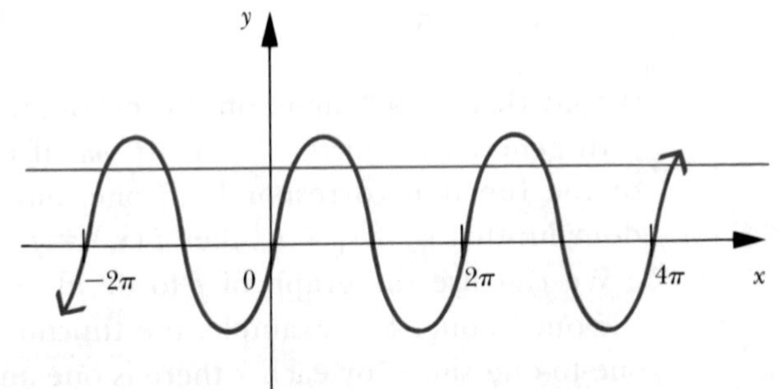

In fact, an examination of the graphs of the trigonometric functions in Section 3 indicates that none of the trigonometric functions is one-to-one.

The following property characterizes the existence of the inverse function.

PROPERTY 1

If a function f is one-to-one, then f has an inverse; conversely, if f has an inverse, then f is one-to-one.

Thus, we can examine the graph of the function to determine whether or not the function has an inverse. For example, $f(x) = x^3$ has an inverse, whereas $f(x) = x^2$ has no inverse (Figures 1 and 2).

If $f(x) = 3x + 2$, then we find the value of f^{-1} at a given number y by solving the equations $y = f(x)$, where $y = 3x + 2$, for x in terms of y. We have $x = \frac{1}{3}y - \frac{2}{3}$; or $f^{-1}(y) = \frac{1}{3}y - \frac{2}{3}$. The letter that we use to denote a number in the domain of the inverse function is of no importance whatsoever, so that the last equation can also be written as $f^{-1}(u) = \frac{1}{3}u - \frac{2}{3}$ or $f^{-1}(t) = \frac{1}{3}t - \frac{2}{3}$, or even $f^{-1}(x) = \frac{1}{3}x - \frac{2}{3}$, and it will still define the same function f^{-1}.

Now we will verify that $f[f^{-1}(x)] = f^{-1}[f(x)] = x$ as follows:

$$f[f^{-1}(x)] = f(\tfrac{1}{3}x - \tfrac{2}{3}) = 3(\tfrac{1}{3}x - \tfrac{2}{3}) + 2 = (x - 2) + 2 = x$$

and

$$f^{-1}[f(x)] = f^{-1}(3x + 2) = \tfrac{1}{3}(3x + 2) - \tfrac{2}{3} = (x + \tfrac{2}{3}) - \tfrac{2}{3} = x.$$

Geometrically, the graph of f^{-1} is obtained from the graph of f by reflecting ("flipping") the graph across the line $y = x$ since the numbers in the ordered pairs of the inverse graph are the same numbers as in the ordered pairs of the original function graph except that the order is reversed. For example, $(1, 5) \in f$, whereas $(5, 1) \in f^{-1}$

Figure 4

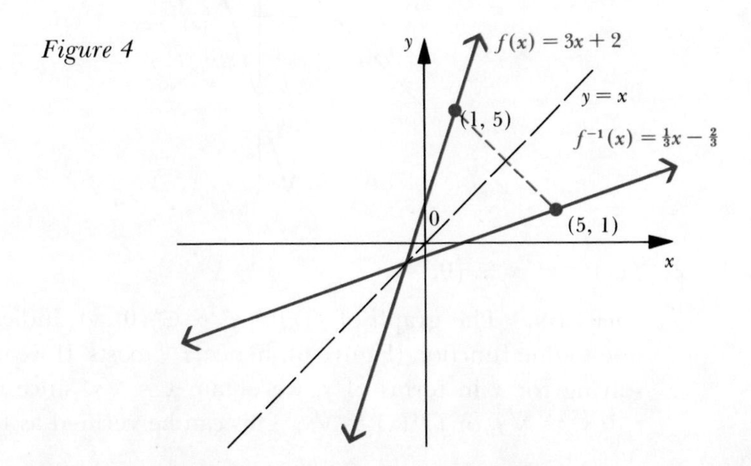

EXAMPLES

Given the function f, find f^{-1} and graph f^{-1} and f on the same coordinate system.

1 $f(x) = 3x - 4$

SOLUTION. f is a one-to-one function (Figure 5); hence, f^{-1} exists. If we let $y = 3x - 4$, then $x = (y + 4)/3$ results from solving for x in terms of y; therefore, $f^{-1}(y) = (y + 4)/3$, or $f^{-1}(x) = (x + 4)/3$. This can be verified as follows:

$$f[f^{-1}(x)] = f\left(\frac{x+4}{3}\right) = 3\left(\frac{x+4}{3}\right) - 4 = x$$

and

$$f^{-1}[f(x)] = f^{-1}(3x - 4) = \frac{3x - 4 + 4}{3} = x$$

After graphing $f(x) = 3x - 4$ and $f^{-1}(x) = (x + 4)/3$ on the same coordinate system, we again observe that the graph of the inverse function f^{-1} can be obtained by reflecting the graph of f across the line $y = x$ (Figure 5).

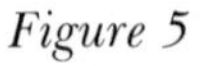

Figure 5

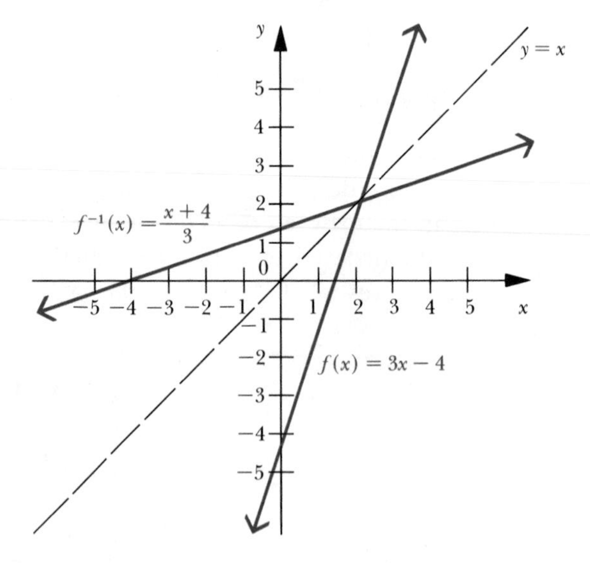

2 $f(x) = x^2$, $x \in [0, \infty)$

SOLUTION. The graph of $f(x) = x^2$, $x \in [0, \infty)$, indicates that f is a one-to-one function (Figure 6); hence, f^{-1} exists. If we let $y = x^2$, then, solving for x in terms of y, we obtain $x = \sqrt{y}$, since $x \in [0, \infty)$, so $f^{-1}(y) = \sqrt{y}$, or $f^{-1}(x) = \sqrt{x}$. This can be verified as follows:

$$f[f^{-1}(x)) = f(\sqrt{x}) = (\sqrt{x})^2 = x$$

and

$$f^{-1}[f(x)] = f^{-1}(x^2) = \sqrt{x^2} = x \qquad \text{for } x \in [0, \infty)$$

The graph of f^{-1} is a reflection of the graph of f across $y = x$ (Figure 6).

Figure 6

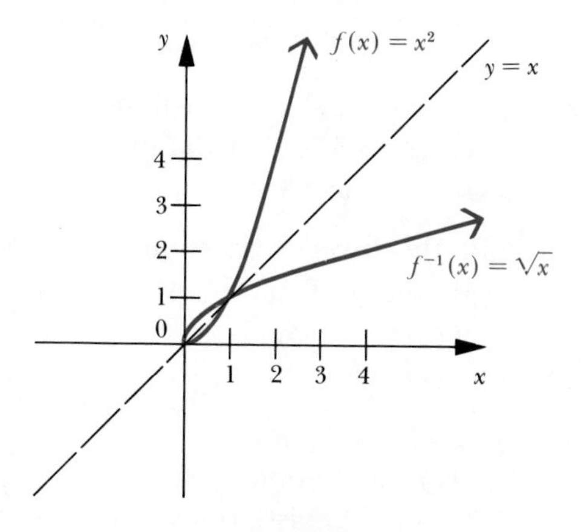

Notice here that $f(x) = x^2$ (Figure 2 on page 166) does not have an inverse. However, if we restrict the domain to construct the function $f(x) = x^2$, $x \in [0, \infty)$ (Figure 6), we get an invertible function. This is a preview of the process that will be used when we construct the inverses of the trigonometric functions.

3 Prove that every increasing function has an inverse.

PROOF. Assume that f is an increasing function. If x_1 and x_2 are different members of the domain of f and $x_1 < x_2$, then, since f is increasing, $f(x_1) < f(x_2)$ (Figure 7). This means that no two different ordered pairs of the function f have the same second members. In other words, each member of the range is the image of one and only one member of the domain; that is, f is one-to-one, from which we can conclude by Property 1 that f^{-1} exists.

Figure 7

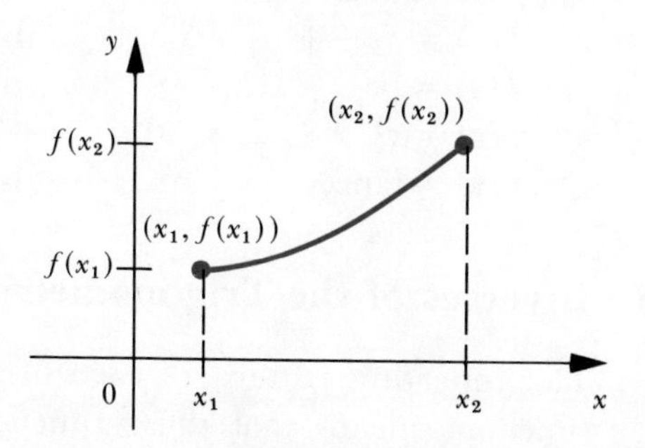

PROBLEM SET 3

1 Verify that $g = f^{-1}$ for each of the following pairs of functions.

a) $f = \{(5, -3), (6, -2), (3, 4)\}$, $g = \{(-3, 5), (-2, 6), (4, 3)\}$

b) $f(x) = 2x - 3$ and $g(x) = \dfrac{x+3}{2}$

c) $f(x) = 2x - 1$ and $g(x) = \dfrac{x+1}{2}$

d) $f(x) = \dfrac{1}{x}$ and $g(x) = \dfrac{1}{x}$, $x \neq 0$

2 Graph each pair of functions in Problem 1 on the same coordinate system. Observe that the graph of f^{-1} is a reflection of the graph of f across the line $y = x$.

3 Find the inverse function f^{-1} of each of the following functions. Verify that $f[f^{-1}(x)] = x$ and $f^{-1}[f(x)] = x$.

a) $f(x) = 1 - 3x$ b) $f(x) = x^3 - 1$

c) $f(x) = 3/x$ d) $f(x) = \frac{2}{3}x + \frac{1}{2}$

4 Let $f(x) = x^3 + 4$.

a) Is f one-to-one?

b) Does f^{-1} exist? If so, find it.

c) If f^{-1} exists, find $f^{-1}(4)$, $f^{-1}(0)$, and $f^{-1}(-4)$.

5 Let $f(x) = -3x + 7$.

a) Is f a decreasing function?

b) Does f^{-1} exist? If so, find it.

c) If f^{-1} exists, find $f^{-1}(0)$ and $f^{-1}(1)$ and sketch the graph of f^{-1}.

6 Let $f(x) = x^3 + x$.

a) Is f an increasing function?

b) Does f^{-1} exist?

7 a) Verify that every decreasing function has an inverse.

b) Explain why no periodic function is invertible.

8 Examine the graphs of each of the following functions to determine whether or not f^{-1} exists. If f^{-1} exists, find it and graph f^{-1} on the same coordinate system as f.

a) $f(x) = 3x - 1$ b) $f(x) = -2x + 5$

c) $f(x) = |x|$ d) $f(x) = 5 \sin 2x$

e) $f(x) = x^2$, $x \in (-\infty, 0]$ f) $f(x) = -2 \cos x$

g) $f(x) = \tan x$ h) $f(x) = \frac{1}{2} \sin (x/2 + 1)$

5 Inverses of the Trigonometric Functions

The repeating pattern of each of the graphs of the trigonometric functions means that these functions are periodic. By drawing a horizontal line through the graph of any one of the trigonometric functions, we can observe that this horizontal line intersects the graph "infinitely" often.

For example, if $f(x) = \cos x$ and the horizontal line $y = \frac{1}{2}$ are both graphed on the same coordinate system, it can be seen that this line intersects the graph of f in more than one point (Figure 1).

Figure 1

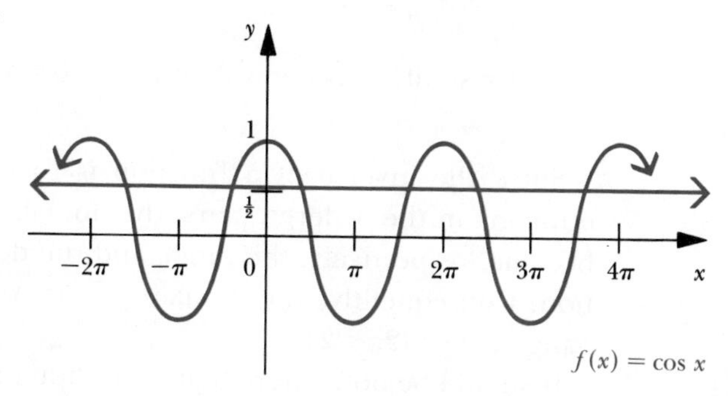

This means that none of the trigonometric functions is invertible. What we will do is to restrict the domains of the trigonometric functions to construct new functions that are invertible.

5.1 Inverses of the Sine and Cosine Functions

We will begin by constructing the function $f(x) = \text{Sin } x$ (notice that a capital letter S is used here to distinguish this function from the sine function) by restricting the domain of $f(x) = \sin x$ to the interval $[-\pi/2, \pi/2]$. The range is $[-1, 1]$ (Figure 2). (Why?) Since $f(x) = \text{Sin } x$ is increasing in the interval $[-\pi/2, \pi/2]$, it has an inverse. This inverse is called the Arcsine function.

Figure 2

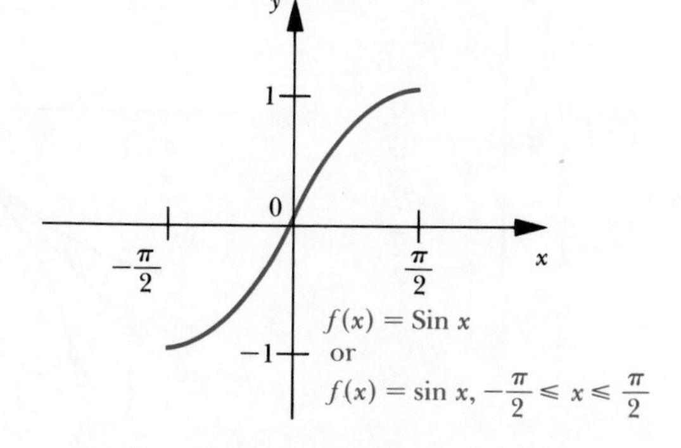

Definition

The *Arcsine function,* denoted by $y = \text{Arcsin } x$ or by $y = \sin^{-1} x$, is the inverse of the function $f(x) = \text{Sin } x$.

$y = \text{Arcsin } x$ is equivalent to $x = \sin y$ with $-\pi/2 \leq y \leq \pi/2$.

For example,

$$\frac{\sqrt{3}}{2} = \sin \frac{\pi}{3} \quad \text{is equivalent to} \quad \frac{\pi}{3} = \text{Arcsin } \frac{\sqrt{3}}{2}$$

$$\frac{1}{\sqrt{2}} = \sin \frac{\pi}{4} \quad \text{is equivalent to} \quad \frac{\pi}{4} = \text{Arcsin } \frac{1}{\sqrt{2}}$$

$$0 = \sin 0 \quad \text{is equivalent to} \quad 0 = \sin^{-1} 0.$$

Since the inverse of a function is formed by interchanging the numbers in the ordered pairs, the domain and range of $f(x) = \text{Sin } x$ become, respectively, the range and the domain of the inverse function. Consequently, the domain of $y = \text{Arcsin } x$ is $[-1, 1]$ and the range is $[-\pi/2, \pi/2]$.

It should be noted here that even though we write $\sin^2 x = (\sin x)^2$, we never use $\sin^{-1} x = (\sin x)^{-1}$, since $(\sin x)^{-1}$ is $1/\sin x$, which is not the same as Arcsin x. The problem arises because we are using the "exponent" "$^{-1}$" ambiguously—it denotes both reciprocals and inverse functions.

Geometrically, the graph of the inverse sine function $y = \text{Arcsin } x$ can be obtained by "reflecting" the graph of the function $f(x) = \text{Sin } x$ about the line whose equation is $y = x$ (Figure 3).

Figure 3

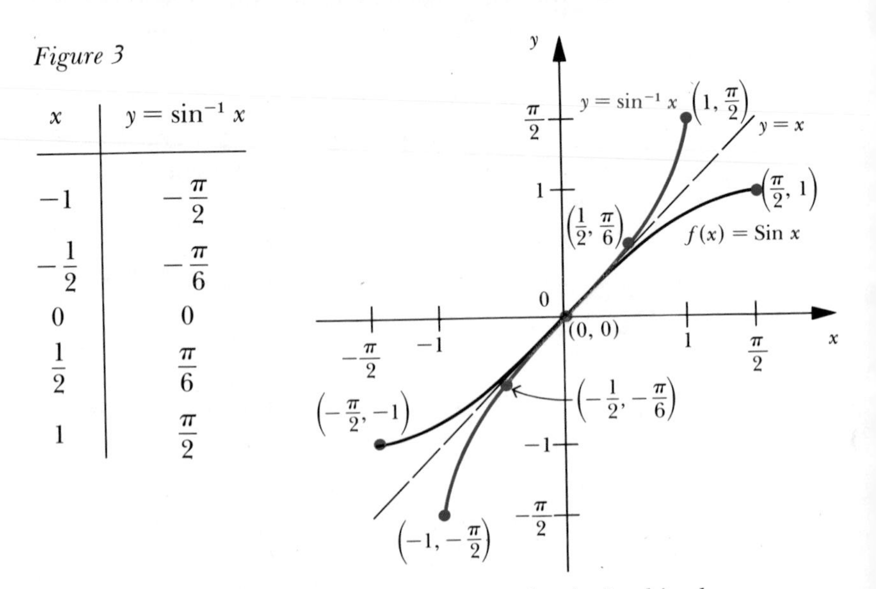

x	$y = \sin^{-1} x$
-1	$-\frac{\pi}{2}$
$-\frac{1}{2}$	$-\frac{\pi}{6}$
0	0
$\frac{1}{2}$	$\frac{\pi}{6}$
1	$\frac{\pi}{2}$

The inverse of the cosine function can be derived in the same way as the inverse sine function was derived. First, we can construct $g(x) = \text{Cos } x$ (notice the use of capital C) from $y = \cos x$ by restricting the domain to the interval $[0, \pi]$. The range is $[-1, 1]$ (Figure 4). (Why?) Since $g(x) = \text{Cos } x$ is a *decreasing* function, it has an inverse called the Arccosine function.

Figure 4

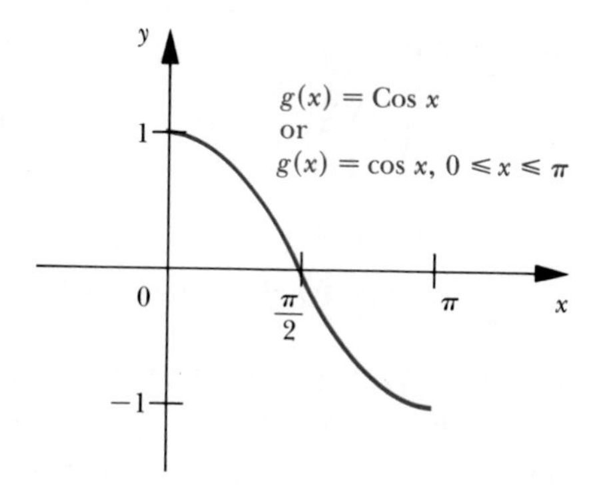

Definition

The *Arccosine function,* denoted by $y = \text{Arccos } x$ or by $y = \cos^{-1} x$, is the inverse of the function $g(x) = \text{Cos } x$.

$y = \text{Arccos } x$ is equivalent to $x = \cos y$ with $0 \leqslant y \leqslant \pi$.

For example,

$$-1 = \cos \pi \quad \text{is equivalent to} \quad \pi = \text{Arccos } (-1)$$

$$-\frac{1}{\sqrt{2}} = \cos \frac{3\pi}{4} \quad \text{is equivalent to} \quad \frac{3\pi}{4} = \cos^{-1}\left(-\frac{1}{\sqrt{2}}\right)$$

$$0 = \cos \frac{\pi}{2} \quad \text{is equivalent to} \quad \frac{\pi}{2} = \cos^{-1} 0$$

The domain of $y = \text{Arccos } x$ is $[-1, 1]$ and the range is $[0, \pi]$.

The graph of $g(x) = \cos^{-1} x$ can be obtained from the graph of $y = \text{Cos } x$ by reflecting the latter graph across the line $y = x$ as illustrated in Figure 5.

Figure 5

x	$y = \cos^{-1} x$
-1	π
$-\frac{1}{2}$	$\frac{2\pi}{3}$
0	$\frac{\pi}{2}$
$\frac{1}{2}$	$\frac{\pi}{3}$
1	0

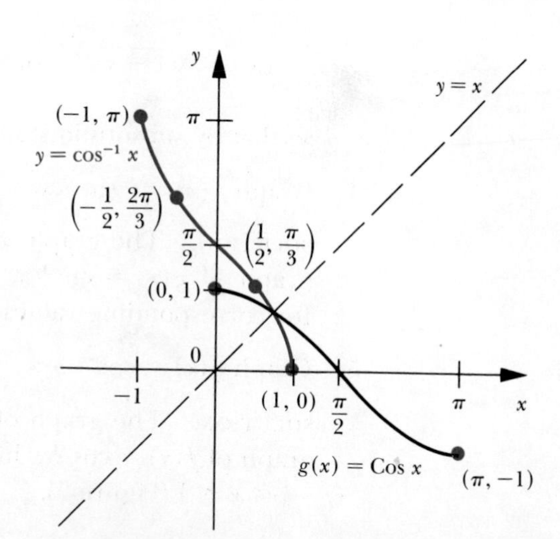

EXAMPLES

1 Find each of the following values.

a) $\text{Arcsin } \frac{1}{2}$ b) $\text{Arccos } \dfrac{\sqrt{2}}{2}$

c) $\sin^{-1}\left(\dfrac{-\sqrt{2}}{2}\right)$ d) $\tan\left[\cos^{-1}\left(-\dfrac{\sqrt{3}}{2}\right)\right]$

SOLUTION

a) Let $\text{Arcsin } \frac{1}{2} = t$; then $\sin t = \frac{1}{2}$, where $t \in [-\pi/2, \pi/2]$, so that $t = \pi/6$.
b) Let $\text{Arccos } \sqrt{2}/2 = t$; then $\cos t = \sqrt{2}/2$, where $t \in [0, \pi]$, so that $t = \pi/4$.
c) Let $\sin^{-1}(-\sqrt{2}/2) = t$; then $\sin t = -\sqrt{2}/2$, where $t \in [-\pi/2, \pi/2]$ so that $t = -\pi/4$.
d) Let $\cos^{-1}(-\sqrt{3}/2) = t$; then $\cos t = -\sqrt{3}/2$, where $t \in [0, \pi]$ so that $t = 5\pi/6$. Thus $\tan[\cos^{-1}(-\sqrt{3}/2)] = \tan(5\pi/6) = -\sqrt{3}/3$.

2 Use Table III to find each of the following values.

a) $\sin^{-1} 0.8016$ b) $\sin^{-1}(-0.8016)$ c) $\cos^{-1} 0.2675$

SOLUTION

a) Let $\sin^{-1} 0.8016 = t$; then $\sin t = 0.8016$, where $t \in [-\pi/2, \pi/2]$. From Table II we find that $t = 0.93$, so that $\sin^{-1} 0.8016 = 0.93$.
b) From part (a) we have $\sin^{-1} 0.8016 = 0.93$. Consequently, $\sin^{-1}(-0.8016) = -0.93$.
c) Let $\cos^{-1} 0.2675 = t$; then $\cos t = 0.2675$, where $t \in [0, \pi]$. From Table II we find that $t = 1.30$, so that $\cos^{-1} 0.2675 = 1.30$.

3 Show that $\sin^{-1} x = \cos^{-1}\sqrt{1-x^2}$ for $0 \leq x \leq 1$.

PROOF. Let $t = \sin^{-1} x$, then $\sin t = x$, where $t \in [-\pi/2, \pi/2]$. Since $\cos t = \sqrt{1-\sin^2 t}$ in quadrant I, we have

$$\cos t = \sqrt{1-x^2} \qquad \text{or} \qquad t = \cos^{-1}\sqrt{1-x^2}$$

so that by substitution, $\sin^{-1} x = \cos^{-1}\sqrt{1-x^2}$.

4 Graph $f(x) = \frac{1}{2}\sin^{-1} x$.

SOLUTION. The graph of $f(x) = \frac{1}{2}\sin^{-1} x$ can be obtained from the graph of $g(x) = \sin^{-1} x$. The ordinates of f will be $\frac{1}{2}$ of those of g for corresponding values of x (Figure 6).

5 Graph $g(x) = \cos^{-1} 3x$.

SOLUTION. The graph of $g(x) = \cos^{-1} 3x$ can be obtained from the graph of $f(x) = \cos^{-1} x$ by noting that $-1 \leq 3x \leq 1$ when $-\frac{1}{3} \leq x \leq \frac{1}{3}$ (Figure 7).

Figure 6

x	$f(x)$
-1	$-\frac{\pi}{4}$
$-\frac{1}{2}$	$-\frac{\pi}{12}$
0	0
$\frac{1}{2}$	$\frac{\pi}{12}$
1	$\frac{\pi}{4}$

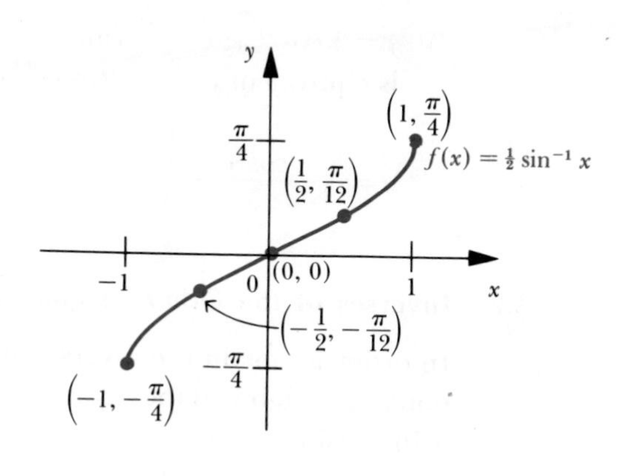

Figure 7

x	$g(x)$
$-\frac{1}{3}$	π
0	$\frac{\pi}{2}$
$\frac{1}{3}$	0

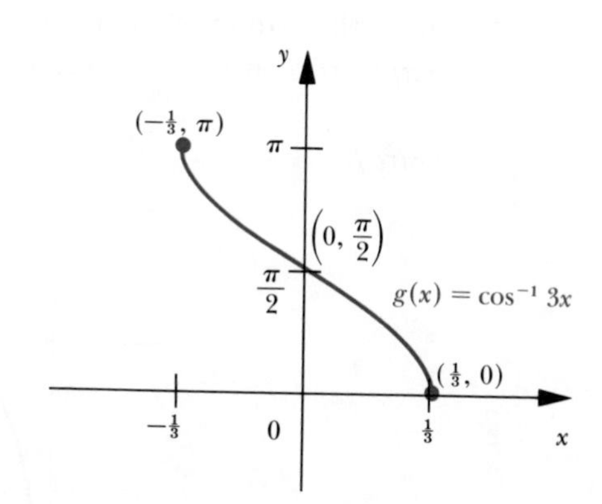

6 Express x in terms of y for each of the following equations.
a) $y = \cos^{-1} 5x$
b) $y = 3 \cos^{-1} 5x$
c) $y = 5 + 3 \cos^{-1} 5x$
d) $y = 3 \cos^{-1} (5x - 2)$

SOLUTION

a) $y = \cos^{-1} 5x$ is equivalent to $5x = \cos y$, so

$$x = \tfrac{1}{5} \cos y$$

b) $y = 3 \cos^{-1} 5x$ can be written as $y/3 = \cos^{-1} 5x$, which is equivalent to $5x = \cos (y/3)$, so that

$$x = \frac{1}{5} \cos \left(\frac{y}{3}\right)$$

c) $y = 5 + 3 \cos^{-1} 5x$ can be written as $(y - 5)/3 = \cos^{-1} 5x$. The latter equation is equivalent to $5x = \cos [(y - 5)/3]$, so that

$$x = \frac{1}{5} \cos \left(\frac{y - 5}{3}\right)$$

d) $y = 3\cos^{-1}(5x - 2)$ can be written as $y/3 = \cos^{-1}(5x - 2)$, which is equivalent to $5x - 2 = \cos(y/3)$, so that

$$x = \frac{2 + \cos(y/3)}{5}$$

5.2 Inverses of the Other Trigonometric Functions

In order to obtain the inverse tangent function, we restrict the function $f(x) = \tan x$ to the open interval $(-\pi/2, \pi/2)$, where the function is increasing.

Construct $f(x) = \operatorname{Tan} x$ (notice the use of capital T) with the domain $(-\pi/2, \pi/2)$. The range is the set of real numbers (Figure 8). Since the function f is an increasing function in the interval $(-\pi/2, \pi/2)$, it has an inverse which we call the Arctangent function.

Figure 8

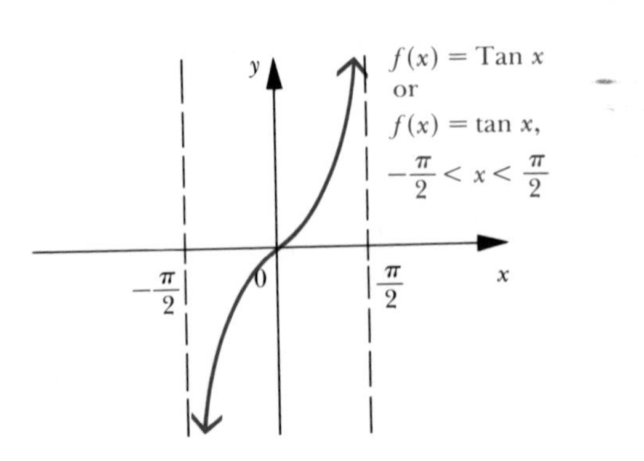

Definition

The *Arctangent function,* denoted by $y = \operatorname{Arctan} x$ or by $y = \tan^{-1} x$, is the inverse function of $f(x) = \operatorname{Tan} x$.

$y = \tan^{-1} x$ is equivalent to $x = \tan y$

with $-\pi/2 < y < \pi/2$.

For example,

$1 = \tan \frac{\pi}{4}$	is equivalent to	$\frac{\pi}{4} = \tan^{-1} 1$
$0 = \tan 0$	is equivalent to	$0 = \operatorname{Arctan} 0$
$-1 = \tan\left(-\frac{\pi}{4}\right)$	is equivalent to	$-\frac{\pi}{4} = \operatorname{Arctan}(-1)$

The domain of Arctangent is the set of real numbers and the range is $(-\pi/2, \pi/2)$.

Figure 9

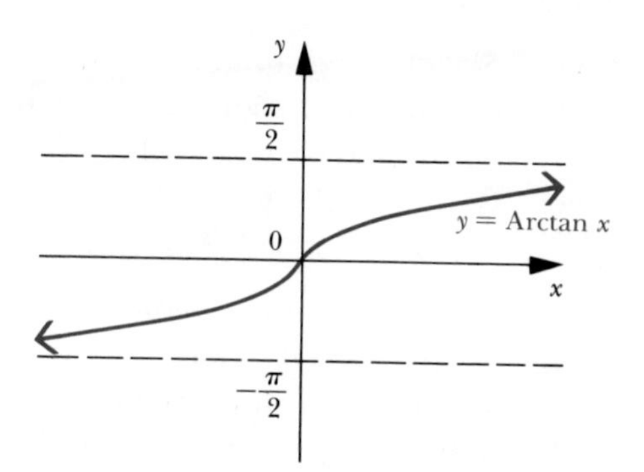

The graph of $y = \text{Arctan } x$ is displayed in Figure 9 (see Problem 4).

The derivations of the inverses for the cotangent, secant, and cosecant functions are assigned in Problems 5, 6, and 7.

PROBLEM SET 4

1 Fill in the blanks in each of the following cases.

a) If $y = \tan^{-1} x$, then $x = \tan y$ and __________ $< y <$ __________.

b) If $y = \cos^{-1} x$, then $x = \cos y$ and __________ $\leqslant y \leqslant$ __________.

c) If $-\pi/2 < x < \pi/2$, then $\tan^{-1} (\tan x) =$ __________ and $\sin^{-1} (\sin x) =$ __________.

d) If $0 \leqslant x \leqslant \pi$, then $\cos^{-1} (\cos x) =$ __________.

2 Find the value of each of the following expressions. Use Table II. if necessary.

a) $\cos^{-1} \frac{\sqrt{2}}{2}$

b) $\sin^{-1} \frac{\sqrt{3}}{2}$

c) $\sin^{-1} \left(-\frac{\sqrt{2}}{2}\right)$

d) $\cos^{-1} \left(-\frac{\sqrt{3}}{2}\right)$

e) $\cos^{-1} 0.8678$

f) $\sin (\sin^{-1} \frac{2}{3})$

g) $\tan^{-1} \sqrt{3}$

h) $\tan^{-1} \left(-\frac{1}{\sqrt{3}}\right)$

i) $\cot^{-1} (-1)$

j) $\cot (\cot^{-1} 3)$

k $\sin^{-1} (0.9819)$

l) $\tan^{-1} (-1.072)$

3 Evaluate each of the following expressions.

a) $\tan (\sin^{-1} \frac{1}{2})$

b) $\sec (\cos^{-1} - \frac{1}{2})$

c) $\sin^{-1} \left(\cos \frac{5\pi}{3}\right)$

d) $\cos^{-1} \left(\sin \frac{4\pi}{3}\right)$

e) $\cos (\cos^{-1} \frac{7}{8})$

f) $\tan (\tan^{-1} \frac{4}{5})$

g) $\cos (\sin^{-1} \frac{1}{3})$

h) $\cot (\tan^{-1} \sqrt{3})$

i) $\sin^{-1} \left(\cos \frac{\pi}{6}\right)$

j) $\cot^{-1} \left[\tan \left(-\frac{\pi}{4}\right)\right]$

4 a) Sketch the graph of $y = \tan x$, $-\pi/2 < x < \pi/2$.

b) On the same coordinate system, plot the points of the graph of $y = \text{Arctan } x$ for $x = -1, -\sqrt{3}, 0, 1$, and $1/\sqrt{3}$.

c) Complete the graph of $y = \text{Arctan } x$ by reflecting the graph of $y = \tan x$, $-\pi/2 < x < \pi/2$, across $y = x$.

5 *The Inverse Cotangent Function*

a) Assume $f(x) = \text{Cot } x$ is the function $f(x) = \cot x$, $0 < x < \pi$. Graph f. Explain why f has an inverse.

b) What is the inverse of f? Indicate the domain and range of f^{-1}.

c) Graph f^{-1}.

d) State the definition of the inverse cotangent function.

6 *The Inverse Secant Function*

a) Let $f(x) = \text{Sec } x$ be the function $f(x) = \sec x$, $0 \leqslant x < \pi/2$ or $\pi/2 < x \leqslant \pi$. Graph f. Explain why f has an inverse.

b) What is the inverse of f? Indicate the domain and range of f^{-1}.

c) Graph f^{-1}.

d) State the definition of the inverse secant function.

7 *The Inverse Cosecant Function*

a) Let $f(x) = \text{Csc } x$ be the function $f(x) = \csc x$, $-\pi/2 \leqslant x < 0$ or $0 < x \leqslant \pi/2$. Graph f. Explain why f has an inverse.

b) What is the inverse of f? Indicate the domain and range of f^{-1}.

c) Graph f^{-1}.

d) State the definition of the inverse cosecant function.

8 Show that the following statements are true.

a) $\sin^{-1} 1 - \sin^{-1} (-\frac{1}{2}) + \tan^{-1} 1 = 17\pi/12$

b) $\cot^{-1} x = \cos^{-1} \dfrac{x}{\sqrt{1 + x^2}}$

9 Express x in terms of y for each of the following equations.

a) $y = \cos^{-1} 2x$

b) $y = 2 \sin^{-1} 3x$

c) $y = 5 - \tan^{-1} 4x$

d) $y = \cos^{-1} (2x + 1)$

REVIEW PROBLEM SET

1 Use the sine and cosine functions to graph a cycle of each of the following functions. Indicate the amplitude, the period, and the zeros for the cycle.

a) $f(x) = 5 \sin x$

b) $f(x) = \frac{1}{3} \cos x$

c) $f(x) = \cos 4x$

d) $f(x) = \sin \frac{1}{3}x$

e) $f(x) = \sin (-2x)$

f) $f(x) = \cos (-\pi x)$

2 Use the sine and cosine functions as an aid to graph a cycle of the following functions. Indicate the amplitude, the period, and the phase shift.

a) $f(x) = 2 \sin (x - \pi/6)$
b) $f(x) = 4 \sin (x - 1)$
c) $f(x) = \frac{1}{2} \cos (2x + \pi/8)$
d) $f(x) = 3 \sin [(x/2) - 1]$
e) $f(x) = 3 \sin [(x - \pi)/3]$
f) $f(x) = \frac{2}{3} \cos [(\pi/4) + 3x]$

3 Use the graphs of the sine function and the cosine function to obtain the graph of $f(x) = 2 \sin x + \cos x$.

4 Sketch the graphs of each of the following functions on the interval $[0, 2\pi)$. What is the fundamental period in each case?

a) $f(x) = 2 \tan 3x$
b) $f(x) = \csc 2x$
c) $f(x) = \frac{1}{3} \sec [x - (\pi/2)]$
d) $f(x) = \cot (2x + \pi)$

5 Use the graph of each of the following functions to decide whether or not the function has an inverse. If the inverse exists, find it, graph it, and determine the domain and range.

a) $f(x) = -5x + 1$
b) $f(x) = \sin x$
c) $f(x) = \sin x,\ 0 \leqslant x \leqslant \pi/2$
d) $f(x) = -\tan x$
e) $f(x) = \cos x,\ \pi \leqslant x \leqslant 2\pi$

6 Find the values of each of the following expressions without the use of tables.

a) $\tan^{-1} 1$
b) $\cos^{-1} (-\frac{1}{2})$
c) $\sec^{-1} \sqrt{2}$
d) $\cot^{-1} (-\sqrt{3})$
e) $\csc^{-1} 2$
f) $\sin^{-1} (-1)$

7 Use Table II to find the values of the following expressions.

a) $\sin^{-1} 0.4969$
b) $\tan^{-1} 0.5334$
c) $\cos^{-1} (-0.6294)$
d) $\cos^{-1} 0.8419$
e) $\sin^{-1} (-0.9959)$
f) $\tan^{-1} (-0.3203)$

8 Evaluate each of the following expressions without the use of tables.

a) $\cos (\sin^{-1} \frac{7}{25})$
b) $\sin (\tan^{-1} \frac{3}{4})$
c) $\sin [\cos^{-1} (-\frac{24}{25})]$
d) $\tan (\cos^{-1} \frac{5}{13})$
e) $\tan [\sin^{-1} (-\frac{3}{4})]$
f) $\tan^{-1} [\cot (4\pi/9)]$

9 Express x in terms of y for each of the following equations.

a) $y = \sin^{-1} 3x$
b) $y = 2 \tan^{-1} (x + 3)$
c) $y = 3 + \sec^{-1} 4x$
d) $2y = \pi - \cos^{-1} (2x - 1)$

CHAPTER 5

Trigonometric Identities and Equations

CHAPTER 5

Trigonometric Identities and Equations

1 Introduction

An equation that is true for all values for which both sides are defined is called an *identity*. The following relations are called *trigonometric identities.*

$$\tan t = \frac{\sin t}{\cos t} \qquad \cot t = \frac{\cos t}{\sin t} \qquad \sec t = \frac{1}{\cos t} \qquad \csc t = \frac{1}{\sin t} \qquad \text{and}$$

$$\cot t = \frac{1}{\tan t}.$$

Note that these identities hold for the trigonometric functions as well as the circular functions.

In this chapter we will examine identities that involve one, two, or more real numbers or multiple of real numbers. Throughout this chapter, even though the proofs of the identities are given in terms of real numbers s and t, the results are also true if s and t are interpreted as angles. In other words, s and t represent real numbers or angles.

2 Trigonometric Identities Involving One Number or Angle

The trigonometric identities derived in Section 3 of Chapter 3 were proved by use of the definition of the circular functions on page 107. The following is a complete list of our previous results.

1 $\tan t = \dfrac{\sin t}{\cos t}$ 2 $\cot t = \dfrac{\cos t}{\sin t}$

3 $\sec t = \dfrac{1}{\cos t}$ 4 $\csc t = \dfrac{1}{\sin t}$

5 $\cot t = \dfrac{1}{\tan t}$ 6 $\sin^2 t + \cos^2 t = 1$

7	$\cos(-t) = \cos t$	8	$\sin(-t) = -\sin t$
9	$\tan(-t) = -\tan t$	10	$\cot(-t) = -\cot t$
11	$\sec(-t) = \sec t$	12	$\csc(-t) = -\csc t$
13	$\sin t = \sin(t + 2\pi)$	14	$\cos t = \cos(t + 2\pi)$
15	$\tan t = \tan(t + 2\pi)$	16	$\cot t = \cot(t + 2\pi)$
17	$\sec t = \sec(t + 2\pi)$	18	$\csc t = \csc(t + 2\pi)$

In this section we will develop new identities called *Pythagorean identities,* using those already given. If t is any real number or angle, then we have the following.

2.1 Pythagorean Identities

1 $\sin^2 t + \cos^2 t = 1$ (repeating identity 6 above)

2 $\tan^2 t + 1 = \sec^2 t$

PROOF

$$\tan^2 t + 1 = \frac{\sin^2 t}{\cos^2 t} + 1 = \frac{\sin^2 t + \cos^2 t}{\cos^2 t} = \frac{1}{\cos^2 t} = \sec^2 t$$

3 $\cot^2 t + 1 = \csc^2 t$

PROOF

$$\cot^2 t + 1 = \frac{\cos^2 t}{\sin^2 t} + 1 = \frac{\cos^2 t + \sin^2 t}{\sin^2 t} = \frac{1}{\sin^2 t} = \csc^2 t$$

EXAMPLES

1 If $\tan t = -\frac{4}{3}$ and t is in quadrant II, find each of the following values.

a) $\sec t$ b) $\cot t$ c) $\csc t$

SOLUTION

a) Using identity 2 of Section 2.1 together with the fact that the secant is negative in quadrant II, we have

$$\sec t = -\sqrt{\tan^2 t + 1} = -\sqrt{(-\tfrac{4}{3})^2 + 1} = -\sqrt{\tfrac{16}{9} + 1}$$
$$= -\sqrt{\tfrac{25}{9}} = -\tfrac{5}{3}$$

b) $\cot t = \dfrac{1}{\tan t} = \dfrac{1}{-\frac{4}{3}} = -\dfrac{3}{4}$

c) Since the cosecant is positive in quadrant II,

$$\csc t = \sqrt{\cot^2 t + 1} = \sqrt{\left(-\frac{3}{4}\right)^2 + 1} = \sqrt{\frac{9}{16} + 1} = \sqrt{\frac{9+16}{16}}$$
$$= \sqrt{\frac{25}{16}} = \frac{5}{4}$$

2 If $W(t)$ is in quadrant II and $\tan t = -\frac{5}{12}$, find
a) $\sec t$, b) $\cot t$, c) $\csc t$, d) $\cos t$, e) $\sin t$.

SOLUTION a) Since $\sec t$ is negative in quadrant II, we have,

$$\sec t = -\sqrt{1 + \tan^2 t} = -\sqrt{1 + \left(-\frac{5}{12}\right)^2} = -\sqrt{\frac{144+25}{144}}$$
$$= -\frac{13}{12}$$

b) $\cot t = \dfrac{1}{\tan t} = \dfrac{1}{-\frac{5}{12}} = -\dfrac{12}{5}$

c) $\csc t = \sqrt{1 + \cot^2 t} = \sqrt{1 + \left(-\dfrac{12}{5}\right)^2} = \sqrt{\dfrac{25+144}{25}} = \dfrac{13}{5}$

d) $\cos t = \dfrac{1}{\sec t} = \dfrac{1}{-\frac{13}{12}} = -\dfrac{12}{13}$

e) $\sin t = \dfrac{1}{\csc t} = \dfrac{1}{\frac{13}{5}} = \dfrac{5}{13}$

3 Use identities to express the other five functions of θ in terms of x, if θ is in quadrant I.
a) $\csc \theta = x$ b) $\sec \theta = x/4$

SOLUTION

a) Using the identity $\sin t = \dfrac{1}{\csc t}$ and replacing t by θ, we have

$$\sin \theta = \frac{1}{\csc \theta} = \frac{1}{x}$$

Since θ is in quadrant I, the signs of the values of the trigonometric functions are positive, so that

$$\cos \theta = \sqrt{1 - \sin^2 \theta} = \sqrt{1 - \frac{1}{x^2}} = \sqrt{\frac{x^2 - 1}{x^2}} = \frac{\sqrt{x^2 - 1}}{x}$$

$$\tan\theta = \frac{\sin\theta}{\cos\theta} = \frac{\frac{1}{x}}{\frac{\sqrt{x^2-1}}{x}} = \frac{1}{\sqrt{x^2-1}}$$

$$\cot\theta = \frac{\cos\theta}{\sin\theta} = \frac{\frac{\sqrt{x^2-1}}{x}}{\frac{1}{x}} = \sqrt{x^2-1}$$

$$\sec\theta = \frac{1}{\cos\theta} = \frac{1}{\frac{\sqrt{x^2-1}}{x}} = \frac{x}{\sqrt{x^2-1}}$$

b) $\cos\theta = \dfrac{1}{\sec\theta} = \dfrac{1}{\frac{x}{4}} = \dfrac{4}{x}$

$$\sin\theta = \sqrt{1-\cos^2\theta} = \sqrt{1-\frac{16}{x^2}} = \frac{\sqrt{x^2-16}}{x}$$

$$\tan\theta = \frac{\sin\theta}{\cos\theta} = \frac{\frac{\sqrt{x^2-16}}{x}}{\frac{4}{x}} = \frac{\sqrt{x^2-16}}{4}$$

$$\cot\theta = \frac{\cos\theta}{\sin\theta} = \frac{\frac{4}{x}}{\frac{\sqrt{x^2-16}}{x}} = \frac{4}{\sqrt{x^2-16}}$$

$$\csc\theta = \frac{1}{\sin\theta} = \frac{1}{\frac{\sqrt{x^2-16}}{x}} = \frac{x}{\sqrt{x^2-16}}$$

4 Use identities to simplify each of the following expressions.

a) $\sin^2 315° + \cos^2 315°$ b) $\tan^2(\theta + 60°) + 1$

c) $\csc^2[(\theta/3) + 120°] - \cot^2[(\theta/3) + 120°]$

SOLUTION

a) Since $\sin^2\theta + \cos^2\theta = 1$, then $\sin^2 315° + \cos^2 315° = 1$.

b) Since $\tan^2\theta + 1 = \sec^2\theta$, then $\tan^2(\theta + 60°) + 1 = \sec^2(\theta + 60°)$.

c) Since $\cot^2\theta + 1 = \csc^2\theta$, $\csc^2\theta - \cot^2\theta = 1$, so that $\csc^2(\theta/3 + 120°) - \cot^2(\theta/3 + 120°) = 1$.

5 Express each of the following expressions in terms of $\sin\theta$.

a) $\dfrac{\sec\theta}{\tan\theta + \cot\theta}$ b) $(\tan\theta + \cot\theta)\cot\theta$

c) $\dfrac{\sin\theta}{1-\cos^2\theta}$ d) $\dfrac{\sec^2\theta - 1}{\sin^2\theta}$

SOLUTION

a) $$\frac{\sec\theta}{\tan\theta+\cot\theta}=\frac{\dfrac{1}{\cos\theta}}{\dfrac{\sin\theta}{\cos\theta}+\dfrac{\cos\theta}{\sin\theta}}=\frac{\dfrac{1}{\cos\theta}}{\dfrac{\sin^2\theta+\cos^2\theta}{\sin\theta\cos\theta}}=\frac{\dfrac{1}{\cos\theta}}{\dfrac{1}{\sin\theta\cos\theta}}$$
$$=\frac{\sin\theta\cos\theta}{\cos\theta}=\sin\theta$$

b) $$(\tan\theta+\cot\theta)\cot\theta=\left(\frac{1}{\cot\theta}+\cot\theta\right)\cot\theta=1+\cot^2\theta$$
$$=\csc^2\theta=\frac{1}{\sin^2\theta}$$

c) Since $\sin^2\theta+\cos^2\theta=1$, we have

$$\frac{\sin\theta}{1-\cos^2\theta}=\frac{\sin\theta}{\sin^2\theta+\cos^2\theta-\cos^2\theta}=\frac{\sin\theta}{\sin^2\theta}=\frac{1}{\sin\theta}$$

d) $$\frac{\sec^2\theta-1}{\sin^2\theta}=\frac{\dfrac{1}{\cos^2\theta}-1}{\sin^2\theta}=\frac{\dfrac{1-\cos^2\theta}{\cos^2\theta}}{\sin^2\theta}=\frac{1-\cos^2\theta}{\sin^2\theta\cos^2\theta}=\frac{\sin^2\theta}{\sin^2\theta\cos^2\theta}$$
$$=\frac{1}{\cos^2\theta}=\frac{1}{1-\sin^2\theta}$$

2.2 Proving Trigonometric Identities

The identities already discussed can be used to prove other identities, and, consequently, to change an expression involving trigonometric functions into a different but equivalent form which is more suitable for the purpose at hand.

The following suggestions are useful in proving identities.

1 Simplify one side until it is the same as the other side.

2 Simplify each side separately until both are the same.

3 To simplify a side of an identity, we suggest the following steps:
   a) If the side contains a sum of fractions, it may be helpful to combine them into a single fraction.
   b) If the numerator of a fraction has more than one term, it may be helpful to split the fraction into a sum of fractions.

4 If in doubt, write the expressions in terms of sines and cosines and simplify.

EXAMPLES

1 Prove each of the following identities.

a) $\sec^2 \theta \csc^2 \theta = \sec^2 \theta + \csc^2 \theta$

b) $1 - \tan^4 \theta = 2 \sec^2 \theta - \sec^4 \theta$

c) $\dfrac{\sin \theta}{\csc \theta - \cot \theta} = 1 + \cos \theta$

d) $\dfrac{1 + \tan^2 \theta}{\csc^2 \theta} = \tan^2 \theta$

PROOF

a)
$$\begin{aligned}\sec^2 \theta \csc^2 \theta &= (1 + \tan^2 \theta)(1 + \cot^2 \theta)\\ &= 1 + \cot^2 \theta + \tan^2 \theta + \tan^2 \theta \cot^2 \theta\\ &= 1 + \cot^2 \theta + \tan^2 \theta + 1\\ &= \csc^2 \theta + \sec^2 \theta\end{aligned}$$

b)
$$\begin{aligned}2 \sec^2 \theta - \sec^4 \theta &= \sec^2 \theta\,(2 - \sec^2 \theta)\\ &= \sec^2 \theta\,[2 - (\tan^2 \theta + 1)]\\ &= (1 + \tan^2 \theta)(1 - \tan^2 \theta)\\ &= 1 - \tan^4 \theta\end{aligned}$$

c)
$$\begin{aligned}\frac{\sin \theta}{\csc \theta - \cot \theta} &= \frac{\sin \theta}{\dfrac{1}{\sin \theta} - \dfrac{\cos \theta}{\sin \theta}}\\ &= \frac{\sin^2 \theta}{1 - \cos \theta}\\ &= \frac{1 - \cos^2 \theta}{1 - \cos \theta}\\ &= \frac{(1 - \cos \theta)(1 + \cos \theta)}{1 - \cos \theta}\\ &= 1 + \cos \theta\end{aligned}$$

d)
$$\begin{aligned}\frac{1 + \tan^2 \theta}{\csc^2 \theta} &= \frac{\sec^2 \theta}{\csc^2 \theta}\\ &= \frac{\dfrac{1}{\cos^2 \theta}}{\dfrac{1}{\sin^2 \theta}}\\ &= \frac{\sin^2 \theta}{\cos^2 \theta}\\ &= \tan^2 \theta\end{aligned}$$

PROBLEM SET 1

1 Use identities to express the values of the other five functions of θ in terms of x, if θ is in the first quadrant for each of the following.

a) $\cot\theta = x$ b) $\sec\theta = x$

2 a) Express each of the other five trigonometric functions in terms of $\sin\theta$ if θ is in the first quadrant.
b) Express each of the other five trigonometric functions in terms of $\cot\theta$ if θ is in the third quadrant.

3 Use identities to simplify each of the following expressions.
a) $\dfrac{1-\tan^2\theta}{1+\tan^2\theta}$ b) $\dfrac{1+\cot^2\theta}{\sec^2\theta}$
c) $(1-\cos\theta)(1+\cos\theta)$
d) $\sec^2\theta\,(\csc^2\theta-1)(\sin\theta+1)-\csc\theta$
e) $2\sin\theta\cos\theta\csc\theta$ f) $1+\dfrac{\sin^3\theta}{\csc\theta}-\dfrac{\cos^3\theta}{\sec\theta}$
g) $\dfrac{1}{1-\cos\theta}+\dfrac{1}{1+\cos\theta}$ h) $\dfrac{\sec^2\theta-\tan^2\theta}{\csc^2\theta}$

4 Express each of the following in terms of the sine and/or cosine; then simplify each expression as much as possible.
a) $\dfrac{\cos t-1}{\sec t-1}$ b) $\cos\theta+\tan\theta\sin\theta$
c) $\sin^4 t-\cos^4 t$ d) $\cos t\tan t$
e) $(\csc\theta+\cot\theta)^2$ f) $\dfrac{1-\cos^2\theta}{\sin\theta}$
g) $\tan t-\csc t\sec t\,(1-2\cos^2 t)$
h) $\sin\theta\cot\theta+\cos^2\theta\sec\theta$

5 Prove each of the following identities.
a) $\sec^2 t+\tan^2 t+1=2\sec^2 t$ b) $\sin t\cot t=\cos t$
c) $\tan^2 t-\tan^2 t\sin^2 t=\sin^2 t$ d) $\cot t-\cos t=\cot t\,(1-\sin t)$
e) $\dfrac{\sin\theta}{\tan\theta}+\dfrac{\cos\theta}{\cot\theta}=\cos\theta+\sin\theta$ f) $(\cot^2\theta-\cos^2\theta)^2=\cos^4\theta\cot^4\theta$
g) $\dfrac{\sec^2 t-1+\tan^2 t\cos t}{\cos t+\cos^2 t}=\tan^2 t\sec t$
h) $\dfrac{\tan\theta\,(\csc^2\theta-1)}{\sin\theta+\cot\theta\cos\theta}=\cos\theta$
i) $(\tan t+\cot t)^2=\sec^2 t\csc^2 t$ j) $\sin^2\theta+\dfrac{1-\tan^2\theta}{\sec^2\theta}=\cos^2\theta$
k) $\dfrac{1+\sec t}{\csc t}=\sin t+\tan t$ l) $\dfrac{\cot t}{1+\cot^2 t}=\sin t\cos t$
m) $2\csc\theta-\cot\theta\cos\theta=\sin\theta+\csc\theta$
n) $\sin^4\theta-\cos^4\theta=2\sin^2\theta-1$
o) $2\sin^2\theta\,(\tan^2\theta+1)+1=\sec^2\theta+\tan^2\theta$
p) $\dfrac{\cos\theta}{1+\sin\theta}-\dfrac{1-\sin\theta}{\cos\theta}=0$
q) $2\cot\theta-\cot\theta\cos^2\theta=\sin\theta\cos\theta+\cot\theta$

r) $\csc^4 \theta - 2 \cot^2 \theta = 1 + \cot^4 \theta$ s) $\csc t + \cot t = \dfrac{1}{\csc t - \cot t}$

t) $\dfrac{\sin \theta}{1 - \cos \theta} = \csc \theta + \cot \theta$

3 Difference and Sum Identities

In this section we shall derive some of the standard formulas or identities that enable us to express the cosine or sine of a sum or difference of two numbers (or two angles) in terms of cosines and sines of the two numbers (or the two angles). For example, suppose that $\cos (t + s)$ were to be computed, using the values $\cos t$, $\cos s$, $\sin t$, and $\sin s$. The first response might be to say the $\cos (t + s) = \cos t + \cos s$; however, this is not the case in general. For example,

$$\cos \left(\frac{\pi}{2} + \frac{\pi}{2}\right) = \cos \pi = -1 \neq \cos \frac{\pi}{2} + \cos \frac{\pi}{2} = 0$$

and

$$\cos (45° + 45°) = \cos 90° = 0 \neq \cos 45° + \cos 45° = \sqrt{2}$$

Although the proofs of the following theorems are given in terms of real numbers s and t, remember that the results are also true if s and t are interpreted as angles.

THEOREM 1

If s and t are real numbers, then
a) $\cos (t - s) = \cos t \cos s + \sin t \sin s$
b) $\cos (t + s) = \cos t \cos s - \sin t \sin s$

PROOF

a) For convenience, we shall illustrate the case where $t > s > 0$ (Figure 1). However, the results are true for all possible values of s and t.

By construction, the length of the arc $\widehat{P_0P_3}$ is equal to the length of the arc $\widehat{P_1P_2}$ (why?), it follows from geometry that the chord $\overline{P_0P_3}$ has the same length as the chord $\overline{P_1P_2}$. Using the distance formula, we have

$$\begin{aligned}\overline{P_0P_3}^2 &= [\cos (t - s) - 1]^2 + [\sin (t - s) - 0]^2 \\ &= \cos^2 (t - s) - 2 \cos (t - s) + 1 + \sin^2 (t - s)\end{aligned}$$

Figure 1

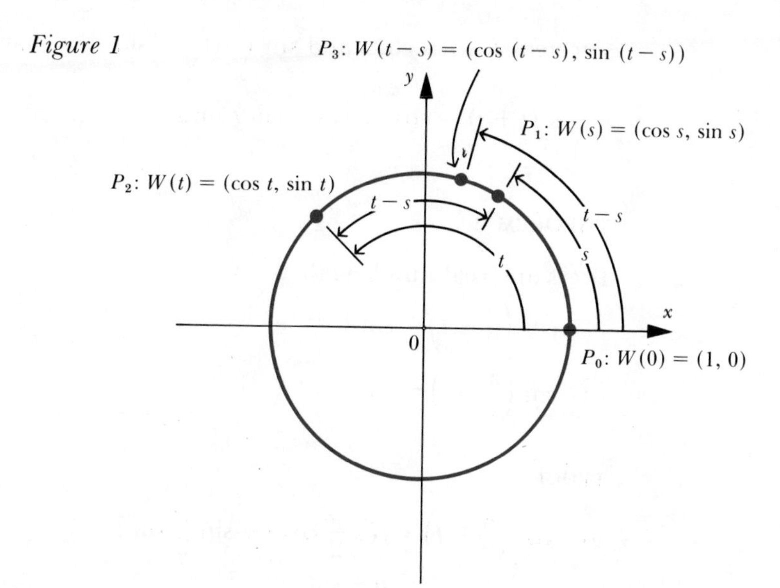

and

$$\begin{aligned}\overline{P_1P_2}^2 &= (\cos t - \cos s)^2 + (\sin t - \sin s)^2 \\ &= \cos^2 t - 2\cos t \cos s + \cos^2 s + \sin^2 t - 2 \sin t \sin s + \sin^2 s\end{aligned}$$

Using the fact that

$$\begin{aligned}\cos^2 (t-s) + \sin^2 (t-s) &= 1 \\ \cos^2 s + \sin^2 s &= 1 \\ \cos^2 t + \sin^2 t &= 1\end{aligned}$$

and equating $\overline{P_1P_2}^2$ with $\overline{P_0P_3}^2$, we get

$$2 - 2\cos(t-s) = 2 - 2\cos t \cos s - 2 \sin t \sin s$$

so that

$$-2\cos(t-s) = -2\cos t \cos s - 2\sin t \sin s$$

or

$$\cos(t-s) = \cos t \cos s + \sin t \sin s$$

b) $\cos(t+s) = \cos[t-(-s)]$

$$= \cos t \cos(-s) + \sin t \sin(-s)$$

since $\cos(-s) = \cos s$ and $\sin(-s) = -\sin s$, we have,

$$\cos(t + s) = \cos t \cos s - \sin t \sin s$$

THEOREM 2

If t is any real number, then

a) $\cos\left(\frac{\pi}{2} - t\right) = \sin t$

b) $\sin\left(\frac{\pi}{2} - t\right) = \cos t$

PROOF

a)
$$\begin{aligned} \cos\left(\frac{\pi}{2} - t\right) &= \cos\frac{\pi}{2}\cos t + \sin\frac{\pi}{2}\sin t \\ &= 0 + \sin t \\ &= \sin t \end{aligned}$$

b)
$$\begin{aligned} \sin\left(\frac{\pi}{2} - t\right) &= \cos\left[\frac{\pi}{2} - \left(\frac{\pi}{2} - t\right)\right] \qquad \text{(Why?)} \\ &= \cos\left(\frac{\pi}{2} - \frac{\pi}{2} + t\right) \\ &= \cos t \end{aligned}$$

THEOREM 3

If t and s are any real numbers, then

a) $\sin(s - t) = \sin s \cos t - \sin t \cos s$

b) $\sin(s + t) = \sin s \cos t + \sin t \cos s$

PROOF

a)
$$\begin{aligned} \sin(s - t) &= \cos\left[\frac{\pi}{2} - (s - t)\right] \\ &= \cos\left[\left(\frac{\pi}{2} - s\right) + t\right] \\ &= \cos\left(\frac{\pi}{2} - s\right)\cos t - \sin\left(\frac{\pi}{2} - s\right)\sin t \end{aligned}$$

Using Theorem 2, we have

$$\sin(s - t) = \sin s \cos t - \cos s \sin t$$

b)
$$\begin{aligned} \sin(s + t) &= \sin[s - (-t)] \\ &= \sin s \cos(-t) - \sin(-t)\cos s \\ &= \sin s \cos t + \sin t \cos s \qquad \text{(Why?)} \end{aligned}$$

THEOREM 4

If t and s are any real numbers, then

a) $\tan(s+t) = \dfrac{\tan s + \tan t}{1 - \tan s \tan t}$

b) $\tan(s-t) = \dfrac{\tan s - \tan t}{1 + \tan s \tan t}$

PROOF

a) $\tan(s+t) = \dfrac{\sin(s+t)}{\cos(s+t)} = \dfrac{\sin s \cos t + \sin t \cos s}{\cos s \cos t - \sin s \sin t}$

Dividing both the numerator and the denominator of the right-hand side by $\cos s \cos t$, we get

$$\tan(s+t) = \frac{\dfrac{\sin s \cos t}{\cos s \cos t} + \dfrac{\sin t \cos s}{\cos s \cos t}}{\dfrac{\cos s \cos t}{\cos s \cos t} - \dfrac{\sin s \sin t}{\cos s \cos t}}$$

Simplifying the above expression, we have

$$\tan(s+t) = \frac{\tan s + \tan t}{1 - \tan s \tan t}$$

b)
$$\begin{aligned}\tan(t-s) &= \tan[t + (-s)] \\ &= \frac{\tan t + \tan(-s)}{1 - \tan t \tan(-s)}\end{aligned}$$

Using $\tan(-s) = -\tan s$ and $\tan(-t) = -\tan t$, we have

$$\tan(t-s) = \frac{\tan t - \tan s}{1 + \tan t \tan s}$$

EXAMPLES

1 Use the identities to find the exact value of $\sin(7\pi/12)$.

SOLUTION

$$\frac{7\pi}{12} = \frac{\pi}{4} + \frac{\pi}{3}$$

so that

$$\begin{aligned}\sin\frac{7\pi}{12} &= \sin\left(\frac{\pi}{4}+\frac{\pi}{3}\right)\\ &= \sin\frac{\pi}{4}\cos\frac{\pi}{3}+\sin\frac{\pi}{3}\cos\frac{\pi}{4}\\ &= \frac{\sqrt{2}}{2}\cdot\frac{1}{2}+\frac{\sqrt{3}}{2}\cdot\frac{\sqrt{2}}{2}\\ &= \frac{\sqrt{2}}{4}+\frac{\sqrt{6}}{4}\\ &= \frac{\sqrt{2}+\sqrt{6}}{4}\end{aligned}$$

2 Use the fact that $195° = 150° + 45°$ and the identities to evaluate

a) $\cos 195°$ b) $\tan 195°$

SOLUTION

a) $\cos 195° = \cos\,(150° + 45°)$

$$\begin{aligned} &= \cos 150° \cos 45° - \sin 150° \sin 45°\\ &= -\frac{\sqrt{3}}{2}\cdot\frac{\sqrt{2}}{2}-\frac{1}{2}\cdot\frac{\sqrt{2}}{2}\\ &= -\frac{\sqrt{6}}{4}-\frac{\sqrt{2}}{4} = -\frac{\sqrt{6}+\sqrt{2}}{4}\end{aligned}$$

b) $\tan 195° = \tan\,(150° + 45°)$

$$\begin{aligned} &= \frac{\tan 150° + \tan 45°}{1-\tan 150° \tan 45°}\\ &= \frac{-\dfrac{\sqrt{3}}{3}+1}{1+\dfrac{\sqrt{3}}{3}\cdot 1}\\ &= \frac{-\sqrt{3}+3}{3+\sqrt{3}}\\ &= \frac{(3-\sqrt{3})(3-\sqrt{3})}{(3+\sqrt{3})(3-\sqrt{3})}\\ &= \frac{9-6\sqrt{3}+3}{9-3}\\ &= \frac{12-6\sqrt{3}}{6}\\ &= 2-\sqrt{3}\end{aligned}$$

3 Use the identities to simplify each of the following expressions.

a) $\cos\left(\frac{\pi}{2}+t\right)$ b) $\sin\left(\frac{\pi}{2}+t\right)$

c) $\sin 27° \cos 18° + \cos 27° \sin 18°$

d) $\cos 25° \cos 35° - \sin 25° \sin 35°$

e) $\cos\,(s-t)\cos t - \sin\,(s-t)\sin t$

SOLUTION

a) $\cos\left(\frac{\pi}{2}+t\right)=\cos\frac{\pi}{2}\cos t-\sin\frac{\pi}{2}\sin t=0-\sin t=-\sin t$

b) $\sin\left(\frac{\pi}{2}+t\right)=\sin\frac{\pi}{2}\cos t+\cos\frac{\pi}{2}\sin t=\cos t+0=\cos t$

c) $\sin 27°\cos 18°+\cos 27°\sin 18°=\sin(27°+18°)=\sin 45°=\frac{\sqrt{2}}{2}$

d) $\cos 25°\cos 35°-\sin 25°\sin 35°=\cos(25°+35°)=\cos 60°=\frac{1}{2}$

e) $\cos(s-t)\cos t-\sin(s-t)\sin t=\cos[(s-t)+t]=\cos s$

4 Given $\sin\theta=\frac{3}{5}$, θ in quadrant II, and given $\cos\phi=-\frac{12}{13}$, ϕ in quadrant III, find

a) $\sin(\theta+\phi)$ b) $\cos(\theta+\phi)$

c) $\tan(\theta+\phi)$ d) $\tan(\theta-\phi)$

e) $\cos^2(\theta+\phi)+\sin^2(\theta+\phi)$

SOLUTION. Since $\sin\theta=\frac{3}{5}$ and θ is in quadrant II, $r=5$ and $y=3$, so that $x=-4$ and the point $(-4, 3)$ is on the terminal side of θ (Figure 2). Therefore, $\cos\theta=x/r=-\frac{4}{5}$. Similarly, since $\cos\phi=-\frac{12}{13}$ and ϕ is in quadrant III, $r=13$ and $x=-12$, so that $y=-5$ (Figure 2). Hence, $\sin\phi=y/r=-\frac{5}{13}$. Also,

$$\tan\theta=\frac{\sin\theta}{\cos\theta}=\frac{\frac{3}{5}}{-\frac{4}{5}}=-\frac{3}{4}$$

and

$$\tan\phi=\frac{\sin\phi}{\cos\phi}=\frac{-\frac{5}{13}}{-\frac{12}{13}}=\frac{5}{12}$$

Figure 2

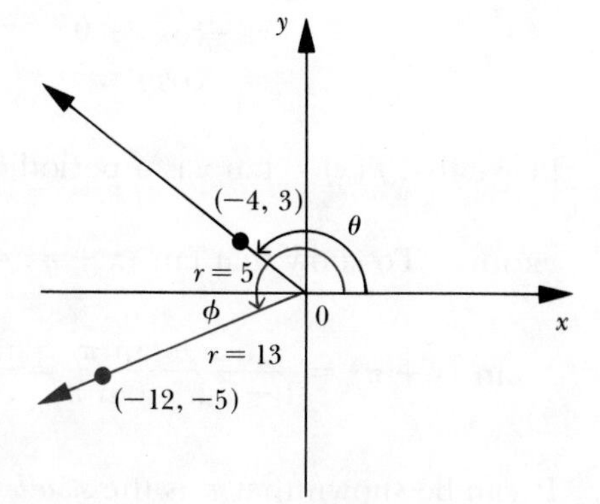

a) $$\begin{aligned}\sin(\theta+\phi)&=\sin\theta\cos\phi+\sin\phi\cos\theta\\&=\left(\tfrac{3}{5}\right)\left(-\tfrac{12}{13}\right)+\left(-\tfrac{5}{13}\right)\left(-\tfrac{4}{5}\right)\\&=-\tfrac{36}{65}+\tfrac{20}{65}\\&=-\tfrac{16}{65}\end{aligned}$$

b) $\cos(\theta + \phi) = \cos\theta\cos\phi - \sin\theta\sin\phi$
$= (-\frac{4}{5})(-\frac{12}{13}) - (\frac{3}{5})(-\frac{5}{13})$
$= \frac{48}{65} + \frac{15}{65}$
$= \frac{63}{65}$

c) $\tan(\theta + \phi) = \dfrac{\sin(\theta + \phi)}{\cos(\theta + \phi)} = \dfrac{-\frac{16}{65}}{\frac{63}{65}} = -\dfrac{16}{63}$

d) $\tan(\theta - \phi) = \dfrac{\tan\theta - \tan\phi}{1 + \tan\theta\tan\phi} = \dfrac{-\frac{3}{4} - \frac{5}{12}}{1 + (-\frac{3}{4})(\frac{5}{12})} = \dfrac{-\frac{14}{12}}{1 - \frac{5}{16}} = -\dfrac{56}{33}$

e) $\cos^2(\theta + \phi) + \sin^2(\theta + \phi) = (\frac{63}{65})^2 + (-\frac{16}{65})^2$
$= \frac{3969}{4225} + \frac{256}{4225}$
$= \frac{4225}{4225}$
$= 1$

5 Verify each of the following identities.

a) $\sin(t + \pi) = -\sin t$ b) $\sin(t - \pi) = -\sin t$
c) $\cos(t + 180°) = -\cos t$ d) $\cos(t - 180°) = -\cos t$

SOLUTION

a) $\sin(t + \pi) = \sin t\cos\pi + \sin\pi\cos t$
$= -\sin t + 0$
$= -\sin t$

b) $\sin(t - \pi) = \sin t\cos\pi - \sin\pi\cos t$
$= -\sin t - 0$
$= -\sin t$

c) $\cos(t + 180°) = \cos t\cos 180° - \sin t\sin 180°$
$= -\cos t - 0$
$= -\cos t$

d) $\cos(t - 180°) = \cos t\cos 180° + \sin t\sin 180°$
$= -\cos t + 0$
$= -\cos t$

6 Prove that $f(x) = \tan x$ is a periodic function of period π.

PROOF. To show that $\tan(x + \pi) = \tan x$, we have

$$\tan(x + \pi) = \frac{\tan x + \tan\pi}{1 - \tan x\tan\pi} = \frac{\tan x + 0}{1 - 0} = \tan x$$

It can be shown that π is the *smallest* period of the tangent function, and hence is its fundamental period.

7 a) Use Theorem 4 to derive a formula for $\cot(t + s)$ in terms of $\cot t$ and $\cot s$.

b) Prove that $g(x) = \cot x$ is a periodic function of period π.

SOLUTION

a) $$\cot(t+s) = \frac{1}{\tan(t+s)} = \frac{1}{\dfrac{\tan t + \tan s}{1 - \tan t \tan s}}$$

$$= \frac{1 - \tan t \tan s}{\tan t + \tan s} = \frac{1 - \dfrac{1}{\cot t} \cdot \dfrac{1}{\cot s}}{\dfrac{1}{\cot t} + \dfrac{1}{\cot s}}$$

$$= \frac{\dfrac{\cot t \cot s - 1}{\cot t \cot s}}{\dfrac{\cot s + \cot t}{\cot t \cot s}} = \frac{\cot t \cot s - 1}{\cot s + \cot t}$$

b) To show that $\cot(x+\pi) = \cot x$, we have

$$\cot(x+\pi) = \frac{1}{\tan(x+\pi)} = \frac{1}{\tan x} = \cot x$$

It can also be shown that π is the *smallest* period of the cotangent function, and hence is its fundamental period.

8 Prove the following identities.
a) $\sin(s+t) + \sin(s-t) = 2 \sin s \cos t$
b) $\sin(s+t) - \sin(s-t) = 2 \cos s \sin t$
c) $\cos(s-t) + \cos(s+t) = 2 \cos s \cos t$
d) $\cos(s-t) - \cos(s+t) = 2 \sin s \sin t$

PROOF

a) $$\begin{aligned}\sin(s+t) + \sin(s-t) &= \sin s \cos t + \sin t \cos s + \sin s \cos t \\ &\quad - \sin t \cos s \\ &= 2 \sin s \cos t\end{aligned}$$

b) $$\begin{aligned}\sin(s+t) - \sin(s-t) &= \sin s \cos t + \sin t \cos s - \sin s \cos t \\ &\quad + \sin t \cos s \\ &= 2 \sin t \cos s\end{aligned}$$

c) $$\begin{aligned}\cos(s-t) + \cos(s+t) &= \cos s \cos t + \sin s \sin t + \cos s \cos t \\ &\quad - \sin s \sin t \\ &= 2 \cos s \cos t\end{aligned}$$

d) $$\begin{aligned}\cos(s-t) - \cos(s+t) &= \cos s \cos t + \sin s \sin t - \cos s \cos t \\ &\quad + \sin s \sin t \\ &= 2 \sin s \sin t\end{aligned}$$

9 Express $\cos 3\theta$ in terms of $\cos \theta$.

SOLUTION

$$\begin{aligned}\cos 3\theta &= \cos(\theta + 2\theta) \\ &= \cos \theta \cos 2\theta - \sin \theta \sin 2\theta\end{aligned}$$

$$= \cos\theta \cos(\theta+\theta) - \sin\theta \sin(\theta+\theta)$$
$$= \cos\theta (\cos\theta\cos\theta - \sin\theta\sin\theta) - \sin\theta(\sin\theta\cos\theta + \sin\theta\cos\theta)$$
$$= \cos\theta(\cos^2\theta - \sin^2\theta) - \sin\theta(2\sin\theta\cos\theta)$$

Using $\sin^2\theta = 1 - \cos^2\theta$, we have

$$\begin{aligned}\cos 3\theta &= \cos\theta[\cos^2\theta - (1-\cos^2\theta)] - 2\sin^2\theta\cos\theta\\ &= \cos\theta(\cos^2\theta - 1 + \cos^2\theta) - 2\sin^2\theta\cos\theta\\ &= \cos\theta(2\cos^2\theta - 1) - 2\cos\theta(1-\cos^2\theta)\\ &= 2\cos^3\theta - \cos\theta - 2\cos\theta + 2\cos^3\theta\\ &= 4\cos^3\theta - 3\cos\theta\end{aligned}$$

10 Show that $\cos^{-1} x = \pi/2 - \sin^{-1} x$ for $-1 \leqslant x \leqslant 1$.

SOLUTION

$$\cos^{-1} x = \pi/2 - \sin^{-1} x \qquad \text{if} \qquad \sin^{-1} x + \cos^{-1} x = \pi/2$$

This latter equation is true if

$$\sin(\sin^{-1} x + \cos^{-1} x) = \sin \pi/2 = 1$$

Expanding, we have

$$\begin{aligned}&\sin(\sin^{-1} x)\cos(\cos^{-1} x) + \sin(\cos^{-1} x)\cos(\sin^{-1} x)\\ &= x \cdot x + \sqrt{1-x^2}\sqrt{1-x^2} \qquad \text{(see Example 2)}\\ &= x^2 + 1 - x^2 = 1\end{aligned}$$

Hence,

$$\cos^{-1} x = \pi/2 - \sin^{-1} x$$

11 Show that $\tan^{-1} 2 + \tan^{-1} 3 = -\pi/4$.

SOLUTION

$\tan^{-1} 2 + \tan^{-1} 3 = -\pi/4$ is true if

$$\tan[\tan^{-1} 2 + \tan^{-1} 3] = \tan\left(-\frac{\pi}{4}\right) = -1$$

Expanding, we have

$$\frac{\tan(\tan^{-1} 2) + \tan(\tan^{-1} 3)}{1 - \tan(\tan^{-1} 2)\tan(\tan^{-1} 3)} = \frac{2+3}{1-(2)(3)} = \frac{5}{-5} = -1$$

Hence

$$\tan^{-1} 2 + \tan^{-1} 3 = -\frac{\pi}{4}$$

PROBLEM SET 2

1 Use the identities to express each of the following expressions as a single function evaluated at some angle.

a) $\sin 33° \cos 27° + \cos 33° \sin 27°$
b) $\sin 35° \cos 25° + \cos 35° \sin 25°$
c) $\cos 55° \cos 25° - \sin 55° \sin 25°$
d) $\sin 2\theta \cos \theta + \cos 2\theta \sin \theta$
e) $\cos (t - s) \cos s - \sin (t - s) \sin s$
f) $\dfrac{\tan 32° + \tan 43°}{1 - \tan 32° \tan 43°}$

2 Use the identities to express each of the following expressions as one trigonometric function of θ.

a) $\sin [(5\pi/2) - \theta]$
b) $\cos [(5\pi/2) - \theta]$
c) $\tan (\pi - \theta)$
d) $\cot (90° - \theta)$
e) $\cos (270° + \theta)$
f) $\sec [(\pi/2) + \theta]$
g) $\csc (90° - \theta)$
h) $\sin (270° - \theta)$
i) $\cot (10\pi - \theta)$
j) $\tan (3\pi - \theta)$
k) $\cos [(11\pi/2) - \theta]$
l) $\sin [(7\pi/2) + \theta]$
m) $\cos (-2\pi - \theta)$
n) $\sin (3\pi - \theta)$

3 Use the identities to find the exact value of each of the following expressions. [*Hint:* $5\pi/12 = (\pi/4) + (\pi/6)$.]

a) $\sin (5\pi/12)$ b) $\cos (5\pi/12)$ c) $\tan (5\pi/12)$

4 Use the fact that $255° = 225° + 30°$ together with the identities to find the exact values of the following expressions.

a) $\sin 255°$ b) $\cos 255°$ c) $\tan 255°$

5 If $\sin t = \frac{12}{13}$ and $\cos s = -\frac{4}{5}$, use the identities to evaluate each of the following expressions, where $\pi/2 < t < \pi$ and $\pi/2 < s < \pi$.

a) $\sin (t + s)$
b) $\cos (t + s)$
c) $\sin (t - s)$
d) $\cos (t - s)$
e) $\tan (t + s)$
f) $\tan (t - s)$

6 Find the values of the indicated expressions by using the identities.

a) $\cos (t + s)$ if $\sin t = \frac{12}{13}$ and $\sin s = \frac{3}{5}$, where $0 < t < \pi/2$ and $0 < s < \pi/2$.
b) $\sin (t + s)$ if $\cos t = -\frac{5}{13}$ and $\cos s = \frac{12}{13}$, where $\pi/2 < t < \pi$ and $0 < s < \pi/2$.
c) $\cos (t - s)$ if $\tan t = -\frac{15}{8}$ and $\cos s = -\frac{4}{5}$, where $\pi/2 < t < \pi$ and $\pi < s < 3\pi/2$.
d) $\tan (t - s)$ if $\tan t = -\frac{12}{5}$ and $\sec s = -\frac{5}{3}$, where $\pi/2 < t < \pi$ and $\pi/2 < s < \pi$.

7 Prove each of the following identities.

a) $\sin(30° + t) = \frac{1}{2}\cos t + \frac{\sqrt{3}}{2}\sin t$

b) $\cos(30° + t) = \frac{\sqrt{3}}{2}\cos t - \frac{1}{2}\sin t$

c) $\tan(45° + t) = \frac{1 + \tan t}{1 - \tan t}$

d) $\sin(45° + t) + \cos(45° + t) = \sqrt{2}\cos t$

e) $\sin(60° + t) - \cos(30° + t) = \sin t$

f) $\cos\left(\frac{\pi}{4} - t\right) = \frac{1}{\sqrt{2}}(\cos t + \sin t)$

g) $\sin\left(t - \frac{\pi}{3}\right) = \frac{1}{2}(\sin t - \sqrt{3}\cos t)$

h) $\cos(90° + t - s) = \cos t \sin s - \sin t \cos s$

8 Find a formula for $\cot(t - s)$ in terms of $\cot t$ and $\cot s$.

9 Use the results of Example 8 to prove each of the following identities. [*Hint:* Using Example 8, let $s = u + v$ and $t = u - v$, so that $u = (s + t)/2$ and $v = (s - t)/2$. Substitute for u and v in terms of s and t.]

a) $\sin s + \sin t = 2\sin\left(\frac{s+t}{2}\right)\cos\left(\frac{s-t}{2}\right)$

b) $\sin s - \sin t = 2\cos\left(\frac{s+t}{2}\right)\sin\left(\frac{s-t}{2}\right)$

c) $\cos s + \cos t = 2\cos\left(\frac{s+t}{2}\right)\cos\left(\frac{s-t}{2}\right)$

d) $\cos t - \cos s = 2\sin\left(\frac{s+t}{2}\right)\sin\left(\frac{s-t}{2}\right)$

10 Show that the following statements are true.

a) $\tan^{-1}\frac{1}{3} + \tan^{-1} 3 = \pi/2$

b) $\sin^{-1}\frac{56}{65} = \sin^{-1}\frac{3}{5} + \cos^{-1}\frac{12}{13}$

c) $2\tan^{-1}\frac{1}{3} = \pi/4 - \tan^{-1}\frac{1}{7}$

d) $\tan^{-1}\frac{4}{7} - \tan^{-1}\frac{1}{5} = \tan^{-1}\frac{1}{3}$

4 Double- and Half-Angle Identities

In this section we shall obtain identities involving expressions of the form $2t$ and $t/2$. These identities are called *double and half-angle identities* and are derived from the identities for the sum of two angles or numbers.

4.1 Double-Angle Identities

THEOREM 1

Let t be any real number, then

$$\sin 2t = 2 \sin t \cos t$$

PROOF. Using the identity $\sin (u + v) = \sin u \cos v + \sin v \cos u$, and replacing both u and v by t, we have

$$\sin 2t = \sin (t + t) = \sin t \cos t + \sin t \cos t = 2 \sin t \cos t$$

THEOREM 2

Let t be any real number, then

$$\cos 2t = \cos^2 t - \sin^2 t = 2 \cos^2 t - 1 = 1 - 2 \sin^2 t$$

PROOF. Using the identity $\cos (u + v) = \cos u \cos v - \sin u \sin v$ and replacing both u and v by t, we have

$$\cos 2t = \cos (t + t) = \cos t \cos t - \sin t \sin t = \cos^2 t - \sin^2 t$$

The other forms may be obtained by using the identity $\sin^2 t + \cos^2 t = 1$. Thus substituting $1 - \cos^2 t$ for $\sin^2 t$, we have

$$\cos 2t = \cos^2 t - \sin^2 t = \cos^2 t - (1 - \cos^2 t) = 2 \cos^2 t - 1$$

Similarly, if we substitute $1 - \sin^2 t$ for $\cos^2 t$, we obtain

$$\cos 2t = \cos^2 t - \sin^2 t = 1 - \sin^2 t - \sin^2 t = 1 - 2 \sin^2 t$$

THEOREM 3

Let t be any real number; then

$$\tan 2t = \frac{2 \tan t}{1 - \tan^2 t}$$

PROOF. Using the identity

$$\tan (u + v) = \frac{\tan u + \tan v}{1 - \tan u \tan v}$$

and replacing both u and v by t, we have

$$\tan 2t = \tan (t + t) = \frac{\tan t + \tan t}{1 - \tan t \tan t} = \frac{2 \tan t}{1 - \tan^2 t}$$

4.2 Half-Angle Identities

In order to derive the half-angle identities, we make use of the identies $\cos 2t = 2\cos^2 t - 1$ and $\cos 2t = 1 - 2\sin^2 t$.

THEOREM 1

Let s be any real number; then

a) $\cos^2 \frac{s}{2} = \frac{1 + \cos s}{2}$

b) $\sin^2 \frac{s}{2} = \frac{1 - \cos s}{2}$

PROOF. Using the identity $\cos 2t = 2\cos^2 t - 1$ and substituting s for $2t$, we have

a) $\cos s = 2\cos^2 (s/2) - 1$, so that $\cos^2 (s/2) = (\cos s + 1)/2$

b) Using the identity $\cos 2t = 1 - 2\sin^2 t$ and substituting s for $2t$, we obtain

$$\cos s = 1 - 2\sin^2 (s/2), \text{ so that } \sin^2 (s/2) = (1 - \cos s)/2$$

Notice that by taking the square root of both sides of the above identities, we have

$$\cos \frac{s}{2} = \pm\sqrt{\frac{1 + \cos s}{2}} \quad \text{and} \quad \sin \frac{s}{2} = \pm\sqrt{\frac{1 - \cos s}{2}}$$

where the choice of sign is determined by the quadrant in which the terminal side of angle $s/2$ lies.

EXAMPLES

1 Express each of the following trigonometric functions in terms of functions with the indicated values.

a) $\sin 80°$ in terms of functions of $40°$.
b) $\cos 100°$ in terms of functions of $50°$.
c) $\sin 10x$ in terms of functions of $5x$.
d) $\cos 8x$ in terms of functions of $4x$.

SOLUTION

a) $\sin 80° = 2\sin 40° \cos 40°$, since $\sin 2t = 2\sin t \cos t$
b) $\cos 100° = \cos^2 50° - \sin^2 50°$, since $\cos 2t = \cos^2 t - \sin^2 t$
c) $\sin 10x = 2\sin 5x \cos 5x$, since $\sin 2t = 2\sin t \cos t$
d) $\cos 8x = 2\cos^2 4x - 1$, since $\cos 2t = 2\cos^2 t - 1$

2 Express each of the following functions as a single term involving an angle twice as large.

a) $2 \sin 5\theta \cos 5\theta$ b) $\cos^2 3\theta - \sin^2 3\theta$

c) $2 \cos^2 4\theta - 1$ d) $1 - 2 \sin^2 6\theta$

SOLUTION

a) $2 \sin 5\theta \cos 5\theta = \sin 2(5\theta) = \sin 10\theta$

b) $\cos^2 3\theta - \sin^2 3\theta = \cos 2(3\theta) = \cos 6\theta$

c) $2 \cos^2 4\theta - 1 = \cos 2(4\theta) = \cos 8\theta$

d) $1 - 2 \sin^2 6\theta = \cos 2(6\theta) = \cos 12\theta$

3 Use the identities to find the exact values of each of the following expressions without the use of the tables.

a) $\sin 15°$ b) $\cos 15°$ c) $\tan 15°$

SOLUTION

a) $\sin\left(\frac{t}{2}\right) = \pm\sqrt{\frac{1 - \cos t}{2}}$

Hence

$$\sin 15° = \sin\left(\frac{30°}{2}\right) = \sqrt{\frac{1 - \cos 30°}{2}} \qquad (\text{since } \sin 15° > 0)$$

$$= \sqrt{\frac{1 - \frac{\sqrt{3}}{2}}{2}}$$

$$= \sqrt{\frac{2 - \sqrt{3}}{4}}$$

$$= \frac{\sqrt{2 - \sqrt{3}}}{2}$$

b) $\cos\left(\frac{t}{2}\right) = \pm\sqrt{\frac{1 + \cos t}{2}}$

Hence,

$$\cos 15° = \cos\left(\frac{30°}{2}\right) = \sqrt{\frac{1 + \cos 30°}{2}} \qquad (\text{since } \cos 15° > 0)$$

$$= \sqrt{\frac{1 + \frac{\sqrt{3}}{2}}{2}}$$

$$= \frac{\sqrt{2 + \sqrt{3}}}{2}$$

c) $\tan 15° = \dfrac{\sin 15°}{\cos 15°}$

$$= \frac{\dfrac{\sqrt{2-\sqrt{3}}}{2}}{\dfrac{\sqrt{2+\sqrt{3}}}{2}} = \frac{\sqrt{2-\sqrt{3}}}{\sqrt{2+\sqrt{3}}} \cdot \frac{\sqrt{2-\sqrt{3}}}{\sqrt{2-\sqrt{3}}}$$

$$= \sqrt{\frac{(2-\sqrt{3})^2}{(2+\sqrt{3})(2-\sqrt{3})}} = \sqrt{\frac{(2-\sqrt{3})^2}{4-3}} = 2-\sqrt{3}$$

4 Use the identities to simplify each of the following expressions.

a) $\sin 2t/(1+\cos 2t)$ b) $\tan t \sin 2t$

c) $1+\tan\theta \tan 2\theta$

SOLUTION

a)
$$\frac{\sin 2t}{1+\cos 2t} = \frac{2\sin t\cos t}{1+2\cos^2 t-1}$$
$$= \frac{2\sin t\cos t}{2\cos^2 t}$$
$$= \frac{\sin t}{\cos t}$$
$$= \tan t$$

b)
$$\tan t\sin 2t = \frac{\sin t}{\cos t}\cdot 2\sin t\cos t$$
$$= 2\sin^2 t$$

c)
$$1+\tan\theta\tan 2\theta = 1+\tan\theta\,\frac{2\tan\theta}{1-\tan^2\theta}$$
$$= 1+\frac{2\tan^2\theta}{1-\tan^2\theta}$$
$$= \frac{1-\tan^2\theta+2\tan^2\theta}{1-\tan^2\theta}$$
$$= \frac{1+\tan^2\theta}{1-\tan^2\theta}$$
$$= \frac{\sec^2\theta}{1-(\sec^2\theta-1)}$$
$$= \frac{\sec^2\theta}{2-\sec^2\theta}$$
$$= \frac{1}{2\cos^2\theta-1}$$
$$= \frac{1}{\cos 2\theta}$$
$$= \sec 2\theta$$

5 Verify the following identities.

a) $\cos 2t+2\sin^2 t = 1$ b) $\dfrac{2\cos 2t}{\sin 2t-2\sin^2 t} = \cot t+1$

c) $\dfrac{2\tan\theta}{1+\tan^2\theta} = \sin 2\theta$ d) $\dfrac{2\cot\theta}{\csc^2\theta-2} = \tan 2\theta$

SOLUTION

a) $$\cos 2t + 2\sin^2 t = \cos^2 t - \sin^2 t + 2\sin^2 t$$
$$= \cos^2 t + \sin^2 t$$
$$= 1$$

b) $$\frac{2\cos 2t}{\sin 2t - 2\sin^2 t} = \frac{2(\cos^2 t - \sin^2 t)}{2\sin t\cos t - 2\sin^2 t}$$
$$= \frac{2(\cos t - \sin t)(\cos t + \sin t)}{2\sin t\,(\cos t - \sin t)}$$
$$= \frac{\cos t + \sin t}{\sin t}$$
$$= \frac{\cos t}{\sin t} + 1$$
$$= \cot t + 1$$

c) $$\frac{2\tan\theta}{1+\tan^2\theta} = \frac{2\tan\theta}{\sec^2\theta}$$
$$= \frac{2\left(\dfrac{\sin\theta}{\cos\theta}\right)}{\dfrac{1}{\cos^2\theta}}$$
$$= 2\sin\theta\cos\theta$$
$$= \sin 2\theta$$

d) $$\frac{2\cot\theta}{\csc^2\theta - 2} = \frac{2\left(\dfrac{\cos\theta}{\sin\theta}\right)}{\dfrac{1}{\sin^2\theta} - 2}$$
$$= \frac{2\sin\theta\cos\theta}{1 - 2\sin^2\theta}$$
$$= \frac{\sin 2\theta}{\cos 2\theta}$$
$$= \tan 2\theta$$

PROBLEM SET 3

1 Use identities to express

a) $\sin 70°$ in terms of functions of $35°$

b) $\cos 40°$ in terms of functions of $80°$

c) $\sin 110°$ in terms of functions of $55°$

d) $\cos 270°$ in terms of functions of $135°$

2 Use the identities fo find the value of $\sin(t/2)$, $\cos(t/2)$, and $\tan(t/2)$ for each of the following values of t.

a) $\dfrac{4\pi}{3}$ b) $\dfrac{7\pi}{3}$ c) $\dfrac{5\pi}{2}$ d) $\dfrac{3\pi}{2}$

3 Use the half-angle identities to find the exact values of each of the following expressions.

a) $\cos \frac{3\pi}{8}$ b) $\cos \frac{5\pi}{12}$ c) $\sin \left(-\frac{7\pi}{12}\right)$ d) $\cos \left(-\frac{\pi}{8}\right)$

e) $\sin 75°$ f) $\tan 75°$ g) $\tan 165°$ h) $\cos 105°$

4 Use the identities to express each of the following as a single term involving an angle twice as large as the given angle.

a) $2 \sin 3t \cos 3t$ b) $1 - 2 \sin^2 37°$

c) $1 - 2 \sin^2 7t$ d) $2 \cos^2 6t - 1$

e) $2 \cos^2 \frac{t}{2} - 1$ f) $100 \sin 20t \cos 20t$

5 If $\cos t = -\frac{7}{25}$ and $\pi < t < 3\pi/2$, use the identities to find the values of each of the following expressions.

a) $\sin 2t$ b) $\cos 2t$ c) $\tan 2t$ d) $\sin (t/2)$ e) $\cos (t/2)$

6 Given the angles s and t such that $s + t = \pi/2$, prove that

a) $\sin (s - t) = \cos 2t = -\cos 2s$ b) $\cos (s - t) = \sin 2s = \sin 2t$

7 Use the fact that $3\theta = 2\theta + \theta$ to show that $\sin 3\theta = 3 \sin \theta - 4 \sin^3 \theta$.

8 Prove each of the following identities.

a) $(\sin t + \cos t)^2 = 1 + \sin 2t$

b) $\frac{\cos^4 t - \sin^4 t}{\sin 2t} = \cot 2t$

c) $\tan \theta \cos 2\theta = \sin 2\theta - \tan \theta$

d) $8 \sin^2 \theta \cos^2 \theta = 1 - \cos 4\theta$

e) $\cot 2t = \frac{2 - \sec^2 t}{2 \tan t}$

f) $\frac{\sin 2t + \sin t}{\cos 2t + \cos t + 1} = \tan t$

g) $\csc t - \cot t = \tan (t/2)$

h) $\cot (\theta/2) - 2 \cot \theta \cos^2 (\theta/2) = \sin \theta$

i) $\frac{1}{2} \sin t \tan (t/2) \csc^2 (t/2) = 1$

j) $\frac{\cos t + \sin t}{\cos t - \sin t} = \tan 2t + \sec 2t$

5 Trigonometric Equations

So far we have dealt only with identities, that is, equations true for all real numbers or angles for which the functions are defined. Now we consider conditional equations, that is, equations that are true only for some values of the variable.

In particular, we shall use the properties of the trigonometric functions, together with the trigonometric identities, to develop techniques to find solutions to equations involving trigonometric or circular expressions. To solve a trigonometric equation, we find all real numbers or angles that satisfy the equation. Such a value is called a *solution of the equation,* and, as before, the set of all solutions is called the *solution set of the equation.*

For example, the equation $\sin t = 1$ is true for $t = \pi/2$ since $\sin(\pi/2) = 1$. But $\sin t = \sin(t + 2\pi n)$, where n is any integer, because the sine function is a periodic function of period 2π. Therefore, $\sin t = 1$ is true for $t = \pi/2, 5\pi/2, 9\pi/2, \ldots$. The solution set of the equation can be expressed as the set

$$\left\{t \mid t = \frac{(4n+1)\pi}{2},\ n \in I\right\}.$$

EXAMPLES

Solve each of the following equations.

1 $\cos t = \sqrt{3}/2$ for $0 \leq t < 2\pi$

SOLUTION. We know that $\cos(\pi/6) = \sqrt{3}/2$. Here $\pi/6$ is the reference angle (Figure 1) and, since the cosine is also positive in quadrant IV, we have $\cos(11\pi/6) = \sqrt{3}/2$. Thus, the solution set of the equation $\cos t = \sqrt{3}/2$, $0 \leq t < 2\pi$ is $\{\pi/6, 11\pi/6\}$.

Figure 1

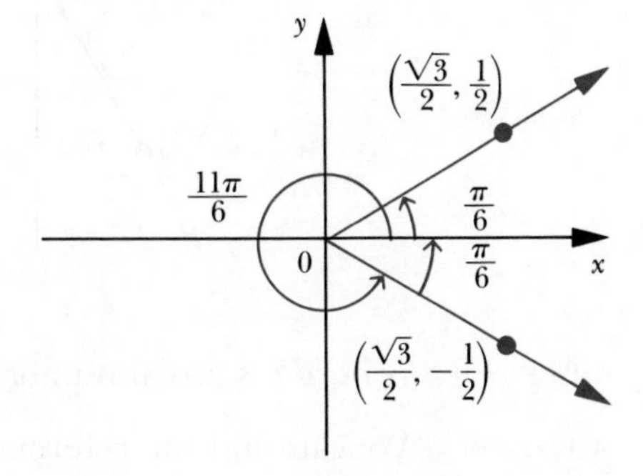

2 $\sin t = \sqrt{2}/2$ for $0 \leq t < 2\pi$

SOLUTION

$\sin(\pi/4) = \sqrt{2}/2$

Here $\pi/4$ is the reference angle (Figure 2) and, since the sine is also positive in quadrant II, we have $\sin(3\pi/4) = \sqrt{2}/2$. Hence, the solution set of the equation $\sin t = \sqrt{2}/2$, $0 \leq t < 2\pi$, is $\{\pi/4, 3\pi/4\}$.

Figure 2

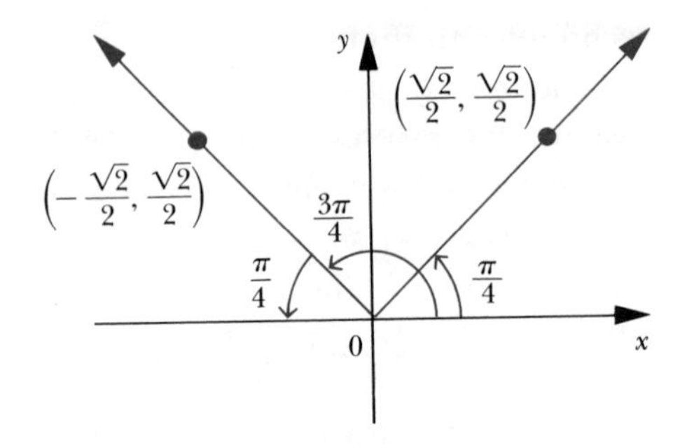

3 $2\cos\theta + 1 = 0$ for $0 \leq \theta < 360°$

SOLUTION. $2\cos\theta + 1 = 0$, so that $2\cos\theta = -1$ or $\cos\theta = -\frac{1}{2}$. Here $60°$ is the reference angle since $\cos 60° = \frac{1}{2}$. The cosine is negative in quadrants II and III so that $\theta = 180° - 60° = 120°$ or $\theta = 180° + 60° = 240°$ (Figure 3). Thus, the solution set is $\{120°, 240°\}$.

Figure 3

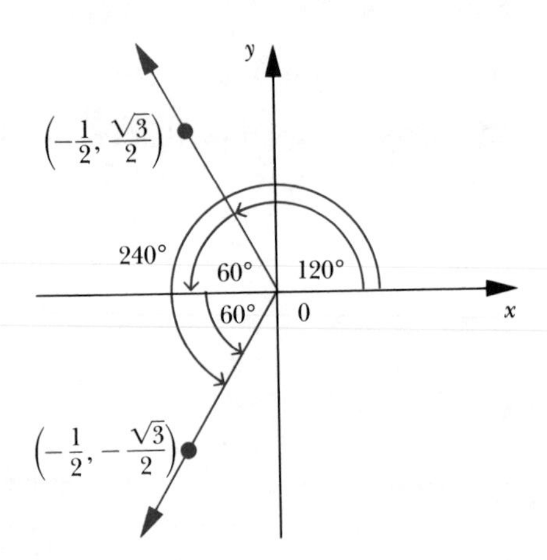

4 $\sin t = -0.8$, where t is a real number (use $\pi = 3.142$)

SOLUTION. We can find the reference angle t_R by using linear interpolation to solve $\sin t_R = 0.8$.

$$0.01\left[\begin{array}{l} \sin 0.92 = 0.7956 \\ d\left[\begin{array}{l} \sin t \quad\;\; = 0.8000 \\ \sin 0.93 = 0.8016 \end{array}\right] 0.0016 \end{array}\right] 0.0060$$

$$\frac{d}{0.01} = \frac{0.0016}{0.006}$$

so that

$$d = 0.003 \qquad \text{(approximately)}$$

Hence,

$$t_R = 0.927 \qquad \text{(approximately)}$$

Since the sine is negative in quadrants III and IV (Figure 4) we have

$$t \in \{t \mid t = 4.069 + 2\pi k,\ k \in I\} \cup \{t \mid t = 5.357 + 2\pi k,\ k \in I\}$$

Figure 4

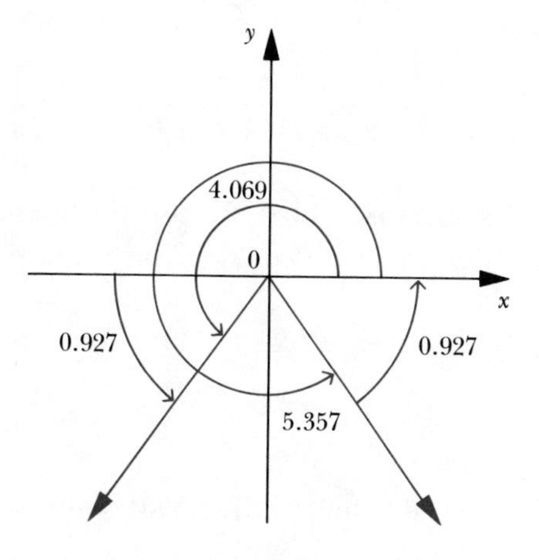

5 $2 \sin^4 t - 9 \sin^2 t + 4 = 0$, for $0 \leqslant t < 2\pi$

SOLUTION. This equation can be factored as

$$(2 \sin^2 t - 1)(\sin^2 t - 4) = 0$$

The solution set, then, is

$$\{t \mid 2 \sin^2 t - 1 = 0\} \cup \{t \mid \sin^2 t - 4 = 0\} \qquad \text{for } 0 \leqslant t < 2\pi$$

$$\{t \mid \sin^2 t = \tfrac{1}{2}\} \cup \{t \mid \sin^2 t = 4\} = \left\{t \mid \sin t = \frac{\pm\sqrt{2}}{2}\right\} \cup \{t \mid \sin t = \pm 2\} \qquad \text{where } 0 \leqslant t < 2\pi$$

Since the sine function has range $[-1, 1]$, $\sin t = \pm 2$ has no solution; therefore, the solution set obtained from the equation $\sin t = \pm\sqrt{2}/2$ is

$$\left\{\frac{\pi}{4}, \frac{3\pi}{4}, \frac{5\pi}{4}, \frac{7\pi}{4}\right\} \qquad \text{(Figure 5)}$$

Figure 5

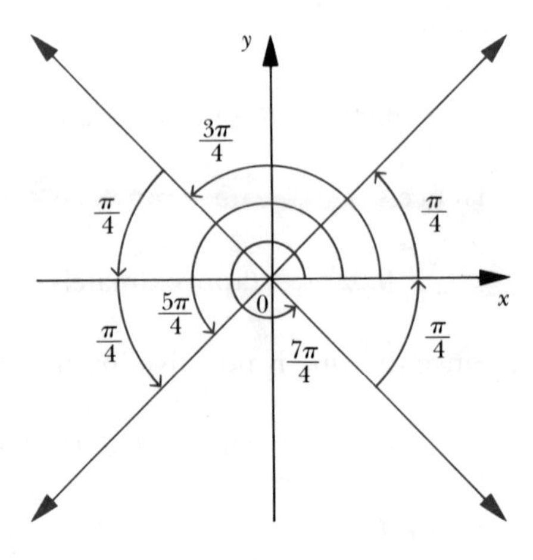

6 $\sin 2t - 2 \sin t = 0$, for $0 \leqslant t < 2\pi$

SOLUTION. First we use the identity for $\sin 2t$ to rewrite the equation as $2 \sin t \cos t - 2 \sin t = 0$, so that the solution set is

$$\begin{aligned} \{t \mid 2 \sin t (\cos t - 1) = 0\} &= \{t \mid \sin t = 0\} \cup \{t \mid \cos t - 1 = 0\} \\ &= \{0, \pi\} \cup \{0\} \\ &= \{0, \pi\} \qquad \text{for } 0 \leqslant t < 2\pi \end{aligned}$$

7 $\cos 2\theta + \sin^2 \theta = 1$, for $0° \leqslant \theta < 360°$

SOLUTION. Since $\cos 2\theta = \cos^2 \theta - \sin^2 \theta$,

$$\cos 2\theta + \sin^2 \theta = \cos^2 \theta$$

so the solution set is

$$\{\theta \mid \cos^2 \theta = 1\} = \{\theta \mid \cos \theta = \pm 1\} = \{0°, 180°\} \qquad \text{for } 0° \leqslant \theta < 360°$$

8 $\sin t + \cos t = 1$, where t is a real number

SOLUTION. Since $\sin t + \cos t = 1$, $\sin t = 1 - \cos t$. Now, square both sides (realizing that we must later check for extraneous roots) to get

$$\sin^2 t = (1 - \cos t)^2$$

but,

$$\sin^2 t = 1 - \cos^2 t = (1 - \cos t)(1 + \cos t)$$

so that

$$(1 - \cos t)(1 + \cos t) = (1 - \cos t)^2$$

or

$$(1 - \cos t)(1 + \cos t) - (1 - \cos t)^2 = 0$$

$$(1 - \cos t)[(1 + \cos t) - (1 - \cos t)] = 0$$

from which we get

$$2(1 - \cos t)(\cos t) = 0$$

or, dividing by 2,

$$(1 - \cos t)(\cos t) = 0$$

But this is equivalent to saying that either

$$1 - \cos t = 0 \qquad \text{or} \qquad \cos t = 0$$

that is, either

$$\cos t = 1 \qquad \text{or} \qquad \cos t = 0$$

The solution set for $\cos t = 1$ is

$$\{t \mid t = 2\pi k,\ k \in I\}$$

and for $\cos t = 0$ is

$$\left\{t \mid t = \pm\frac{\pi}{2} + 2\pi k,\ k \in I\right\}$$

So all solutions to the original equation must be in the union of these two sets. Checking back in the original equation, we find the solution set to be

$$\{t \mid t = 2\pi k,\ k \in I\} \cup \left\{t \mid t = \frac{\pi}{2} + 2\pi k,\ k \in I\right\}$$

Notice here that $-(\pi/2) + 2\pi k$ are extraneous roots.

We can also check this solution by looking at the graph on page 156. To solve $\sin x + \cos x = 1$ is equivalent to finding those values of x for which this graph crosses the line $y = 1$. We see that this occurs at $x = 0$ and $x = \pi/2$ and again every $\pm 2\pi$ units away from these points.

PROBLEM SET 4

1 Solve each of the following trigonometric equations for $0 \leq t < 2\pi$.

a) $\sin t = -\frac{1}{2}$
b) $\sin t = \sqrt{3}/2$
c) $\tan t = 1$
d) $\cot t = -\sqrt{3}$
e) $\sec t = \sqrt{2}$
f) $\csc t = -2/\sqrt{3}$
g) $\cos t = 0.4176$
h) $\sin t = 0.8134$
i) $\tan t = 0.6696$
j) $\cot t = 6.617$

2 Solve each of the following trigonometric equations for $0 \leq t < 2\pi$.

a) $2 \sin t - 1 = 0$
b) $4 \cos^2 \theta - 3 = 0$
c) $\sin 2t = \sin t$
d) $\sin 2t = \sqrt{2} \cos t$
e) $\cos 2t = 2 \sin^2 t$
f) $\sqrt{3} \sec t = 2 \tan t$
g) $\cos 2t = 1 - \sin t$
h) $\sin 2t = \sqrt{2} \cos t$

3 Solve each of the following trigonometric equations for $0 \leq \theta < 360°$.

a) $\sin \theta = -\sqrt{3}/2$
b) $\tan 2\theta = 1$
c) $\sec \theta = -\sqrt{2}$
d) $\cos 2\theta - \cos \theta = 0$
e) $\sin 4\theta - \sin 2\theta = 0$
f) $\sin (45° + \theta) + \sin (45° - \theta) = 1$

4 Solve for s and t if $0 \leq s \leq \pi/2$ and $0 \leq t \leq \pi/2$ when $\tan (s - t) = 1$ and $\sin (s + t) = 1$.

5 Solve each of the following equations for t if $0 \leq t < 2\pi$.

a) $2 \cos t - \sin t = 1$
b) $\sin^2 t - 2 \sin t + 1 = 0$
c) $2 - \sin t = 2 \cos^2 t$
d) $4 \cos^2 t - 5 \sin t \cot t - 6 = 0$
e) $2 \sin^2 t - \sin t - 1 = 0$
f) $4 \csc t - 8 = 0$
g) $2 \sec t + 4 = 0$
h) $3 \cot^2 t - 1 = 0$
i) $\sin t + \cos t \tan t = 1$

6 Solve each of the following equations for θ if $0 \leq \theta < 360°$.

a) $\cos \theta = -\sqrt{3}/2$
b) $\cos \theta = -0.8880$
c) $2 \cos^2 \theta - \cos \theta = 0$
d) $\tan^2 \theta + 3 \sec \theta - 3 = 0$
e) $\tan^2 \theta - 2 \tan \theta + 1 = 0$
f) $\cot^2 \theta - 5 \cot \theta + 4 = 0$
g) $2 \cos^2 \theta + \cos \theta = 0$
h) $2 \cos^2 \theta - \sin \theta = 1$
i) $\tan \theta - 3 \cot \theta = 0$
j) $\cos 2\theta + \sin 2\theta = 0$

7 Solve each of the following trigonometric equations, where t is a real number.

a) $\sec^2 t - 2 \tan t = 0$
b) $\sqrt{3} \csc^2 t + 2 \csc t = 0$
c) $2 \sin^2 t - \sqrt{3} \sin t = 0$
d) $\sec^2 \theta + 1 - 3 \tan \theta = 0$

e) $9(\cos^2 t + \sin t) = 11$ f) $3 \sin^2 t - 4 \sin t + 1 = 0$
g) $2 \sin^2 t + 3 \sin t - 2 = 0$ h) $\sec^2 t + 5 - 3\sqrt{3} \tan t = 0$

REVIEW PROBLEM SET

1 Use the identities to simplify each of the following expressions.

a) $\sec^2 300° - \tan^2 300°$ b) $\csc^2 315° - \cot^2 315°$

c) $\csc t - \cot t \cos t$ d) $\dfrac{\sin^2 t}{1 - \cos t} - 1$

e) $\tan t \sin t + \cos t$ f) $\dfrac{\tan t(1 + \cot^2 t)}{1 + \tan^2 t}$

2 Use the identities to express each of the following in terms of the sine and /or cosine.

a) $(\sec t + \csc t)^2 \tan t$ b) $1/(1 + \cot^2 t)$
c) $(\sec t + \tan t)^2$ d) $(1/\cot^2 t) - (1/\csc^2 t)$

3 Prove each of the following identities.

a) $\sin t \cot t = \cos t$ b) $\tan t \csc t = \sec t$

c) $\sec t - \cos t = \sin t \tan t$ d) $\dfrac{1 - \sin t}{\cos t} = \dfrac{\cos t}{1 + \sin t}$

e) $\dfrac{1}{1 + \sin t} + \dfrac{1}{1 - \sin t} = 2 \sec^2 t$ f) $\cot t + \tan t = \cot t \sec^2 t$

g) $\cos^2 t - \sin^2 t = \dfrac{1 - \tan^2 t}{1 + \tan^2 t}$ h) $\dfrac{\cot^2 t - 1}{1 + \cot^2 t} = 2 \cos^2 t - 1$

i) $\dfrac{(\sin^2 t - \cos^2 t)^2}{\sin^4 t - \cos^4 t} = 2 \sin^2 t - 1$ j) $\dfrac{\tan t + \sec t}{\cos t - \sec t - \tan t} = -\dfrac{1}{\sin t}$

4 Use the identities to express each of the following expressions as a single expression of one angle.

a) $\sin 25° \cos 20° + \sin 20° \cos 25°$
b) $\cos (\pi/5) \cos (\pi/4) - \sin (\pi/5) \sin (\pi/4)$
c) $\cos 3 \cos 2 + \sin 3 \sin 2$
d) $\dfrac{\tan 37° - \tan 17°}{1 + \tan 37° \tan 17°}$

5 Use the identities to express each of the following as an expression involving a trigonometric function of θ.

a) $\sin (11\pi/2 + \theta)$ b) $\sec (180° - \theta)$
c) $\tan (3\pi/4 + \theta)$ d) $\csc (5\pi + \theta)$
e) $\cos (630° + \theta)$ f) $\cot (135° - \theta)$

6 Use the identities and the fact that $\pi/12 = \pi/3 - \pi/4$ to find the exact value of each of the following expressions.

a) $\sin (\pi/12)$ b) $\cos (\pi/12)$ c) $\tan (\pi/12)$

7 If $\sin t = \frac{4}{5}$ and $\cos s = -\frac{3}{5}$, use the identities to evaluate each of the following expressions, where $\pi/2 < t < \pi$ and $\pi < s < 3\pi/2$.

a) $\sin (t+s)$ b) $\cos (t+s)$ c) $\cos (t-s)$
d) $\tan (t-s)$ e) $\tan (t+s)$ f) $\sin (t-s)$

8 Prove each of the following identities.

a) $\cos (45° + t) = (\sqrt{2}/2)(\cos t - \sin t)$
b) $\sin (150° - t) = \frac{1}{2}(\cos t - \sqrt{3} \sin t)$
c) $\tan (t - 120°) = \dfrac{\sqrt{3} + \tan t}{1 - \sqrt{3} \tan t}$
d) $\sec (60° - t) = \dfrac{2}{\cos t + \sqrt{3} \sin t}$

9 Use the half-angle formulas to find the exact values of each of the following expressions.

a) $\sin (-5\pi/12)$ b) $\cos 75°$ c) $\cot (5\pi/12)$
d) $\tan 105°$ e) $\sec (3\pi/8)$ f) $\csc (\pi/12)$

10 Prove each of the following identities.

a) $\sin \dfrac{t}{2} \cos \dfrac{t}{2} = \dfrac{\sin t}{2}$ b) $\dfrac{1 - \cos 2t}{\sin 2t} = \tan t$
c) $(\sin t - \cos t)^2 = 1 - \sin 2t$ d) $\dfrac{1}{\csc 2\theta - \cot 2\theta} = \cot \theta$
e) $\dfrac{1 + \tan^2 \theta}{2 \tan \theta} = \csc 2\theta$ f) $\dfrac{\sin 3\theta}{\cos \theta} + \dfrac{\cos 3\theta}{\sin \theta} = 2 \cot 2\theta$
g) $\dfrac{\sin 3t}{\sin t} - \dfrac{\cos 3t}{\cos t} = 2$ h) $\tan 3\theta = \dfrac{3 \tan \theta - \tan^3 \theta}{1 - 3 \tan^2 \theta}$

11 Solve each of the following trigonometric equations for $0 \leq t < 2\pi$.

a) $\sin t = -\sqrt{2}/2$ b) $\sec t = -2$
c) $\cot t = -1$ d) $2 \cos t = 1$
e) $\sqrt{3} \tan t - 3 = 0$ f) $\sqrt{3} \sin t - \sin 2t = 0$
g) $\sin t - \cos 2t = 0$ h) $2 \cos^2 t + 2 = 5 \cos t$

12 Show that the following statements are true.

a) $\tan^{-1} \dfrac{1}{2} + \tan^{-1} \dfrac{1}{3} = \dfrac{\pi}{4}$ b) $\sin^{-1} \dfrac{1}{2} + \sin^{-1} \dfrac{\sqrt{3}}{2} = \dfrac{\pi}{2}$

CHAPTER 6

Triangle Trigonometry

CHAPTER 6

Triangle Trigonometry

1 Introduction

This chapter deals with the use of trigonometric functions to find the measures of unknown sides and angles of given triangles. We shall begin by restricting ourselves to right triangles, and then we will solve other triangles by use of the *Law of Cosines* and the *Law of Sines*.

Before we begin we should introduce some of the standard notation used to describe the sides and angles of any triangle. The triangle determined by the points A, B, and C is often denoted $\triangle ABC$. The angle at vertex A is denoted by the Greek letter alpha, α; the angle at vertex B is denoted by beta, β; the angle at vertex C is denoted by gamma, γ. The side opposite α is denoted by a; the side opposite β by b; and the side opposite γ by c (Figure 1). Also, recall from geometry that the sum of the angles in any triangle is 180°. Thus, in Figure 1, we have $\alpha + \beta + \gamma = 180°$.

Figure 1

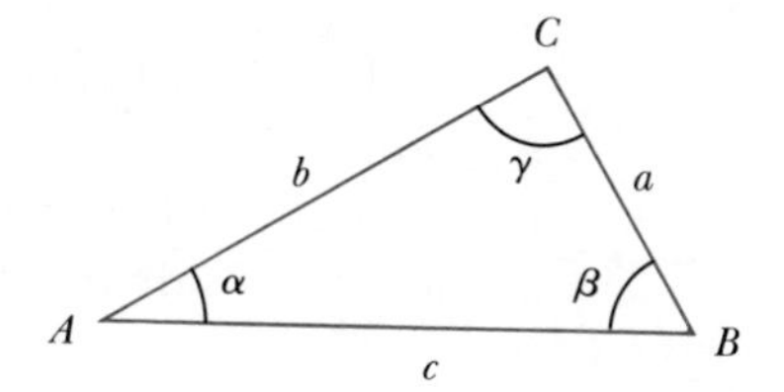

2 Right Triangles

The definitions of the trigonometric functions in Chapter 2 enable us to establish relationships between the sides and angles of right triangles.

Suppose that we are given a right triangle, $\triangle ABC$, with $\gamma = 90°$ and α one of the acute angles (Figure 1). If the right triangle is placed on the coordinate system with α in standard position, then the values of the trigonometric functions of α may be expressed in terms of the sides of the right triangle as follows.

Figure 1

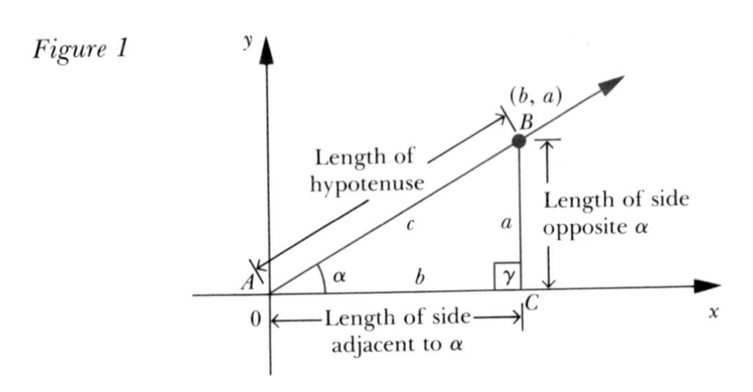

Note that a denotes the length of the side opposite α, b denotes the length of the side adjacent to α, and c denotes the length of the hypotenuse (Figure 1).

Using the definition of the trigonometric functions (Chapter 2 on page 67) with $\theta = \alpha$, $r = c$, $x = b$, and $y = a$, we get

$$\sin \alpha = \frac{a}{c} = \frac{\text{length of side opposite } \alpha}{\text{length of hypotenuse}}$$

$$\cos \alpha = \frac{b}{c} = \frac{\text{length of side adjacent to } \alpha}{\text{length of hypotenuse}}$$

$$\tan \alpha = \frac{a}{b} = \frac{\text{length of side opposite } \alpha}{\text{length of side adjacent to } \alpha}$$

$$\cot \alpha = \frac{b}{a} = \frac{\text{length of side adjacent to } \alpha}{\text{length of side opposite } \alpha}$$

$$\sec \alpha = \frac{c}{b} = \frac{\text{length of hypotenuse}}{\text{length of side adjacent to } \alpha}$$

$$\csc \alpha = \frac{c}{a} = \frac{\text{length of hypotenuse}}{\text{length of side opposite } \alpha}$$

EXAMPLES

1 Determine the lengths of the unknown sides of the right triangle, $\triangle ABC$, if $b = 4$ and $\alpha = 40°$.

Figure 2

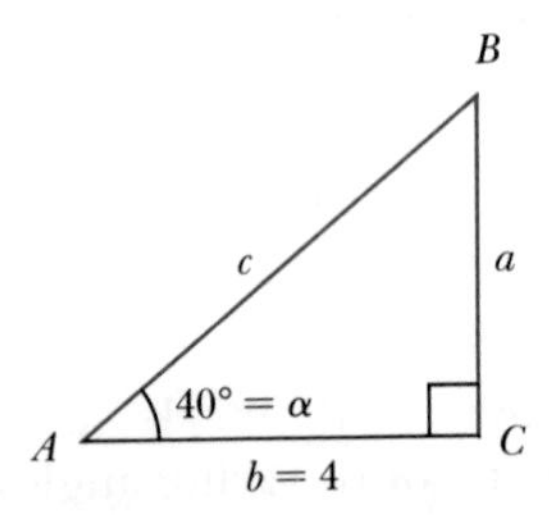

SOLUTION. From Figure 2 and Table I of Appendix B we see that

$$\tan 40^\circ = \frac{a}{b} = \frac{a}{4} = 0.8391$$

so that

$$\begin{aligned} a &= 4(0.8391) \\ &= 3.36 \end{aligned}$$

Also,

$$\cos 40^\circ = \frac{b}{c} = \frac{4}{c} = 0.7660$$

so that

$$c = \frac{4}{0.7660} = 5.22$$

2 Determine the lengths of the unknown sides of the right triangle, $\triangle ABC$, if $b = 5.75$ and $\alpha = 43^\circ$, as illustrated in Figure 3.

Figure 3

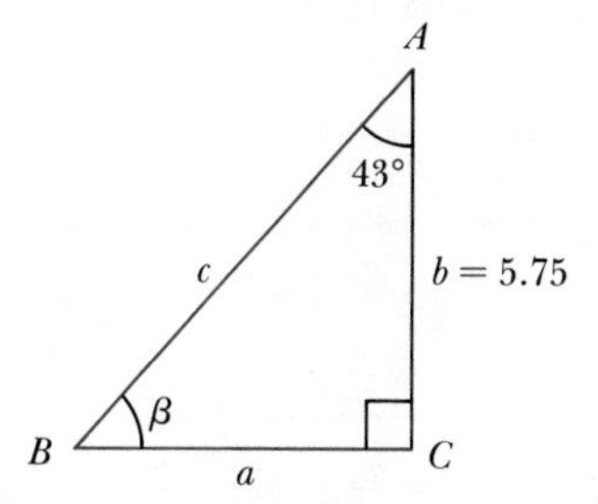

SOLUTION. From Figure 3, we see that $\beta = 90^\circ - 43^\circ = 47^\circ$, so

$$\csc 47^\circ = \frac{c}{5.75}$$

Hence

$$c = 5.75 \csc 47^\circ = 5.75(1.367) = 7.86$$

Since

$$\cot 47^\circ = \frac{a}{5.75}$$

$$a = 5.75 \cot 47^\circ = (5.75)(0.9325) = 5.36$$

3 Determine the measure of the angle α of the right triangle, $\triangle ABC$, if $a = 4$, $b = 3$, and $c = 5$.

Figure 4

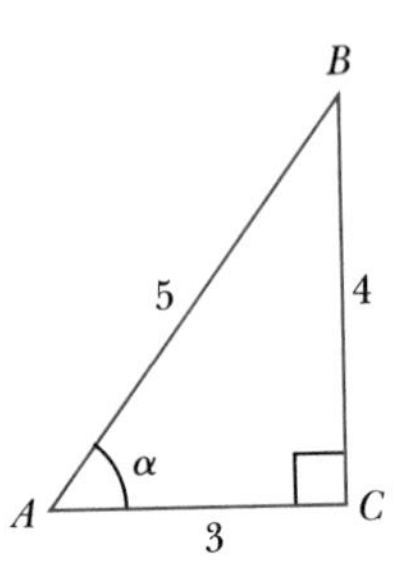

SOLUTION. From Figure 4, we see that

$$\sin \alpha = \tfrac{4}{5} = 0.8$$

so that, by linear interpolation, we get

$$\alpha = 53°8'$$

4 Find the measure of an acute angle α of a right triangle if $\sin \alpha = \cos(\alpha - 22°)$.

Figure 5

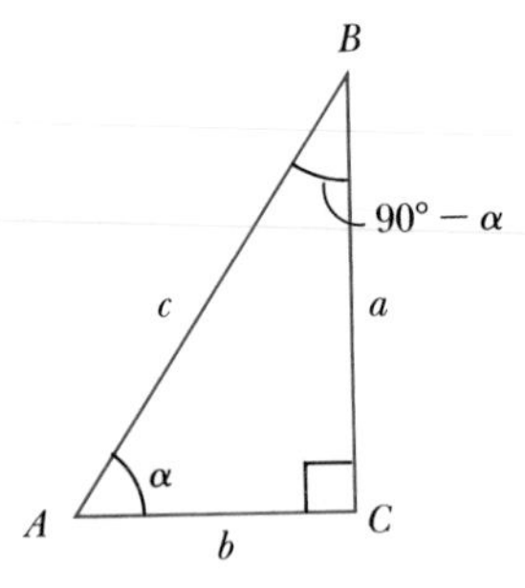

SOLUTION. From Figure 5 we see that

$$\frac{a}{c} = \sin \alpha = \cos(90° - \alpha)$$

so that

$$\cos(90° - \alpha) = \cos(\alpha - 22°)$$

and $90° - \alpha = \alpha - 22°$, that is, $2\alpha = 112°$ or $\alpha = 56°$.

5 Find the degree measures of the two acute angles of a right triangle if the hypotenuse is 100 and side a is 66.91.

Figure 6

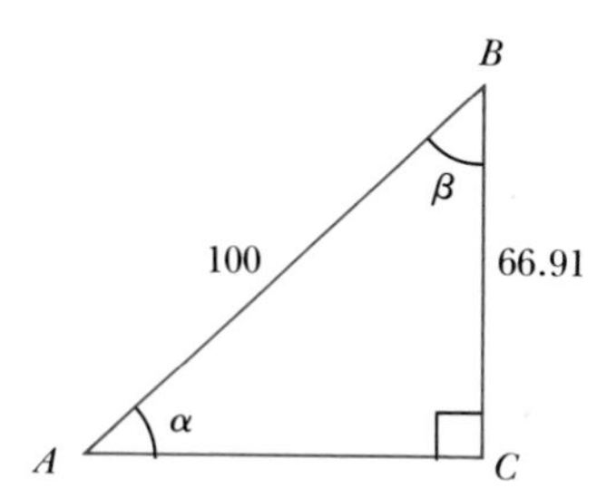

SOLUTION. Using right triangle trigonometry (Figure 6) we have

$$\sin \alpha = \frac{66.91}{100} = 0.6691$$

From Table I we get $\alpha = 42°$ so that $\beta = 90° - \alpha = 90° - 42° = 48°$.

6 A man is standing 150 feet from the foot of a flagpole, which is on his eye level, and the angle of elevation (acute angle formed by the line of sight and a horizontal line passing through the position of the sighting) of the top of the flagpole is 45°. Find the height of the pole.

Figure 7

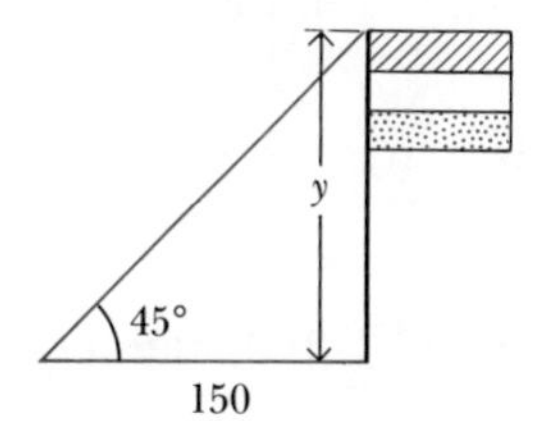

SOLUTION. Let y be the height of the flagpole (Figure 7), then $\tan 45° = y/150 = 1$, so $y = 150$ feet. Therefore, the height of the flagpole is 150 feet.

7 A man observes the angle of elevation of the top of a tower to be 45°. From a point in the same horizontal plane and 100 feet farther away he observes the angle of elevation to be 30°. If the two points of observation and the top of the tower are in the same vertical plane, find the height of the tower.

Figure 8

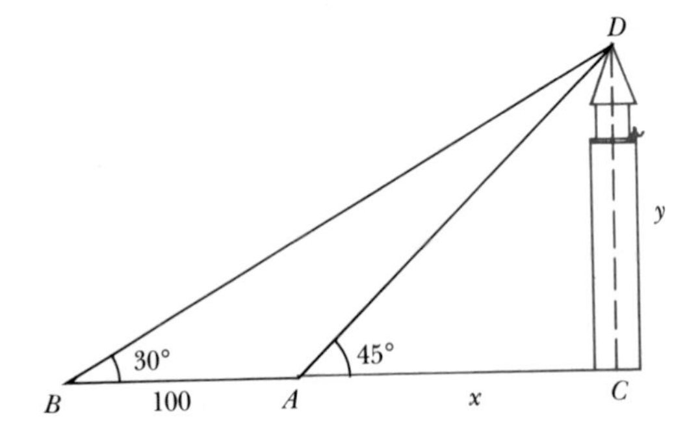

SOLUTION. Let A represent the position of the man initially, B his position after moving 100 feet away, C a point at the foot of the tower, and D a point on top of the tower. Let y feet be the height of the tower and $\overline{AC} = x$ feet (Figure 8). Then

$$\tan 45° = \frac{y}{x}$$

Since $\tan 45° = 1$, we have

$$(1) \quad y = x$$

Also,

$$\tan 30° = \frac{y}{x + 100}$$

so that

$$(2) \quad y = (x + 100) \tan 30° = (x + 100)\left(\frac{\sqrt{3}}{3}\right)$$

Combining Equations (1) and (2) we get

$$y = (y + 100)\,\frac{\sqrt{3}}{3}$$

or

$$y = \frac{\sqrt{3}y}{3} + \frac{100\sqrt{3}}{3}$$

$$y - \frac{\sqrt{3}}{3}\,y = \frac{100\sqrt{3}}{3}$$

$$y\left(1-\frac{\sqrt{3}}{3}\right)=\frac{100\sqrt{3}}{3}$$

$$y=\frac{\left(\frac{100\sqrt{3}}{3}\right)}{\left(1-\frac{\sqrt{3}}{3}\right)}$$

$$=\frac{100\sqrt{3}}{3-\sqrt{3}}=136.59 \text{ feet} \qquad \text{(approximately)}$$

PROBLEM SET 1

1 Use right triangle, $\triangle ABC$, in Figure 9 to answer each of the following.
a) If $c = 6$ and $\alpha = 30°$, find a and b.
b) If $\sin\alpha = \frac{1}{2}$ and $c = 60$, find a and $\cos\beta$.
c) If $\alpha = 34°$ and $c = 200$, find a and b.
d) If $a = 6$ and $b = 8$, find c, $\sin\alpha$ and $\cos\alpha$.
e) If $a = 5$ and $\sin\alpha = \frac{4}{5}$, find b and c.
f) If $c = 50$ and $\tan\alpha = 2$, find a and b, $\sin\alpha$ and $\cos\alpha$.
g) If $c = 10$ and $\sin\alpha = \frac{4}{7}$, find a, b, and $\cos\alpha$.
h) If $\sin\alpha < 0$ and $\cos\alpha = -\frac{3}{5}$, find $\sin\alpha$ and $\tan\alpha$.
i) If $\alpha = 40°$ and $c = 5$, find a and b.

Figure 9

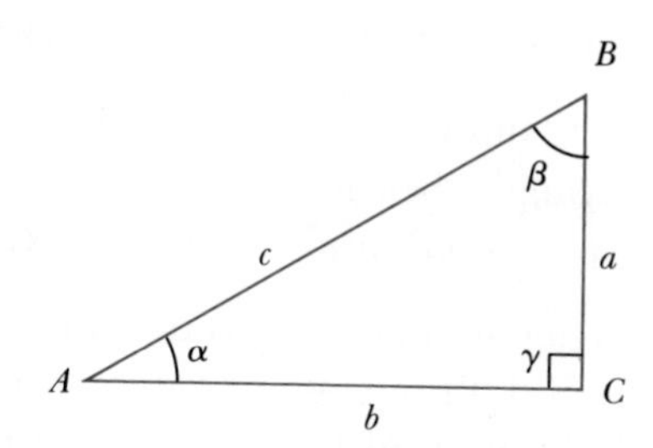

2 Find the acute angles of the right triangle in Figure 9 if the lengths of two of the sides are
a) $a = 3$ and $b = 4$
b) $b = 25$ and $a = 7$
c) $a = 5$ and $c = 12$
d) $b = 17$ and $a = 8$
e) $c = 10\sqrt{2}$ and $b = 10$

3 In each case find the measure of an acute angle α of a right triangle for which the equation is true.
a) $\cos\alpha = \sin(\alpha + 45°)$
b) $\cot 3\alpha = \tan\alpha$
c) $\tan(\alpha/2) = \cot 5\alpha$
d) $\sin(\alpha + 20°) = \cos(\alpha - 20°)$
e) $\sec 2\alpha = \csc(\alpha - 10°)$

4 If the angle of elevation of the top of a tower from a distance 200 feet away on the ground level is 60°, find the height of the tower.

5 From the top of an observation post the angle of depression (acute angle determined by the plane of the observer and the line of sight) of a ship is 30°. If the distance between the observer and the ship is 450 feet, find the height of the observation post.

6 From the top of a tower of height 300 feet, the angles of depression of the top and the bottom of a flag pole standing on a plane level with the base of the tower are observed to be 45° and 60°, respectively. Find the height of the flag pole.

7 From the top of a mountain 2000 feet high, the angles of depression of two boats lying in a lake below are 18° and 14°, respectively. Find the distance between the boats.

8 From an airplane at an altitude of 15,000 feet, the pilot observes the angle of depression of the base of a building to be 20°. How far is the building from a point on the ground directly beneath the airplane?

9 A ladder 14 feet long is leaning against a building. The angle formed by the ladder and the ground is 65°. How far from the building is the foot of the ladder?

10 Points A and B are opposite each other on shores of a straight river that is 3 miles wide. Point C is on the same shores as B but down the river from B. A telephone company wishes to lay a cable underwater from A to C. If $\angle ACB$ is 48°, what is the length of the cable $\overline{AC}$?

11 An island is at point A, 6 miles offshore from the nearest point B on a straight beach. A store is at point C down the beach from B. If $\angle ACB$ is 35°, find the distance from the point C to A.

12 A television antenna stands on top of a house that is located on level ground. From a point 140 feet from the base of the house, the angles of elevation of the top and bottom of the antenna measure 70° and 66°, respectively. How high is the house? How long is the television antenna?

13 The angle of elevation of the top of a hill is 59° from a given point. After moving a distance of 170 feet in a horizontal line away from the point, the angle of elevation was found to be 52°. How high is the hill?

3 Law of Cosines

In Section 2 we used trigonometric functions to determine the unknown parts of *right triangles*. In this section and the next, we shall solve similar problems involving other triangles. In particular, if we are given two sides and the included angle of a triangle or the three sides of a triangle, a unique triangle is determined, and the unknown parts may be found by using the *Law of Cosines.*

THEOREM 1 LAW OF COSINES

In any $\triangle ABC$ (Figure 1)

$$a^2 = b^2 + c^2 - 2bc \cos \alpha$$

Figure 1

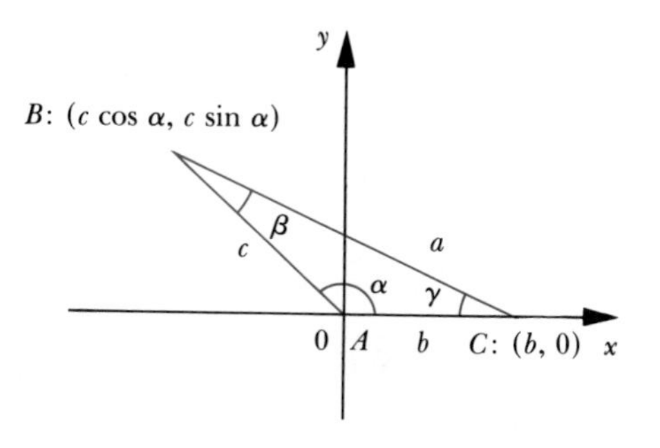

PROOF. First, locate $\triangle ABC$ on a Cartesian coordinate system with angle α in standard position (Figure 1).

B has coordinates $(c \cos \alpha, c \sin \alpha)$ (why?) and C has coordinates $(b, 0)$, so that by the distance formula, we get

$$\begin{aligned} a^2 &= (c \cos \alpha - b)^2 + (c \sin \alpha - 0)^2 \\ &= c^2 \cos^2 \alpha - 2bc \cos \alpha + b^2 + c^2 \sin^2 \alpha \\ &= c^2 (\cos^2 \alpha + \sin^2 \alpha) + b^2 - 2bc \cos \alpha \\ &= c^2 + b^2 - 2bc \cos \alpha = b^2 + c^2 - 2bc \cos \alpha \end{aligned}$$

Note: Any letters may be used to denote the known and unknown parts of a triangle, so the Law of Cosines can also be expressed for $\triangle ABC$ as

$$b^2 = a^2 + c^2 - 2ac \cos \beta$$

or

$$c^2 = a^2 + b^2 - 2ab \cos \gamma$$

EXAMPLES

Use the Law of Cosines to find the indicated part of each of the following triangles.

1 In $\triangle ABC$, $b = 4$, $c = 3$, and $\alpha = 60°$. Find a.

Figure 2

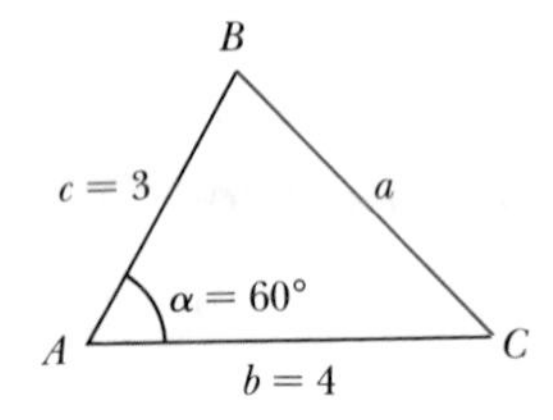

SOLUTION. Using the Law of Cosines (Figure 2) we have

$$\begin{aligned} a^2 &= b^2 + c^2 - 2bc \cos \alpha \\ &= 16 + 9 - 2(4)(3) \cos 60° \\ &= 25 - 24(\tfrac{1}{2}) \\ &= 25 - 12 \\ &= 13 \end{aligned}$$

so that $a = \sqrt{13}$.

2 In $\triangle ABC$, $a = 7.2$, $b = 4$, and $c = 6$. Find β.

Figure 3

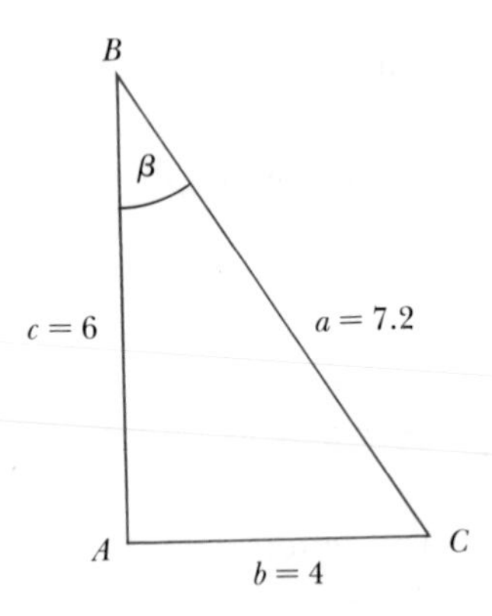

SOLUTION. Using the Law of Cosines (Figure 3) we have

$$b^2 = a^2 + c^2 - 2ac \cos \beta$$
$$16 = 51.84 + 36 - 2(7.2)(6) \cos \beta$$

so

$$\cos \beta = \frac{71.84}{86.40} = 0.8315 \qquad \text{(approximately)}$$

Using Table I and linear interpolation we find that β is approximately 33°45′.

3 In $\triangle ABC$, $b = 5$, $c = 2$, and $\alpha = 120°$. Find a.

Figure 4

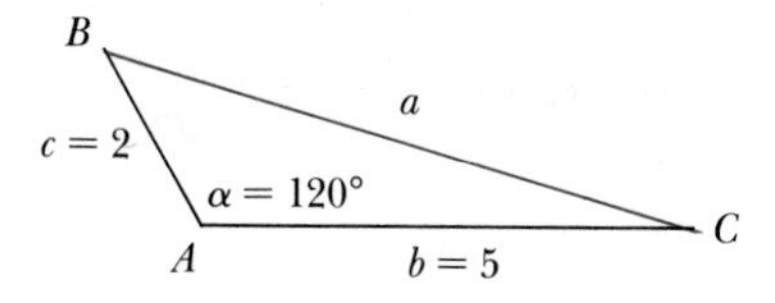

SOLUTION. Using the Law of Cosines (Figure 4) we have

$$\begin{aligned} a^2 &= b^2 + c^2 - 2bc \cos \alpha \\ &= 5^2 + 2^2 - 2(5)(2) \cos 120° \\ &= 25 + 4 - 20(-\tfrac{1}{2}) \\ &= 29 + 10 \\ &= 39 \end{aligned}$$

so that $a = \sqrt{39}$.

4 In $\triangle ABC$, $a = 7$, $b = 4$, and $c = 5$. Find α to the nearest degree.

Figure 5

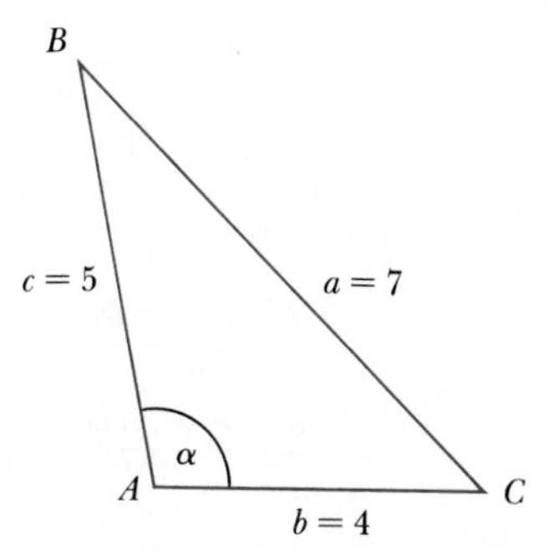

SOLUTION. Using the Law of Cosines (Figure 5) we have

$$\begin{aligned} a^2 &= b^2 + c^2 - 2bc \cos \alpha \\ 7^2 &= 4^2 + 5^2 - 2(4)(5) \cos \alpha \\ 49 &= 16 + 25 - 40 \cos \alpha \\ 49 &= 41 - 40 \cos \alpha \end{aligned}$$

so that

$$\cos \alpha = -\tfrac{8}{40} = -0.2$$

Since $\cos \alpha$ is negative, $90° < \alpha < 180°$. From Table I, we find that

$$\cos 78° = 0.2 \qquad \text{(approximately)}$$

so that $\alpha = 180° - 78° = 102°$.

5 Use the Law of Cosines to show that

$$1 + \cos \alpha = \frac{(b + c + a)(b + c - a)}{2bc}$$

SOLUTION. Using the Law of Cosines, we have

$$a^2 = b^2 + c^2 - 2bc \cos \alpha$$

so that

$$2bc \cos \alpha = b^2 + c^2 - a^2$$

or

$$\cos \alpha = \frac{b^2 + c^2 - a^2}{2bc}$$

Then

$$\begin{aligned} 1 + \cos \alpha &= 1 + \frac{b^2 + c^2 - a^2}{2bc} \\ &= \frac{2bc + b^2 + c^2 - a^2}{2bc} \\ &= \frac{(b + c)^2 - a^2}{2bc} \\ &= \frac{(b + c + a)(b + c - a)}{2bc} \end{aligned}$$

6 The air speed of an airplane heading north 42° east is 300 miles per hour. If the wind velocity is 50 miles per hour blowing south 25° east, find the ground speed of the airplane.

Figure 6

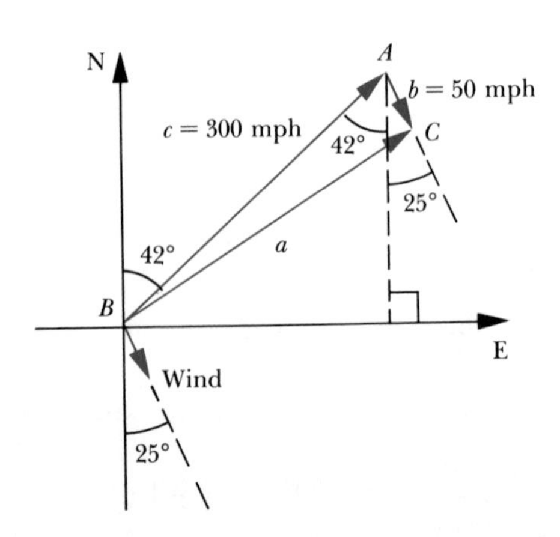

SOLUTION. By geometry, we see that $\angle BAC = 67°$ (Figure 6). Let $\overline{BA} = c = 300$ represent the air speed of the plane and $\overline{AC} = b = 50$ represent the wind velocity and $\overline{BC} = a$ represent the ground speed of the airplane. Using the Law of Cosines, we have

$$\begin{aligned} a^2 &= b^2 + c^2 - 2bc \cos \alpha \\ &= 2500 + 90{,}000 - 2(50)(300) \cos 67° = 80{,}779 \end{aligned}$$

so that

$$a = 284.22 \qquad \text{(approximately)}$$

Thus the ground speed of the airplane is 284.22 miles per hour.

7 The straight-line distance between two cities, A and B, cannot be measured because of a swamp. A surveyor at C is able to measure $\overline{AC} = 5.73$ miles, $\overline{CB} = 8.19$ miles, and $\angle ACB = 113°$. What is the distance from city A to city B?

Figure 7

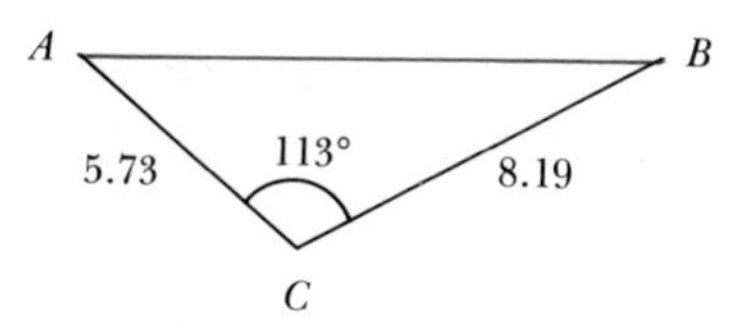

SOLUTION. Using the Law of Cosines (Figure 7), we have

$$\begin{aligned} \overline{AB}^2 &= \overline{AC}^2 + \overline{BC}^2 - 2\overline{AC} \cdot \overline{BC} \cos \alpha \\ &= (5.73)^2 + (8.19)^2 - 2(5.73)(8.19) \cos 113° \\ &= 32.8329 + 67.0761 - (93.8574)(-0.3907) \\ &= 136.58 \end{aligned}$$

Therefore, $\overline{AB} = \sqrt{136.58} = 11.69$ miles.

Another useful formula, called *Hero's formula,* can be derived using the Law of Cosines. Hero's formula expresses the area of a triangle in terms of the lengths of the three sides.

EXAMPLES

1 (Hero's Formula) Prove that the area A of a triangle with sides a, b, and c and semiperimeter $s = \frac{1}{2}(a + b + c)$, is

$$A = \sqrt{s(s-a)(s-b)(s-c)}$$

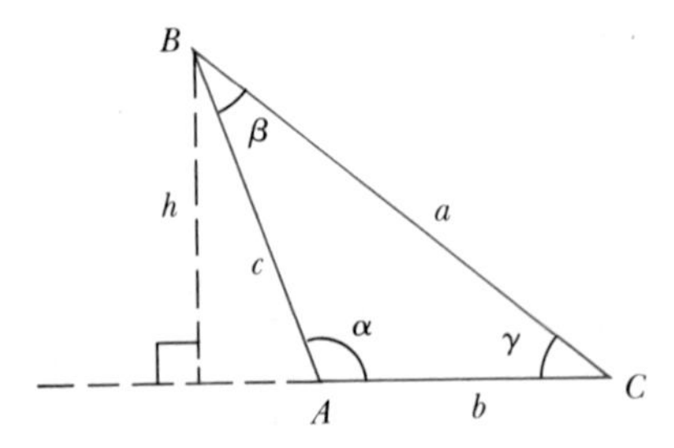

Figure 8

PROOF. The area A of $\triangle ABC$ can be expressed as $A = \frac{1}{2}bh$, where h is the altitude (Figure 8). But since $h/c = \sin(180° - \alpha) = \sin\alpha$ (why)?, we have $h = c \sin\alpha$, from which it follows that $A = \frac{1}{2}bc \sin\alpha$. Now any positive number is the square root of its own square; hence,

$$\begin{aligned} A &= \sqrt{\tfrac{1}{4}b^2c^2 \sin^2\alpha} \\ &= \sqrt{\tfrac{1}{4}b^2c^2(1 - \cos^2\alpha)} \end{aligned}$$

or

$$A = \sqrt{\tfrac{1}{2}bc(1 + \cos\alpha) \cdot \tfrac{1}{2}bc(1 - \cos\alpha)}$$

Using the results of Example 5 on page 228 and Problem 3, Problem Set 2, we have

$$\begin{aligned} A &= \sqrt{\tfrac{1}{2}bc\,\frac{(b+c+a)(b+c-a)}{2bc}\;\tfrac{1}{2}bc\,\frac{(a-b+c)(a+b-c)}{2bc}} \\ &= \sqrt{\left(\frac{a+b+c}{2}\right)\left(\frac{b+c-a}{2}\right)\left(\frac{a-b+c}{2}\right)\left(\frac{a+b-c}{2}\right)} \\ &= \sqrt{\left(\frac{a+b+c}{2}\right)\left(\frac{a+b+c-2a}{2}\right)\left(\frac{a+b+c-2b}{2}\right)\left(\frac{a+b+c-2c}{2}\right)} \\ &= \sqrt{\left(\frac{a+b+c}{2}\right)\left(\frac{a+b+c}{2}-a\right)\left(\frac{a+b+c}{2}-b\right)\left(\frac{a+b+c}{2}-c\right)} \\ &= \sqrt{s(s-a)(s-b)(s-c)} \end{aligned}$$

2 Use Hero's formula to find the area of a triangle if $a = 6$, $b = 8$, and $c = 12$.

SOLUTION

$$s = \frac{a+b+c}{2} = \frac{6+8+12}{2} = 13$$

so that

$$\begin{aligned} A &= \sqrt{s(s-a)(s-b)(s-c)} \\ &= \sqrt{13(13-6)(13-8)(13-12)} \\ &= \sqrt{13(7)(5)(1)} \\ &= \sqrt{455} \\ &= 21.33 \end{aligned}$$

3 Suppose that a room with a triangular floor plan is to be carpeted at a cost of \$9.50 per square yard. What is the cost of carpeting the room if the lengths of the three sides are 10 yards, 15 yards, and 17 yards?

SOLUTION. The area A of the room can be found by using Hero's formula.

$$\begin{aligned} A &= \sqrt{s(s-a)(s-b)(s-c)} \\ &= \sqrt{21(11)(6)(4)} \\ &= \sqrt{5544} \\ &= 74.46 \text{ square yards} \qquad \text{(approximately)} \end{aligned}$$

Since the cost is \$9.50 per square yard, the total cost T is given by

$$T = (\$9.50)(74.46) = \$707.37$$

PROBLEM SET 2

1 Use the Law of Cosines to find the indicated part of the following triangles.

a) a if $c = 3$, $b = 10$, and $\alpha = 60°$
b) c if $a = 10$, $b = 12$, and $\gamma = 35°$
c) c if $a = 250$, $b = 300$, and $\gamma = 58°$
d) γ if $a = 200$, $b = 50$, and $c = 177$
e) γ if $a = 2$, $b = 3$, and $c = 4$
f) b if $a = 4.9$, $c = 6.8$, and $\beta = 122°$

2 Use the Law of Cosines to find the unknown parts of each of the following triangles.

a) $a = 10$, $b = 7$, and $\gamma = 32°$
b) $a = 10$, $b = 7$, and $c = 12$
c) $a = 60$, $c = 30$, and $\beta = 40°$
d) $a = 2\sqrt{61}$, $b = 8$, and $c = 10$
e) $a = 4$, $b = 20$, and $c = 18$
f) $a = 54$, $c = 15$, and $\beta = 95°$

3 Use the Law of Cosines to prove each of the following identities.

a) $1 - \cos \alpha = \dfrac{(a-b+c)(a+b-c)}{2bc}$
b) If $b = c$, then $a^2 = 2b^2(1 - \cos \alpha)$.

4 Use the Law of Cosines to prove that if $\alpha = 90°$ in $\triangle ABC$, then $a^2 = b^2 + c^2$.

5 The lengths of two sides of a parallelogram are 5 inches and 6 inches. If the length of one diagonal is 8.5 inches, find the angles of the parallelogram.

6 A triangular lot has two sides with lengths 200 feet and 80 feet. The angle between the two sides is 120°. Find the length of the third side of the lot.

7 The sides of a large triangular flower bed are 20 feet, 25 feet, and 40 feet. Find the largest angle of the triangle.

8 A railroad crosses a highway at an angle of 60°. A locomotive is 100 feet from the intersection and moving away from it at the rate of 40 miles per hour. A car is 60 feet from the intersection and moving away from it at the rate of 50 miles per hour. What is the distance between them after 3 hours?

9 Use Hero's formula to find the area of the triangle ABC with the given sides.
a) $a = 5$, $b = 8$, and $c = 7$
b) $a = 10$, $b = 12$, and $c = 14$
c) $a = 7.3$, $b = 11.1$, and $c = 15.2$
d) $a = 13$, $b = 15$, and $c = 17$

10 Which triangle has the largest area, a triangle with sides 10, 10, and 15 or a triangle with sides 9, 11, and 12?

4 Law of Sines

In this section we shall develop another formula, called the *Law of Sines,* which also helps us to determine the unknown parts of triangles that are not necessarily right triangles. We will see that the Law of Sines is applicable if we are given any two angles and any side of a triangle, or if we are given two sides and an angle opposite one of them. The first situation yields a unique triangle, whereas the second situation, called the *ambiguous case,* may result in no triangle, one triangle, or two triangles with the given measures, as we shall see later.

THEOREM 1 LAW OF SINES

In any $\triangle ABC$ (Figure 1), we have

$$\frac{\sin \alpha}{a} = \frac{\sin \beta}{b} = \frac{\sin \gamma}{c}$$

Figure 1

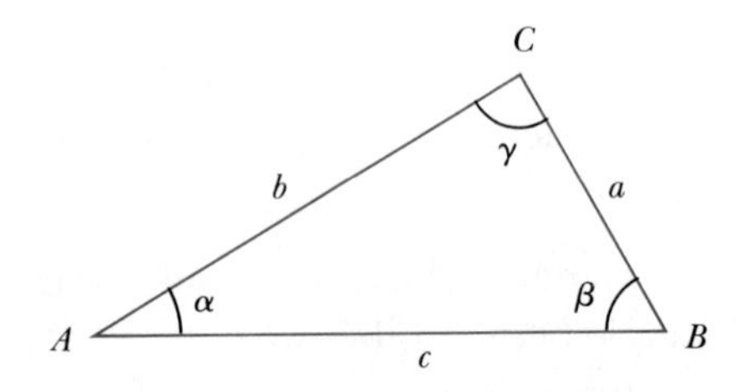

PROOF. Place $\triangle ABC$ in a Cartesian coordinate system so that angle γ is in standard position. (Here we have illustrated the case when γ is an obtuse angle, however the proof will hold for any γ, where $0° < \gamma < 180°$.) Label the parts as illustrated in Figure 2.

Figure 2

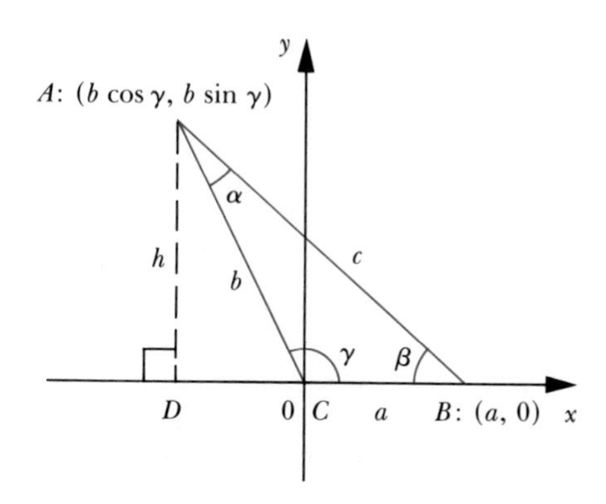

In right triangle BDA,

$$\sin \beta = \frac{h}{c}$$

so that

(1) $c \sin \beta = h$

Also, in right triangle CDA,

$$\sin (180° - \gamma) = \frac{h}{b}$$

so that

$$b \sin (180° - \gamma) = h$$

But since $\sin (180° - \gamma) = \sin \gamma$, we have

(2) $b \sin \gamma = h$

Hence, $c \sin \beta = b \sin \gamma$.

Dividing both sides of the equation by bc we get

$$\frac{\sin \beta}{b} = \frac{\sin \gamma}{c}$$

If we locate the triangle so that the vertex of α is on the x axis, the same proof shows that

$$\frac{\sin \alpha}{a} = \frac{\sin \gamma}{c}$$

Thus

$$\frac{\sin \alpha}{a} = \frac{\sin \beta}{b} = \frac{\sin \gamma}{c}$$

EXAMPLES

1 Use the Law of Sines to determine the indicated parts of the triangle.

a) In $\triangle ABC$, $a = 10$, $\beta = 42°$, and $\gamma = 51°$, find b.

Figure 3

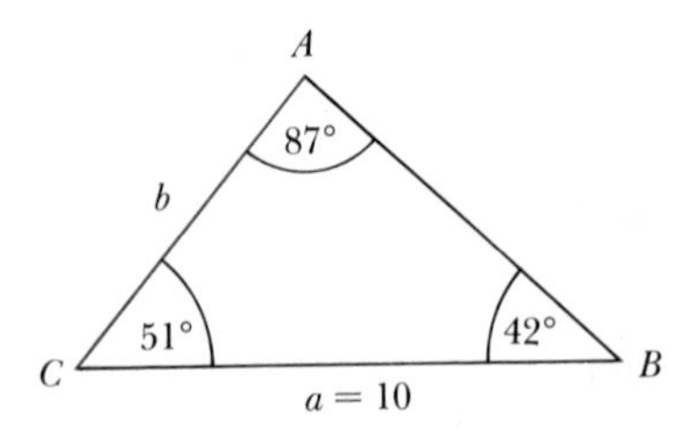

SOLUTION. Since $\alpha + \beta + \gamma = 180°$, we obtain $\alpha = 180° - 42° - 51° = 87°$ (Figure 3).

Using the Law of Sines, we have

$$\frac{\sin 87°}{10} = \frac{\sin 42°}{b}$$

so that

$$b = \frac{10 \sin 42°}{\sin 87°} = \frac{10(0.6691)}{0.9986} = \frac{6.691}{0.9986} = 6.70 \qquad \text{(approximately)}$$

b) In $\triangle ABC$, if $\alpha = 30°$, $\beta = 45°$ and $a = 7$, find b.

Figure 4

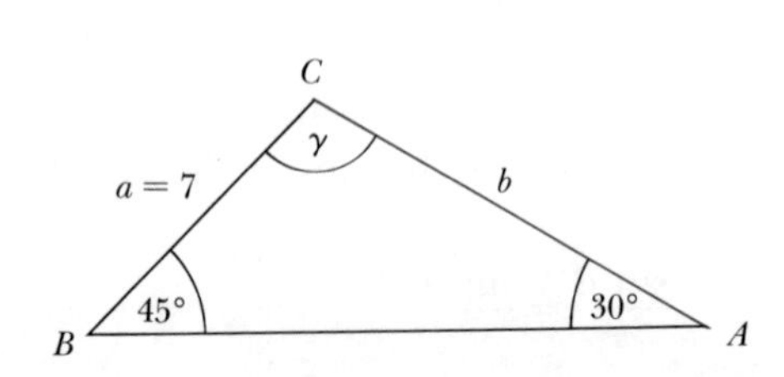

SOLUTION. (See Figure 4.)
Using the Law of Sines, we have

$$\frac{\sin 30°}{7} = \frac{\sin 45°}{b}$$

so that

$$b = \frac{7(\sin 45°)}{\sin 30°} = \frac{7(\sqrt{2}/2)}{\frac{1}{2}} = 7(1.414) = 9.90 \qquad \text{(approximately)}$$

2 A tower 200 feet high is on a cliff on the shore of a lake. From the top of the tower the angle of depression of a ship in the lake is 28°40′, and from the base of the tower the angle of depression of the same ship is 18°20′. Find the height of the cliff and the distance between the ship and the cliff.

Figure 5

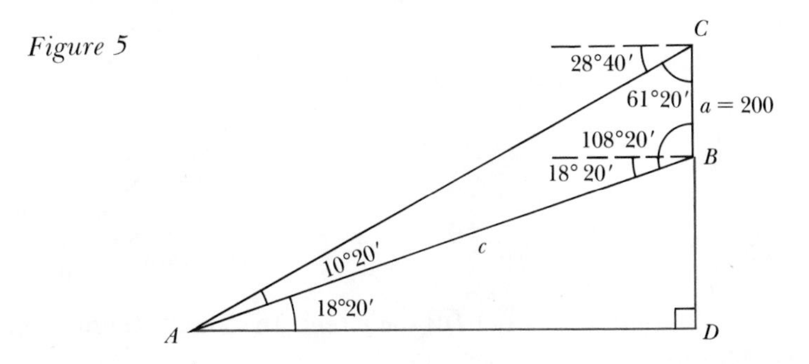

SOLUTION. In Figure 5, let $\overline{BC} = a$ represent the height of the tower, $\overline{BD}$ represents the height of the cliff, $\overline{AD}$ represents the distance between the ship and the cliff, and A represents the position of the ship in the lake. In triangle ABC,

$$\angle ABC = 90° + 18°20′ = 108°20′$$
$$\angle ACB = 90° - 28°40′ = 61°20′$$
$$\text{and } \angle BAC = 180° - (61°20′ + 108°20′) = 10°20′$$

Using the Law of Sines, we have

$$\frac{\sin 61°20′}{c} = \frac{\sin 10°20′}{200}$$

so that

$$c = \frac{200 \sin 61°20′}{\sin 10°20′} = \frac{200(0.8774)}{0.1794} = 978.15$$

In right triangle ADB

$$\overline{BD} = c \sin 18°20' = (978.15)(0.3145) = 307.63$$
$$\overline{AD} = c \cos 18°20' = (978.15)(0.9492) = 928.46$$

so the height of the cliff is 307.63 feet and the distance between the ship and the tower is 928.46 feet.

3 A surveyor wishes to find the distance between the city of Detroit D and the city of Windsor W (the two cities are on the opposite banks of a river). From D a line $\overline{DC} = 550$ feet is laid off and $\angle CDW = 125°40'$ and $\angle DCW = 48°50'$ are measured. Find the distance between the two cities, that is, the length of $\overline{DW}$.

Figure 6

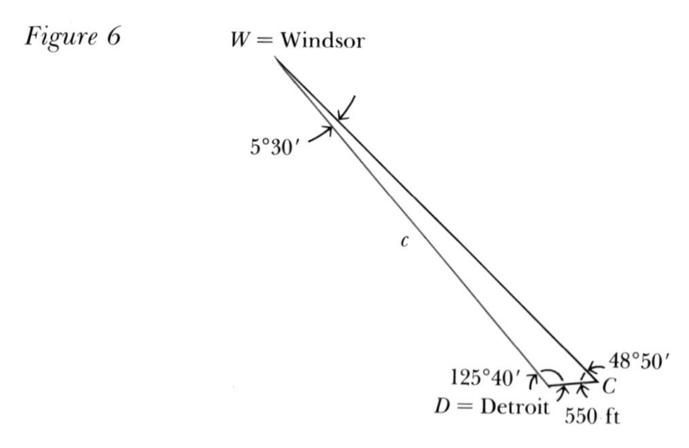

SOLUTION. Let $\overline{DW}$ be c feet. In $\triangle WDC$ (Figure 6),

$$\angle DWC = 180° - (125°40' + 48°50') = 5°30'$$

Using the Law of Sines, we have

$$\frac{\sin 48°50'}{c} = \frac{\sin 5°30'}{550}$$

so that

$$\overline{DW} = c = \frac{550(\sin 48°50')}{\sin 5°30'} = \frac{550(0.7528)}{0.0958} = 4321.92$$

Therefore, the distance is 4321.92 feet.

Logarithms[1] of the trigonometric functions found in Table IV can be used to perform the computations when applying the Law of Sines as illustrated in the following examples.

[1] A brief review of logarithms is included in Appendix A.

EXAMPLES

Use logarithms to solve each of the following problems.

1 In $\triangle ABC$, $a = 12$, $\alpha = 43°$, and $\beta = 100°$. Find c.

Figure 7

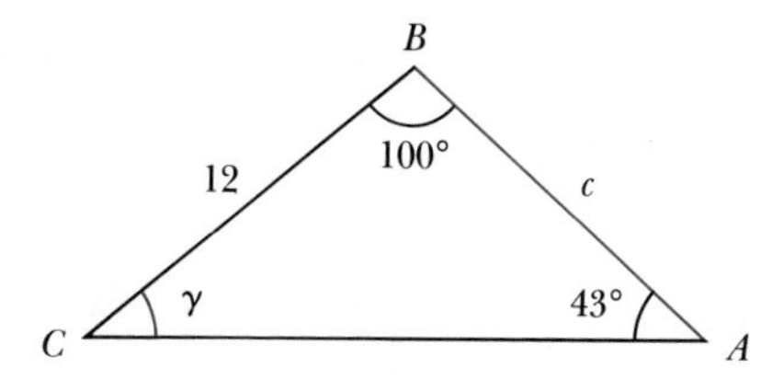

SOLUTION. (See Figure 7). Since $\alpha + \beta + \gamma = 180°$,

$$\gamma = 180° - 143° = 37°$$

Applying the Law of Sines, we have

$$\frac{\sin 37°}{c} = \frac{\sin 43°}{12}$$

so that

$$c = \frac{12 \sin 37°}{\sin 43°}$$

Taking logarithms of each side, we obtain

$$\begin{aligned}\log c &= \log 12 + \log \sin 37° - \log \sin 43° \\ &= (0.0792 + 1) + (9.7795 - 10) - (9.8338 - 10) \\ &= 1.0249\end{aligned}$$

Taking the antilogarithm we get $c = 10.59$ (approximately).

2 In $\triangle ABC$, $\alpha = 105°$, $\beta = 45°$, and $b = 2$. Find c.

Figure 8

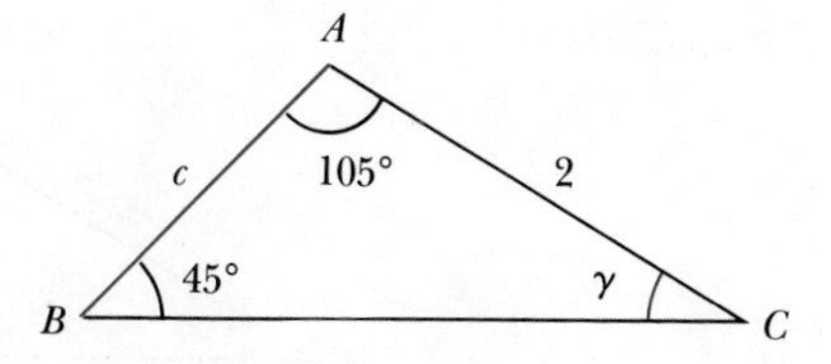

SOLUTION. (See Figure 8.) $\gamma = 180° - 150° = 30°$, so, upon applying the Law of Sines, we obtain

$$\frac{\sin 30°}{c} = \frac{\sin 45°}{2} \quad \text{or} \quad c = \frac{2 \sin 30°}{\sin 45°}$$

Taking logarithms of each side, we get

$$\begin{aligned}\log c &= \log 2 + \log \sin 30° - \log \sin 45° \\ &= 0.3010 + (9.6990 - 10) - (9.8495 - 10) \\ &= 0.1505\end{aligned}$$

Hence, $c = 1.41$ (approximately).

4.1 The Ambiguous Case

If two sides and an angle opposite one of them are given, there may be no triangle possible with the given measures, only one triangle possible, or two triangles possible with these parts. Because of these three possibilities, this case is often referred to as the *ambiguous case.* Before illustrating how the Law of Sines can be used in this case, let us examine how the three possibilities can occur, if the given angle is acute. (See Problem 8 for the case when the angle is obtuse.)

1 If the given parts include side b, acute angle α, and side $a = b \sin \alpha$, then there is one possible (right) triangle (Figure 9).

Figure 9

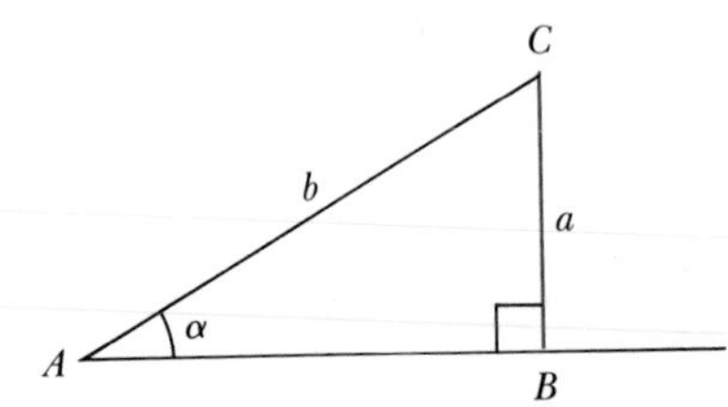

2 If the given parts include side b, acute angle α, and side

$$a < h = b \sin \alpha$$

then there is no possible triangle (Figure 10).

Figure 10

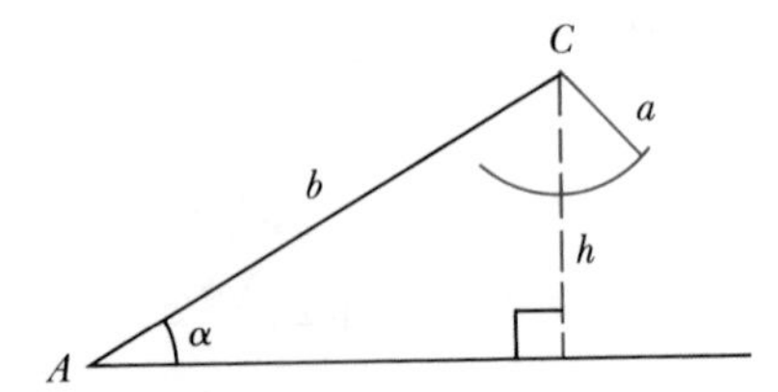

3 If the given parts include acute angle α and $a \geqslant b$, then there is only one possible triangle (Figure 11), either $\triangle ABC$ if $a = b$ or $\triangle AB'C$ if $a > b$.

Figure 11

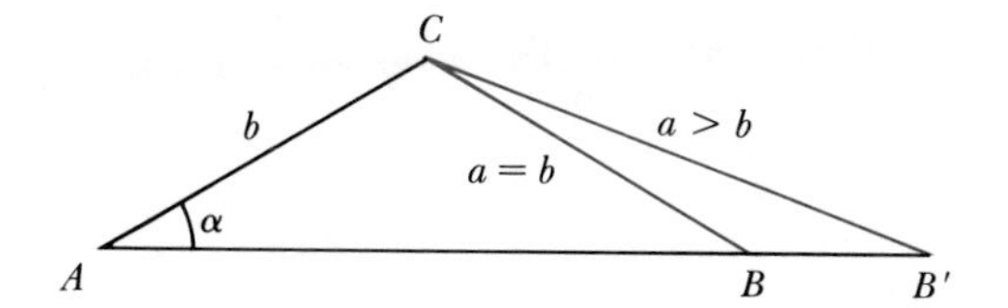

4 If the given parts include acute angle α and $b > a > h = b \sin \alpha$, then there are two possible triangles, $\triangle ABC$ and $\triangle AB_1C$ (Figure 12).

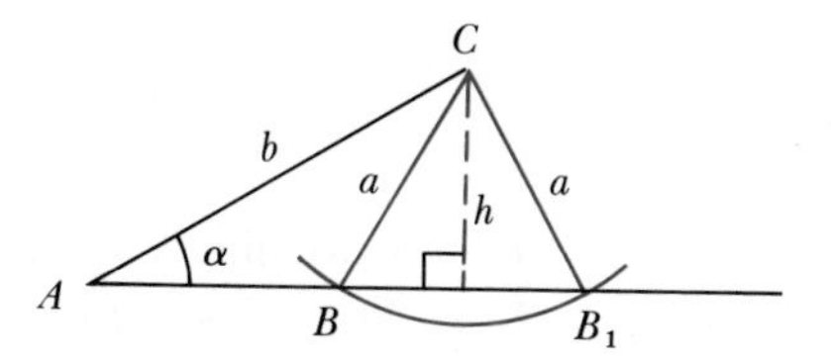

EXAMPLES

1 Is it possible to have a triangle, $\triangle ABC$, such that $a = 10$, $b = 50$, and $\alpha = 22°$? If so, find β.

SOLUTION. We will attempt to find β with the idea in mind that there may be no triangle, one triangle, or two triangles. If we assume that there is at least one possible triangle, then the Law of Sines applies and we get

$$\frac{\sin \beta}{50} = \frac{\sin 22°}{10}$$

so that

$$\sin \beta = \frac{50(0.3746)}{10} = 5(0.3746) = 1.8730$$

But $|\sin \beta| \leq 1$; therefore, no such β exists. Consequently, no triangle exists with these parts (Figure 13).

Figure 13

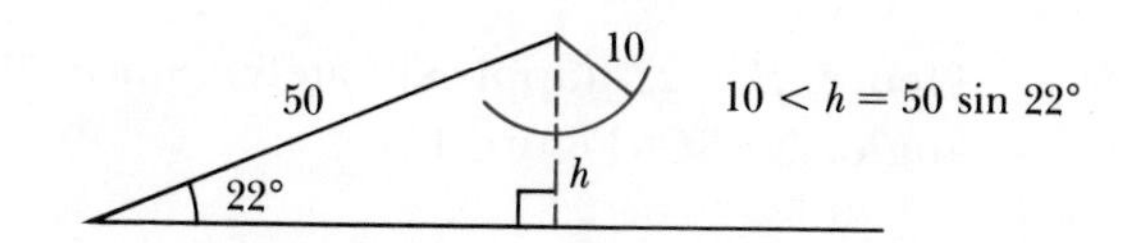

2 Is it possible to have a triangle, $\triangle ABC$, such that $b = 27$, $c = 20$, and $\gamma = 20°$? If so, find β to the nearest 10 minutes.

SOLUTION. If there is at least one such triangle, then, by the Law of Sines,

$$\frac{\sin \beta}{27} = \frac{\sin 20°}{20}$$

or

$$\sin \beta = \frac{27 \sin 20°}{20} = \frac{27}{20}(0.3420) = 0.4617$$

Hence,

$$\beta = 27°30' \qquad \text{or} \qquad \beta = 180° - 27°30' = 152°30'$$

Thus, there are two possible triangles, $\triangle CBA$ and $\triangle CB'A$ (Figure 14). Note that $27 > 20 > 27 \sin 20° = 9.234$.

Figure 14

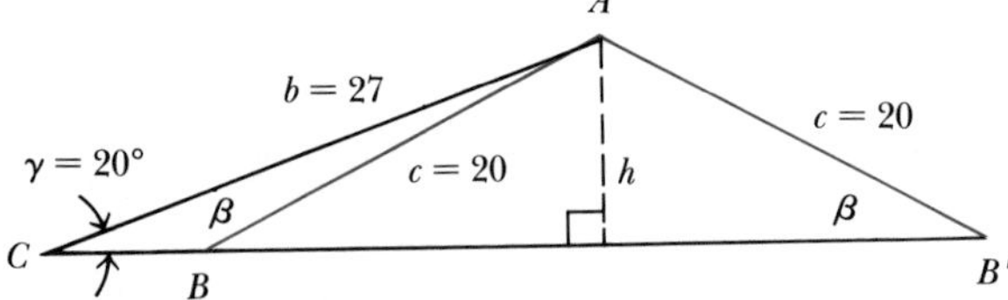

3 Is it possible to have a triangle such that $a = 20$, $b = 15$, and $\alpha = 30°$? If so, find β to the nearest degree.

SOLUTION. If there is at least one such triangle, then, by the Law of Sines, we have

$$\frac{\sin \beta}{15} = \frac{\sin 30°}{20}$$

or

$$\sin \beta = \tfrac{15}{20} \sin 30° = \tfrac{15}{20}(0.5) = 0.3750$$

Hence, $\beta = 22°$ (approximately). Since $20 > 15$, there is only one triangle, $\triangle ABC$ (Figure 15).

Figure 15

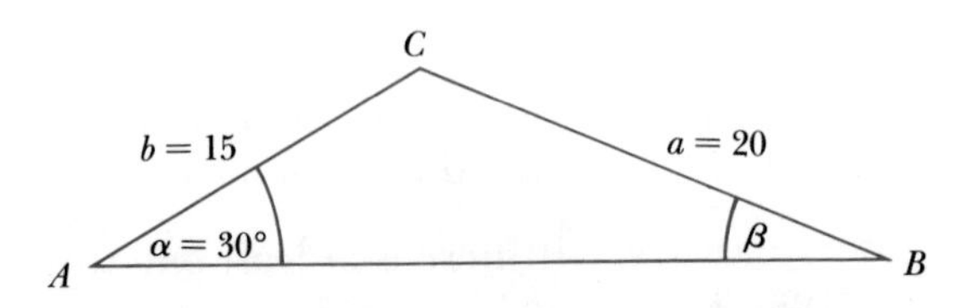

We can use the law of sines to derive another useful formula, called the *Law of Tangents,* which can also be used to find unknown parts of triangles.

EXAMPLES

1 (*Law of Tangents*) Prove that for any $\triangle ABC$ (Figure 16),

$$\frac{a-b}{a+b}=\frac{\tan \frac{1}{2}(\alpha-\beta)}{\tan \frac{1}{2}(\alpha+\beta)}$$

Figure 16

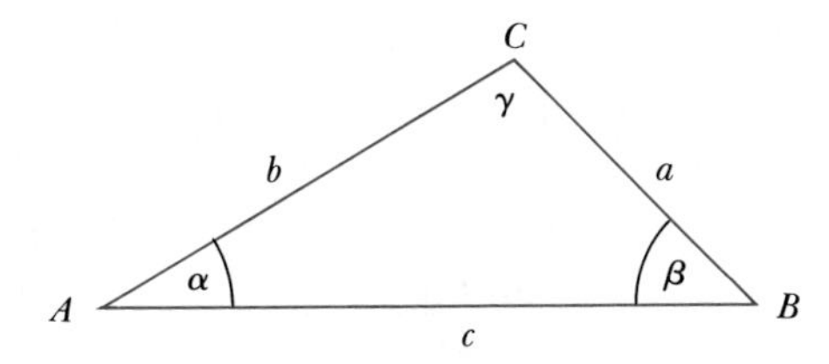

PROOF. Using the Law of Sines (Figure 16), we have

$$\frac{\sin \alpha}{a}=\frac{\sin \beta}{b}$$

so that

$$\text{(1)} \quad \frac{a}{b}=\frac{\sin \alpha}{\sin \beta}$$

Adding 1 to both sides of Equation (1) gives us

$$\frac{a}{b}+1=\frac{\sin \alpha}{\sin \beta}+1$$

or

$$\text{(2)} \quad \frac{a+b}{b}=\frac{\sin \alpha+\sin \beta}{\sin \beta}$$

If we add -1 to both sides of Equation (1), we have

$$\frac{a}{b}-1=\frac{\sin \alpha}{\sin \beta}-1$$

so that

$$(3) \quad \frac{a-b}{b} = \frac{\sin \alpha - \sin \beta}{\sin \beta}$$

Divide the two sides of Equation (3) by the corresponding sides of Equation (2) to obtain

$$\frac{a-b}{a+b} = \frac{\sin \alpha - \sin \beta}{\sin \alpha + \sin \beta}$$

so that

$$\begin{aligned} \frac{a-b}{a+b} &= \frac{2 \cos \frac{1}{2}(\alpha+\beta) \sin \frac{1}{2}(\alpha-\beta)}{2 \sin \frac{1}{2}(\alpha+\beta) \cos \frac{1}{2}(\alpha-\beta)} \\ &= \cot \tfrac{1}{2}(\alpha+\beta) \tan \tfrac{1}{2}(\alpha-\beta) \\ &= \frac{\tan \frac{1}{2}(\alpha-\beta)}{\tan \frac{1}{2}(\alpha+\beta)} \end{aligned}$$

(see Chapter 5, Problem Set 2, Problem 9)

2 For $\triangle ABC$, use the Law of Tangents to find the value of a if $\alpha = 60°$, $\beta = 20°$, and $b = 30$.

Figure 17

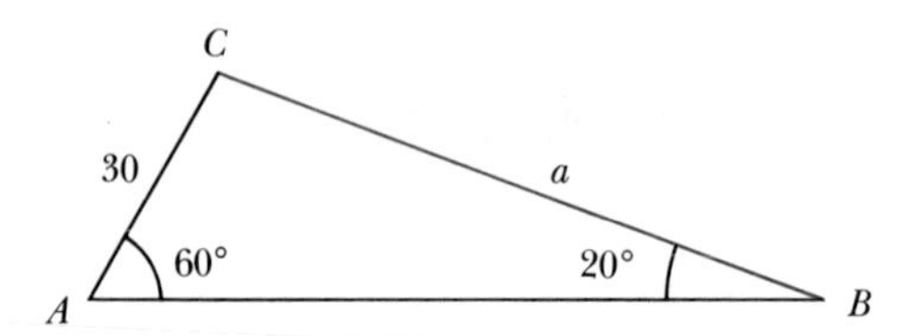

SOLUTION. Using the Law of Tangents (Figure 17) we have

$$\begin{aligned} \frac{a-30}{a+30} &= \frac{\tan \frac{1}{2}(60° - 20°)}{\tan \frac{1}{2}(60° + 20°)} \\ &= \frac{\tan 20°}{\tan 40°} \\ &= \frac{0.3640}{0.8391} \end{aligned}$$

or

$$\frac{a-30}{a+30} = 0.4338$$

so

$$\begin{aligned} a - 30 &= 0.4338(a+30) \\ (1 - 0.4338)a &= 30 + 30(0.4338) \\ a &= 75.97 \end{aligned}$$

PROBLEM SET 3

1 Use the Law of Sines to determine the indicated parts of each of the following triangles. Use logarithms for parts (c), (d), (g), and (h).

a) c and γ if $a = 3$, $b = 5$, and $\beta = 45°$
b) γ if $\alpha = 30°$, $a = 4$, and $c = 7$
c) c and γ if $\alpha = 11°$, $\beta = 72°$, and $a = 10$
d) a and c if $\alpha = 45°$, $\gamma = 60°$, and $b = 10$
e) α if $a = 10$, $b = 26$, and $\beta = 110°$
f) b if $a = 135$, $\alpha = 132°$, and $\beta = 24°$
g) b if $a = 100$, $\beta = 20°$, and $\alpha = 120°$
h) a if $\alpha = 42°$, $\gamma = 61°$, and $b = 52$

2 Use the Law of Sines to determine whether there is one triangle, two triangles, or no triangle in each of the following cases. Find the unknown parts of all possible triangles.

a) $a = 10$, $b = 50$, and $\alpha = 22°$
b) $b = 12$, $c = 10$, and $\beta = 60°$
c) $a = 50$, $c = 60$, and $\gamma = 111°$
d) $b = 480$, $c = 620$, and $\gamma = 55°$
e) $a = 8$, $b = 12$, and $\alpha = 34°$
f) $a = 50.6$, $b = 54.3$, and $\alpha = 59°$
g) $b = 13$, $c = 20$, and $\gamma = 53°$

3 Given $\triangle ABC$, show that $c = b \cos \alpha + a \cos \beta$. Then use the Law of Sines to express b and a in terms of c to obtain the formula $\sin(\alpha + \beta) = \sin \alpha \cos \beta + \sin \beta \cos \alpha$.

4 Use the Law of Sines to show that if $\beta = 2\alpha$, then $\cos \alpha = b/2a$.

5 Two observation posts, A and B, are 600 feet apart. Each observer marks the angle of a shell burst at a point C in enemy territory. If $\angle CAB = 60°$ and $\angle CBA = 73°$, find the range of C from a gun located at B.

6 A surveyor needs to determine the distance from point A to point B. A point C is known to be 5672 feet from B. The distance $\overline{AC}$ is 2431 feet and $\angle BAC$ is 48°. Find the distance $\overline{AB}$.

7 A slide 28 feet long makes an angle of 39° with the ground and is reached by a ladder 18 feet long. Find the angle of inclination of the ladder.

8 Illustrate the possibilities of the ambiguous case if α is an obtuse angle. (*Hint:* There are *two* possibilities.)

9 Use the Law of Tangents to find the unknown parts of $\triangle ABC$ for each of the following cases.

a) $a = 12$, $\beta = 75°$, and $\alpha = 70°$ b) $b = 31$, $\beta = 36°$, and $\gamma = 80°$
c) $a = 20$, $\gamma = 82°$, and $\alpha = 70°$ d) $b = 61$, $\beta = 11°$, and $\gamma = 75°$

10 A tower 150 feet high is on a cliff on a bank of a river. From the top of the tower the angle of depression of a point on the opposite shore of a river is 30° and from the base of the tower the angle of depression of the same point is 20°. Find the width of the river and the height of the cliff.

11 Two men, 500 feet apart, observe a helicopter between them that is in a vertical plane with the men. The respective angles of the elevation of the helicopter are observed by the men to be 74° and 50°. Find the height of the helicopter above the ground.

12 A tree standing vertically on a slope inclined at an angle of 9° to the horizon casts a shadow 50 feet in length up the slope. If the angle that the sun makes with the slope is 65°, find the height of the tree.

13 Two points A and B are 50 feet apart on one bank of a river. A point C is located on the opposite bank so that $\angle CAB$ is 70° and $\angle ABC$ is 80°. How wide is the river?

14 An airplane is sighted simultaneously from two towns that are 5 miles apart. The angles of sighting from the two towns are 20° and 75°. How high above the ground is the airplane?

15 A diagonal of a parallelogram is 16 inches long and forms angles of 43° and 15° with the sides. How long are the two sides of the parallelogram?

16 A guy line from the top of a pole is 40 feet long and forms a 50° angle with the ground. How long is the pole if it is tilted 15° out of line away from the guy line?

REVIEW PROBLEM SET

1 Given the right triangle in Figure 18, find the unknown parts if

a) $a = 10$ and $\beta = 60°$ b) $c = 20$ and $\alpha = 50°$

c) $b = 12$ and $\beta = 48°23'$ d) $a = 8$ and $b = 12$

Figure 18

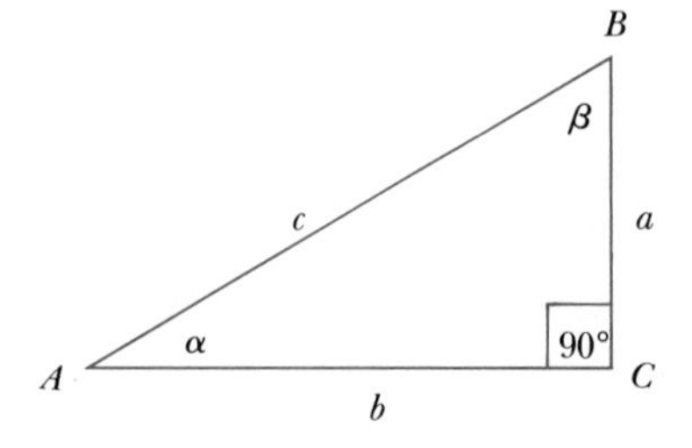

2 A rectangular lot is 100 feet by 150 feet. Find the length of the diagonal and the angles it makes with sides.

3 From the top of a lighthouse 80 feet above the level of the water, the angle of depression of a boat is $32°20'$. How far is the boat from a point at water level directly under the point of observation?

4 At a certain point the angle of elevation of a building is $40°$ and at a point 100 feet farther away in the same horizontal plane its angle of elevation is $30°$. How high is the building?

5 Use the Law of Cosines to find the indicated part of each of the following triangles.

a) c if $a = 5$, $b = 6$, and $\gamma = 30°$

b) b if $a = 11$, $c = 15$, and $\beta = 43°20'$

c) a if $b = 20$, $c = 30$, and $\alpha = 52°$

d) α if $a = 12$, $b = 14$, and $c = 15$

6 Two highways intersect at an angle of $53°$. At a certain point in time, car A is 150 feet from the crossing while car B is 300 feet from the crossing. Find the distance between them at this point in time. (There are two possible solutions.)

7 Use the Law of Sines to find the indicated part of each of the following triangles.

a) b if $\alpha = 62°40'$, $\beta = 79°20'$, and $a = 28$

b) b if $\beta = 81°43'$, $\gamma = 57°51'$, and $c = 100$

c) a if $\alpha = 47°$, $\gamma = 118°$, and $b = 43$

d) c if $\beta = 14°$, $\gamma = 53°$, and $b = 63.7$

8 Find the unknown parts of each of the following triangles or show that there is no possible solution.

a) $b = 50$, $c = 58$, and $\gamma = 57°20'$

b) $a = 147$, $b = 256$, and $\beta = 147°$

c) $a = 49$, $c = 8.7$, and $\alpha = 45°$

d) $b = 5.94$, $c = 7.23$, and $\beta = 38°$

9 Use the Law of Tangents to find the unknown parts of $\triangle ABC$ if $a = 12$, $\alpha = 45°$, and $\beta = 105°$.

10 Use Hero's formula to find the area of $\triangle ABC$ if $a = 13$, $b = 21$, and $c = 30$.

CHAPTER 7

Complex Numbers

CHAPTER 7

Complex Numbers

1 Introduction

The real numbers are used to solve many mathematical problems. However, the real number system does not contain solutions for all equations. For example, consider the equation $x^2 + 1 = 0$. To solve this equation, we need to extend the real number system to a "new" system—the *complex number system.* Our objectives in this chapter are: to review the addition, subtraction, multiplication, and division of complex numbers; to apply the trigonometric functions to represent a complex number in trigonometric or polar form; and to use DeMoivre's theorem to find the powers and the nth roots of a complex number.

2 Complex Numbers

An ordered pair of real numbers (a, b), which shall be denoted by $a + bi$, is called a *complex number.* The symbol i is a constant, denoting $\sqrt{-1}$.

Suppose that $z = a + bi$ and $w = c + di$ are complex numbers, then we have the following rules of equality, addition, and multiplication.

1 *Equality:* $z = w$ if and only if $a = c$ and $b = d$

2 *Addition:* $z + w = (a + c) + (b + d)i$

3 *Multiplication:* $zw = (ac - bd) + (ad + bc)i$

Given a complex number $z = a + bi$, the real number a is called the *real part of* z and the real number b is called the *imaginary part of* z.

It would perhaps be better to identify these numbers as the "non-i part" and the "i part" of the complex numbers, since there is nothing imaginary about the "imaginary" number, but the choice of words "real part" and "imaginary part" is dictated by historical convention. Notice that if $z = a + bi = 0$, then $a = 0$ and $b = 0$. In other words, $z = 0$ means that $z = 0 + 0i$. Furthermore, if the imaginary part of z is 0 (so that $z = a + 0i$), we write $z = a$ and call z a real number [although technically z is the ordered pair $(a, 0)$ on the x axis and not the number a]. We denote the set of complex numbers by C, so in set notation we write

$$C = \{a + bi \mid a \text{ and } b \text{ are real numbers}\}$$

EXAMPLE

1 Find the sum and product of each of the following pairs of complex numbers and identify the real and imaginary parts of the sum.

a) i, i

b) $2 + 3i, 1 - 2i$

c) $1 + 4i, 3 + 5i$

SOLUTION

a) $i + i = 2i$, so the real part of $2i$ is 0 and the imaginary part is 2.

$$i \cdot i = (0 + 1i)(0 + 1i) = (0 \cdot 0 - 1 \cdot 1) + (0 \cdot 1 + 1 \cdot 0)i = -1$$

b) $(2 + 3i) + (1 - 2i) = (2 + 1) + (3 - 2)i = 3 + i$, so the real part of $3 + i$ is 3 and the imaginary part is 1.

$$(2 + 3i)(1 - 2i) = (2 + 6) + (3 - 4)i = 8 - i$$

c) $(1 + 4i) + (3 + 5i) = (1 + 3) + (4 + 5)i = 4 + 9i$, so the real part of $4 + 9i$ is 4 and the imaginary part is 9.

$$(1 + 4i)(3 + 5i) = (3 - 20) + (5 + 12)i = -17 + 17i$$

2 Let $w = 3 + 2i$; find

a) $4w$ b) $-w$ c) $-3w$

SOLUTION

a) $4w = (4 + 0i)(3 + 2i) = (12 + 0) + (0 + 8)i = 12 + 8i$

b) $-w = (-1 + 0i)(3 + 2i) = (-3 + 0) + (0 - 2)i = -3 - 2i$

c) $-3w = (-3 + 0i)(3 + 2i) = (-9 + 0) + (-6 + 0)i = -9 - 6i$

We have in Example 1a) one of the properties of the complex number system that does not hold in the real number system. If $x \in R$, $x^2 \geqslant 0$, whereas it is possible to have $z \in C$ such that $z^2 < 0$, for example, $i^2 = -1 < 0$. That is, there is no order relation in the complex number system.

Positive integral powers of complex numbers have the same meaning in terms of repeated multiplication as with real numbers. For example, $i^3 = i^2 \cdot i = (-1)i = -1$, $i^4 = (i^2)(i^2) = (-1)^2 = 1$, $i^5 = i^4 \cdot i = (1)(i) = i$, and so forth.

EXAMPLE

Write each of the following complex numbers in the form of $a + bi$.

a) $i^4 + 5i^3$ b) $(2 + 3i)^2$

SOLUTION

a) $i^4 + 5i^3 = i^2 \cdot i^2 + 5i^2 \cdot i = (-1)(-1) + 5(-1)i = 1 - 5i$

b) $(2 + 3i)^2 = (2 + 3i)(2 + 3i) = (4 - 9) + (6 + 6)i$
$= -5 + 12i$

The properties of addition and multiplication on C (see Problems 5, 6, and 8 on page 255) are the same as the properties of addition and multiplication on R (these properties are listed in Appendix B), since the operations of addition and multiplication on C are defined in terms of the corresponding operations on R.

2.1 Difference and Quotient of Complex Numbers

Let $z = a + bi$ and $w = c + di$; to *subtract* w from z, we subtract the real part of w from the real part of z and subtract the imaginary part of w from the imaginary part of z. Thus,

$$z - w = (a + bi) - (c + di) = (a - c) + (b - d)i$$

For example, if $z = 2 + 3i$ and $w = 1 - 2i$, then

$$z - w = (2 + 3i) - (1 - 2i) = (2 - 1) + [3 - (-2)]i = 1 + 5i$$

In order to express the quotient of two complex numbers in the form $u + vi$, we first define the *conjugate* of a complex number. If $z = a + bi$, the conjugate of z, written as $\bar{z}$ is $\bar{z} = a - bi$. For example, if $z = 3 + 2i$, then $\bar{z} = 3 - 2i$ and if $z = -1 - 3i$, then $\bar{z} = -1 + 3i$.

The *quotient* of two complex numbers $z = a + bi$ and $w = c + di$, with $w \neq 0$, is given by

$$\frac{z}{w} = \frac{z \cdot \bar{w}}{w \cdot \bar{w}}$$

or equivalently, by

$$\frac{a+bi}{c+di}=\frac{(a+bi)\cdot(c-di)}{(c+di)\cdot(c-di)}=\frac{(ac+bd)+(bc-ad)i}{c^2+d^2}$$
$$=\frac{ac+bd}{c^2+d^2}+\frac{bc-ad}{c^2+d^2}\,i$$

It should be noticed that the result is a complex number in the form $u+vi$ where

$$u=\frac{ac+bd}{c^2+d^2} \qquad \text{and} \qquad v=\frac{bc-ad}{c^2+d^2}$$

For example, if $z=-2+3i$ and $w=1-i$, then

$$\frac{z}{w}=\frac{-2+3i}{1-i}=\frac{(-2+3i)(1+i)}{(1-i)(1+i)}=\frac{-5+i}{1+1}=-\frac{5}{2}+\frac{1}{2}\,i$$

Also, if $z=2+3i$, then

$$\frac{1}{z}=\frac{1}{2+3i}=\frac{1\cdot(2-3i)}{(2+3i)(2-3i)}=\frac{2-3i}{4+9}=\frac{2}{13}-\frac{3}{13}\,i$$

Here, $\frac{2}{13}-\frac{3}{13}i$ is called the *multiplicative inverse* of $z=2+3i$, since

$$z\cdot\frac{1}{z}=(2+3i)\left(\frac{2}{13}-\frac{3}{13}\,i\right)=\left(\frac{4}{13}+\frac{9}{13}\right)+\left(\frac{6}{13}-\frac{6}{13}\right)i=1$$

EXAMPLES

1 Let $z=8-7i$ and $w=-2+i$; find

a) $\bar{z}$ b) $\bar{w}$ c) $z-w$ d) $\overline{z-w}$

SOLUTION

a) $\bar{z}=8+7i$
b) $\bar{w}=-2-i$
c) $z-w=(8-7i)-(-2+i)=[8-(-2)]+(-7-1)i=10-8i$
d) $\overline{z-w}=10+8i$

2 Let $z=2+7i$ and $w=5-2i$; find

a) $\bar{z}$ b) $\bar{w}$ c) $w-z$ d) $\overline{w-z}$
e) zw f) $\overline{zw}$ g) $\bar{z}\cdot\bar{w}$

SOLUTION

a) $\bar{z}=2-7i$
b) $\bar{w}=5+2i$
c) $w-z=(5-2i)-(2+7i)=(5-2)+(-2-7)i=3-9i$

d) $\overline{w - z} = 3 + 9i$

e) $zw = (2 + 7i)(5 - 2i) = (10 + 14) + (35 - 4)i = 24 + 31i$

f) $\overline{zw} = 24 - 31i$

g) $\bar{z} \cdot \bar{w} = (2 - 7i)(5 + 2i) = (10 + 14) + (-35 + 4)i = 24 - 31i$
Note that $\overline{zw} = \bar{z} \cdot \bar{w}$ here (see Problem 4b) on page 255).

3 If $z = a + bi$, show that $z\bar{z}$ is a real number.

SOLUTION. If $z = a + bi$, then $\bar{z} = a - bi$, so that

$$z\bar{z} = (a + bi)(a - bi) = (a^2 + b^2) + (ab - ab)i$$
$$= a^2 + b^2 \in R$$

since a and b are real numbers.

4 If $z = 7 - 4i$, write each of the following in the form of $a + bi$. Also identify the real part and the imaginary part of the result.

a) $z + \bar{z}$ b) $(z - \bar{z})/i$ c) $z\bar{z}$ d) $1/z$

SOLUTION. If $z = 7 - 4i$, then $\bar{z} = 7 + 4i$.

a) $z + \bar{z} = (7 - 4i) + (7 + 4i) = 14$, so that the real part is 14 and the imaginary part is 0.

b) $\dfrac{z - \bar{z}}{i} = \dfrac{(7 - 4i) - (7 + 4i)}{i} = \dfrac{-8i}{i} = -8$, so that the real part is -8 and the imaginary part is 0.

c) $z\bar{z} = (7 - 4i)(7 + 4i) = 49 + 16 = 65$, so that the real part is 65 and the imaginary part is 0.

d) $\dfrac{1}{z} = \dfrac{1 \cdot \bar{z}}{z\bar{z}} = \dfrac{1(7 + 4i)}{(7 - 4i)(7 + 4i)} = \dfrac{7 + 4i}{65} = \dfrac{7}{65} + \dfrac{4}{65}i$, so that the real part is $\frac{7}{65}$ and the imaginary part is $\frac{4}{65}$.

5 Show that if z is a complex number, $z + \bar{z}$ is a real number.

SOLUTION. Let $z = a + bi$; then $\bar{z} = a - bi$, so that

$$z + \bar{z} = (a + a) + (b - b)i$$
$$= 2a \in R$$

6 Find the real numbers x and y such that $x + yi = 1/(3 + 5i)$.

SOLUTION. First, we write $1/(3 + 5i)$ in the form $a + bi$, that is

$$\frac{1}{3 + 5i} = \frac{1(3 - 5i)}{(3 + 5i)(3 - 5i)} = \frac{3 - 5i}{34} = \frac{3}{34} - \frac{5}{34}i$$

so that

$$x + yi = \frac{1}{3 + 5i} = \frac{3}{34} - \frac{5}{34}i$$

Hence

$$x = \frac{3}{34} \quad \text{and} \quad y = -\frac{5}{34}$$

7 Write each of the following quotients in the form of $a + bi$.

a) $\dfrac{2}{3+i}$ b) $\dfrac{2+3i}{1+i}$ c) $\dfrac{2-5i}{-1+4i}$ d) $\dfrac{7-6i}{6-5i}$

SOLUTION

a) $\dfrac{2}{3+i} = \dfrac{2(3-i)}{(3+i)(3-i)} = \dfrac{6-2i}{9+1} = \dfrac{6}{10} - \dfrac{2}{10}i = \dfrac{3}{5} - \dfrac{1}{5}i$

b) $\dfrac{2+3i}{1+i} = \dfrac{(2+3i)(1-i)}{(1+i)(1-i)} = \dfrac{5+i}{1+1} = \dfrac{5}{2} + \dfrac{1}{2}i$

c) $\dfrac{2-5i}{-1+4i} = \dfrac{(2-5i)(-1-4i)}{(-1+4i)(-1-4i)} = \dfrac{(-2-20)+(-8+5)i}{1+16}$

$= \dfrac{-22-3i}{17} = \dfrac{-22}{17} - \dfrac{3}{17}i$

d) $\dfrac{7-6i}{6-5i} = \dfrac{(7-6i)(6+5i)}{(6-5i)(6+5i)} = \dfrac{(42+30)+(-36+35)i}{36+25}$

$= \dfrac{72-i}{61} = \dfrac{72}{61} - \dfrac{1}{61}i$

PROBLEM SET 1

1 Perform the following operations and express the results in the form of $a + bi$.

a) $(2+3i)+(4+5i)$ b) $(-1+2i)+(3+5i)$

c) $(2+i)-(4+3i)$ d) $(-4+2i)-(3+2i)$

e) $(5+2i)+(6-i)-(3+7i)$ f) $(2-3i)(4+6i)$

g) $(-2+3i)(2+5i)$ h) $(4+5i)(-7+2i)$

i) i^3+2i^5+1 j) $(3-i)^2$

k) $i^{14}-3i^5$ l) $(4+i)^3$

m) $(1+2i)^4$ n) $(-1-\sqrt{2}i)^3$

o) $\dfrac{5}{3i}$ p) $-\dfrac{2}{7i^5}$

q) $\dfrac{3}{1+2i}$ r) $\dfrac{5}{1-i}$

s) $\dfrac{3+2i}{1-3i}$ t) $\dfrac{7+4i}{3+5i}$

u) $\dfrac{3+7i}{5i}$ v) $\dfrac{i}{2+5i}$

w) $\dfrac{3}{(1+2i)^2}$ x) $\dfrac{4}{(1-i)^3}$

y) $\dfrac{(\sqrt{3}+i)^2}{2+i}$ z) $\dfrac{2i^4}{(6-i)^2}$

2 Find the real numbers x and y such that each of the following equations holds.

a) $2x + 5i = 8 + 10yi$

b) $(3x - 1) + (y - 1)i = 8 + 5i$

c) $2x + (y - 2)i = x - 2 + (3y + 1)i$

d) $6x + 2i = 12 + 5yi$

3 Find $\bar{z}$, the real part of z, the imaginary part of z, and $1/z$ for each of the following complex numbers.

a) $z = -5 + 7i$ b) $z = 6i$

c) $z = (3 + 2i)^2$ d) $z = 3i$

4 Let $z = a + bi$ and $w = c + di$; verify each of the following statements.

a) $\overline{z + w} = \bar{z} + \bar{w}$

b) $\overline{zw} = \bar{z} \cdot \bar{w}$

c) $\overline{\left(\frac{z}{w}\right)} = \frac{\bar{z}}{\bar{w}}$

d) $\bar{\bar{z}} = z$

e) $z + \bar{z} = 0$ if and only if the real part of z is 0.

5 Write the multiplicative inverse of each of the following complex numbers and check by multiplication.

a) $2i$ b) $-3i$ c) $2 + i$

d) $3 + 5i$ e) $4 + 3i$ f) $1 + 7i$

6 Given the set $A = \{2, -2, i, -3i\}$.

a) Construct a multiplication table.

b) Is the set A closed under multiplication?

c) Is multiplication on A commutative?

d) Is multiplication on A associative?

$\cdot$	2	-2	i	$-3i$
2				
-2				
i				
$-3i$				

e) What is the multiplicative inverse of each element of A?

7 Let z_1 and z_2 be complex numbers; show that

$$\operatorname{Re}\left(\frac{z_1}{z_1 + z_2}\right) + \operatorname{Re}\left(\frac{z_2}{z_1 + z_2}\right) = 1$$

where Re (z) indicates the real part of z.

8 Assume the properties of addition and multiplication on the real numbers R.

a) Prove that C is closed under addition and multiplication.
b) Prove the commutativity of addition and multiplication on C.
c) Prove the associativity of addition and multiplication on C.
d) Prove the distributive properties on C.
e) Prove that the identity elements of addition and multiplication on C are 0 and 1, respectively.

3 Geometric Representation of Complex Numbers

Each ordered pair of real numbers (a, b) can be associated with the complex number $z = a + bi$, and each complex number $z = a + bi$ can be associated with the ordered pair of real numbers (a, b). Because of this one-to-one correspondence between the set of complex numbers and the set of ordered pairs of real numbers, we can use the points in the plane associated with the ordered pairs of real numbers to represent the complex numbers.

For example, the ordered pairs $(2, -3)$, $(5, 2)$, and (e, π) are used to represent complex numbers $z_1 = 2 - 3i$, $z_2 = 5 + 2i$, and $z_3 = e + \pi i$, respectively, as points in the plane (Figure 1).

Figure 1

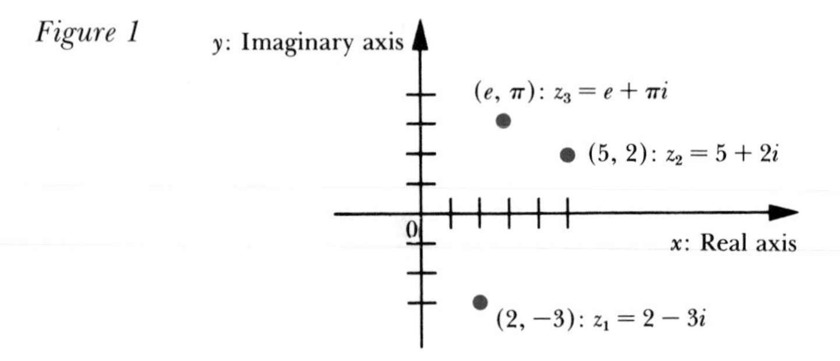

The plane on which the complex numbers are represented is called the *complex plane*. The horizontal axis (x axis) is called the *real axis*, and the vertical axis (y axis) is called the *imaginary axis*. Thus, complex numbers of the form $z = bi$ are represented by points of the form $(0, b)$, that is, by points on the imaginary axis, whereas complex numbers of the form $z = a$ are represented by points of the form $(a, 0)$, that is, by points on the real axis.

3.1 The Modulus of a Complex Number

If $z = a + bi$, then the *absolute value* or *length* or *modulus* of z, written $|z|$ is defined by

$$|z| = \sqrt{a^2 + b^2}$$

The modulus of $z = a + bi$ is the distance between the origin and the point (a, b) (Figure 2). Notice that $|z| = \sqrt{z \cdot \bar{z}}$ (see Problem 8).

Figure 2

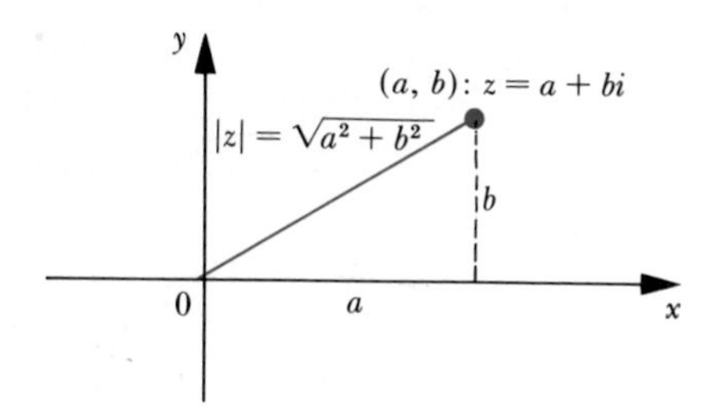

EXAMPLES

1 Let $z = 1 + \sqrt{3}i$. Find $|z|$ and show that $z\bar{z} = |z|^2$.

SOLUTION

$$|z| = \sqrt{1^2 + (\sqrt{3})^2} = \sqrt{1 + 3} = \sqrt{4} = 2$$

Also

$$z\bar{z} = (1 + \sqrt{3}i)(1 - \sqrt{3}i) = 1^2 + (\sqrt{3})^2 = 4 = |z|^2 \qquad \text{(Figure 3)}$$

Figure 3

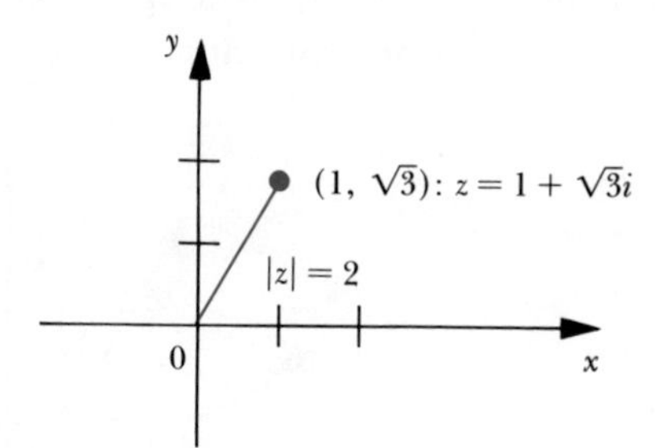

2 Let $z_1 = 4 + 3i$ and $z_2 = \sqrt{3} - i$; find

a) $|z_1|$ b) $|z_2|$ c) $|z_1 z_2|$

SOLUTION

a) $|z_1| = \sqrt{16 + 9} = \sqrt{25} = 5$

b) $|z_2| = \sqrt{(\sqrt{3})^2 + (-1)^2} = \sqrt{4} = 2$

c) $$\begin{aligned} |z_1 z_2| &= |(4 + 3i)(\sqrt{3} - i)| \\ &= |(4\sqrt{3} + 3) + (3\sqrt{3} - 4)i| \\ &= \sqrt{(4\sqrt{3} + 3)^2 + (3\sqrt{3} - 4)^2} = 10 \end{aligned}$$

3 Let $z = 2 + 7i$ and $w = 5 - 2i$; find

a) $|z|$ b) $|w|$ c) $|z||w|$

d) $|zw|$ e) $|z|/|w|$ f) $|z/w|$

SOLUTION

a) $|z| = \sqrt{2^2 + 7^2} = \sqrt{4 + 49} = \sqrt{53}$

b) $|w| = \sqrt{5^2 + (-2)^2} = \sqrt{25 + 4} = \sqrt{29}$

c) $|z||w| = \sqrt{53} \cdot \sqrt{29} = \sqrt{(53)(29)} = \sqrt{1537}$

d) $|zw| = |(2 + 7i)(5 - 2i)| = |24 + 31i| = \sqrt{1537}$

e) $\dfrac{|z|}{|w|} = \dfrac{\sqrt{53}}{\sqrt{29}} = \dfrac{\sqrt{53}\sqrt{29}}{\sqrt{29}\sqrt{29}} = \dfrac{\sqrt{1537}}{29}$

f) $\left|\dfrac{z}{w}\right| = \dfrac{|2 + 7i|}{|5 - 2i|} = \left|\dfrac{(2 + 7i)(5 + 2i)}{29}\right| = \left|\dfrac{-4 + 39i}{29}\right| = \dfrac{\sqrt{(-4)^2 + (39)^2}}{29}$

$= \dfrac{\sqrt{16 + 1521}}{29} = \dfrac{\sqrt{1537}}{29}$

4 Polar Coordinates

We have seen that points in the plane can be referenced or associated with pairs of real numbers by using the Cartesian coordinate system. Before we proceed with the discussion of complex numbers, let us investigate a way of associating pairs of numbers with points in the plane based upon a "grid" composed of concentric circles and rays emanating from the common center ot the circles (Figure 1).

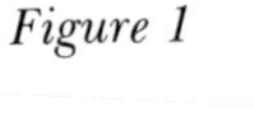

Figure 1

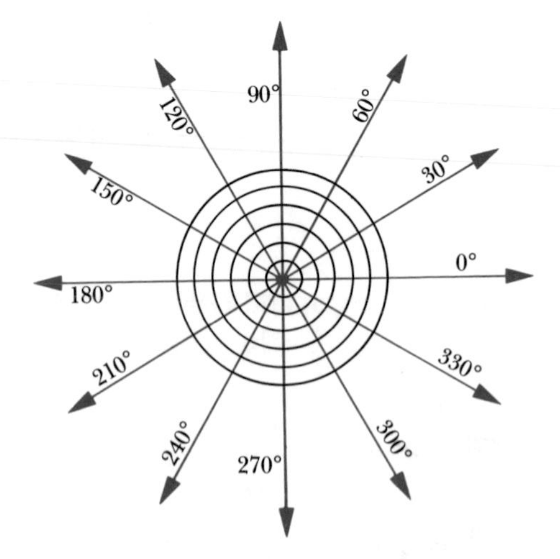

Such a system is called a *polar coordinate system.* The frame of reference for this coordinate system consists of a fixed point O, called the *pole,* and a fixed ray called the *polar axis.* Usually, we already have a Cartesian coordinate system, and we choose O to be the origin and the ray to be the positive x axis. The position of a point P is uniquely determined by r and θ, where θ is any angle in standard position having the ray $\overrightarrow{OA}$ as its initial side and a ray on line OP as its terminal side, and r is the directed distance along the terminal side of θ between P and the pole. The pair (r, θ) is called the *polar coordinates* of P (Figure 2).

Figure 2

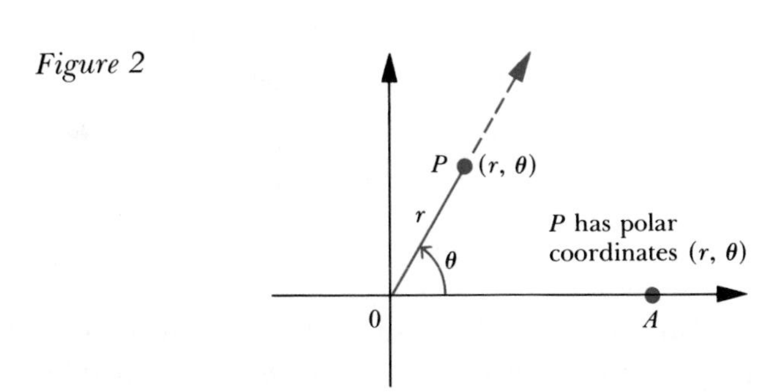

If the angle θ is measured in degrees, then (r, θ) clearly indicates that the ordered pair represents polar coordinates. On the other hand, if θ is given in radians, then the ordered pair of real number (r, θ) is indistinguishable from the notation used in Cartesian coordinates. For example, clearly $(2, 30°)$ represents polar coordinates, whereas $(2, 3)$ could be rectangular or 3 could be the radian measure of an angle (Figures 3*a* and *b*). If the context does not make clear that the coordinates are polar or rectangular, we will assume that they are rectangular coordinates. [In specifying polar coordinates, we use the phrase "plot the polar point (a, b)" as short for "plot the point whose polar coordinates are (a, b)."]

Figure 3

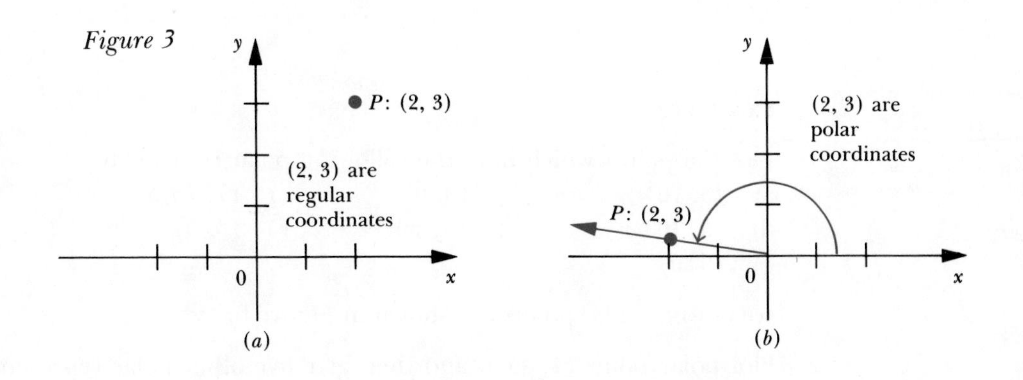

It is important to observe that a polar coordinate system does not establish a one-to-one correspondence between points in a plane and ordered pairs (r, θ). In fact, each point can be represented by infinitely many ordered pairs of numbers. For example, $(2, 30°)$, $(2, 390°)$, and $(2, -330°)$ each represent the same point (Figure 4).

Figure 4

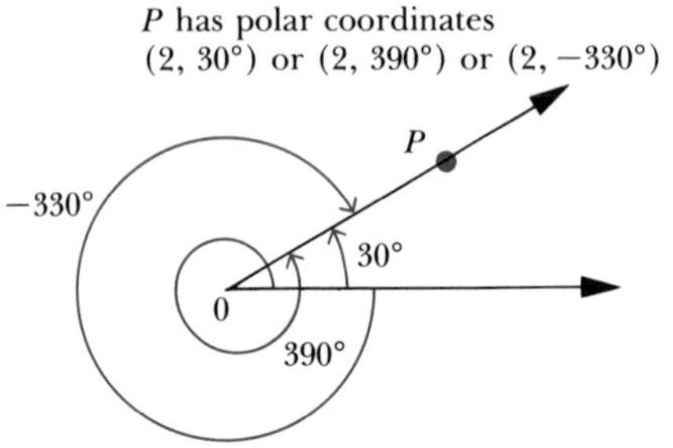

P has polar coordinates $(2, 30°)$ or $(2, 390°)$ or $(2, -330°)$

P

−330°

30°

0

390°

Also, r need not be positive. If $r < 0$, the point (r, θ) is determined by plotting $(|r|, \theta + 180°)$ or $(|r|, \theta + \pi)$, depending on whether θ is measured in degrees or radians. For example, $(-2, 30°)$ is the same as $(2, 210°)$ (Figure 5).

Figure 5

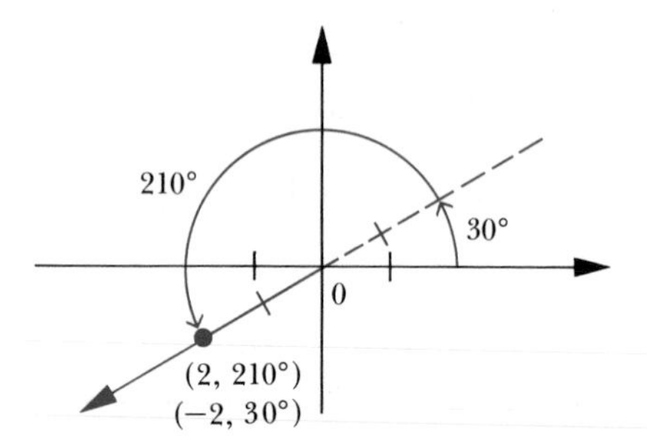

EXAMPLES

1 Plot the points which have the following polar coordinates.

a) $(3, 70°)$ b) $(0, 0)$ c) $(7, 7\pi/5)$
d) $(3, 100°)$ e) (π, π) f) $(3, 5)$
g) $(5, 0)$

SOLUTION. The points are shown in Figure 6.

2 Plot polar point $(4, 45°)$, and then give five other polar representations of the same point.

SOLUTION. $(4, -315°)$, $(4, 405°)$, $(-4, -135°)$, $(-4, 225°)$, and $(4, \pi/4)$ are other polar representations of the same point (Figure 7).

Figure 6

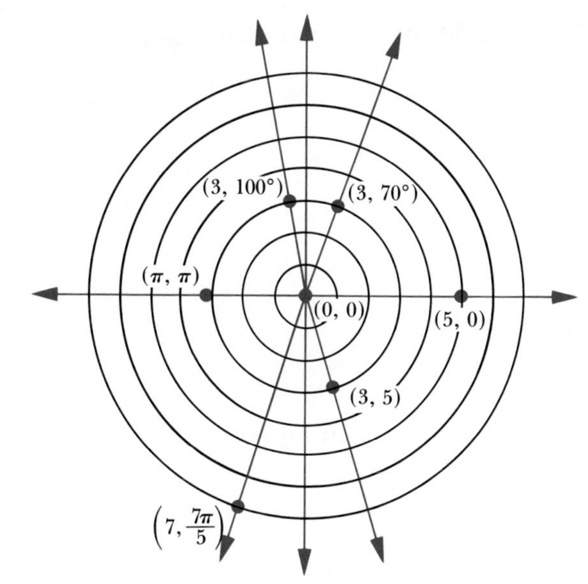

Figure 7

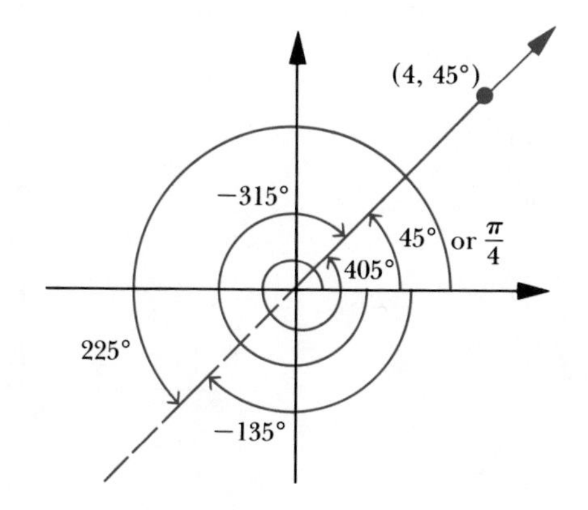

4.1 Conversion of Coordinates

If (r, θ) is a polar representation of a point P, the trigonometric functions can be used to find the rectangular coordinates (x, y) of the same point (Figure 8). We know from trigonometry that $\sin \theta = y/r$ and $\cos \theta = x/r$; hence,

$$y = r \sin \theta \qquad \text{and} \qquad x = r \cos \theta$$

These formulas are often referred to as the *transformation* or *conversion* formulas; they enable us to convert from *polar to rectangular coordinates.*

Figure 8

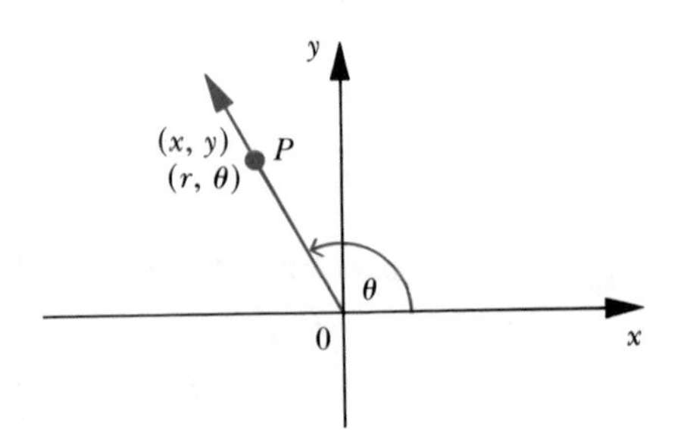

EXAMPLE

Convert the given polar coordinates to rectangular coordinates.

a) $(3, 60°)$ b) $(-2, 180°)$
c) $(4, -150°)$ d) (π, π)

SOLUTION

a) $x = r \cos \theta = 3 \cos 60° = \frac{3}{2}$ and $y = r \sin \theta = 3 \sin 60° = 3\sqrt{3}/2$. Hence, the rectangular coordinates of $(3, 60°)$ are $(\frac{3}{2}, 3\sqrt{3}/2)$ (Figure 9).

Figure 9

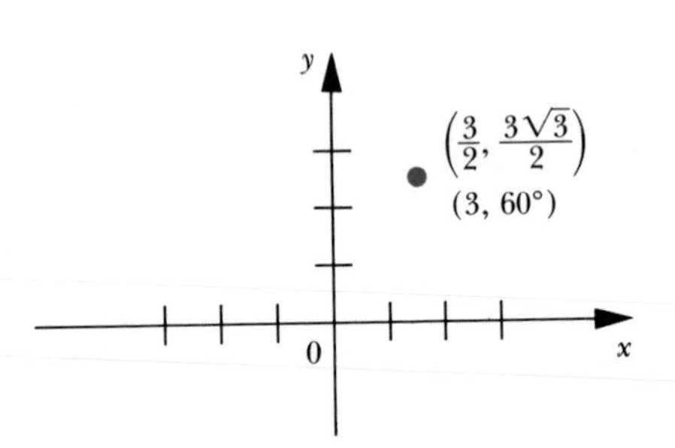

b) $x = r \cos \theta = -2 \cos 180° = 2$ and $y = r \sin \theta = -2 \sin 180° = 0$. Hence, the rectangular coordinates of $(-2, 180°)$ are $(2, 0)$ (Figure 10).

Figure 10

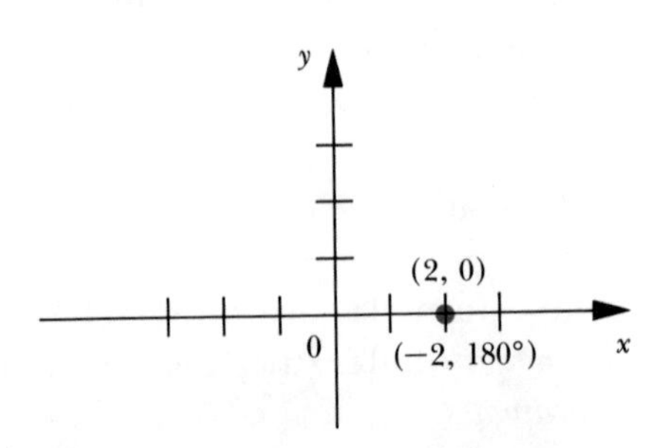

c) $x = r\cos\theta = 4\cos(-150°) = 4(-\sqrt{3}/2) = -2\sqrt{3}$ and $y = r\sin\theta = 4\sin(-150°) = 4(-\frac{1}{2}) = -2$. Hence, the rectangular coordinates of $(4, -150°)$ are $(-2\sqrt{3}, -2)$ (Figure 11).

Figure 11

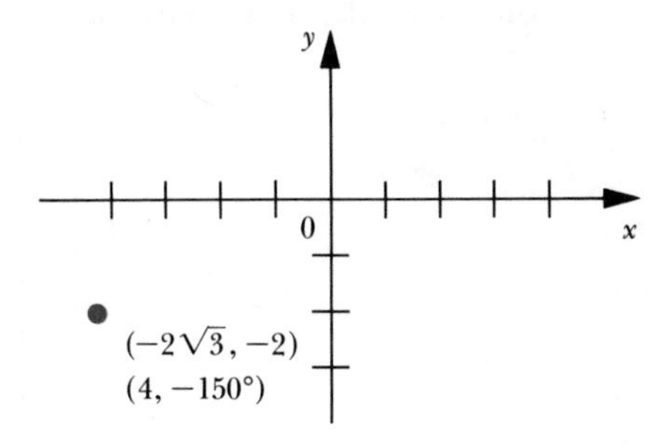

d) $x = r\cos\theta = \pi\cos\pi = -\pi$ and $y = r\sin\theta = \pi\sin\pi = 0$. Hence, the rectangular coordinates of (π, π) are $(-\pi, 0)$ (Figure 12).

Figure 12

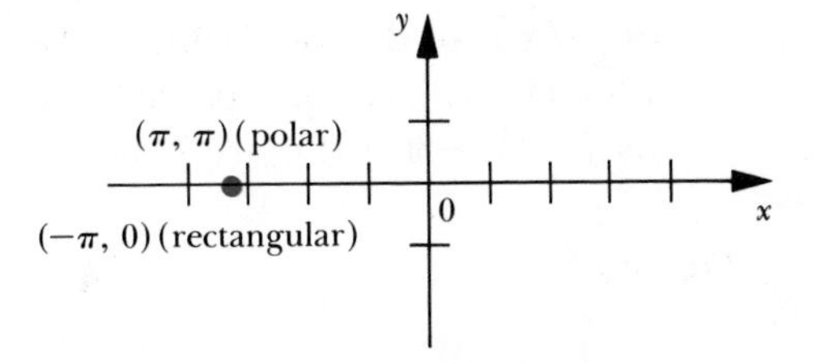

Now, assume that the rectangular coordinates of a point P are given by (x, y). Then $r = \sqrt{x^2 + y^2}$ and $\tan\theta = y/x$ can be used to transform the rectangular coordinates to polar coordinates (Figure 13).

Figure 13

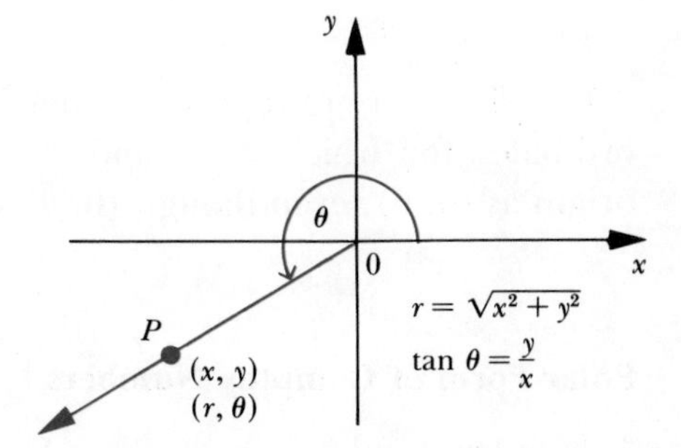

EXAMPLE

Convert the given rectangular coordinates to polar coordinates.

a) $(-1, 1)$ b) $(3, -3/\sqrt{3})$

SOLUTION

a) $r = \sqrt{x^2 + y^2} = \sqrt{(-1)^2 + 1} = \sqrt{2}$, and, since the point is in quadrant II, $\tan \theta = -1$ implies that one value of θ is 135°; hence, one pair of polar coordinates is $(\sqrt{2}, 135°)$ (Figure 14).

Figure 14

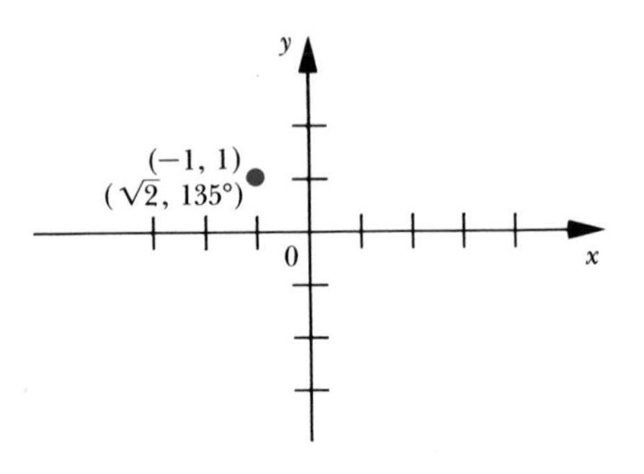

b) $r = \sqrt{3^2 + (-3/\sqrt{3})^2} = \sqrt{9 + 3} = \sqrt{12} = 2\sqrt{3}$ and $\tan \theta = -\sqrt{3}/3 = -1/\sqrt{3}$, so that one value of θ is −30°, since the point is in quadrant IV. Hence, one possible pair of polar coordinates is given by $(2\sqrt{3}, -30°)$ (Figure 15).

Figure 15

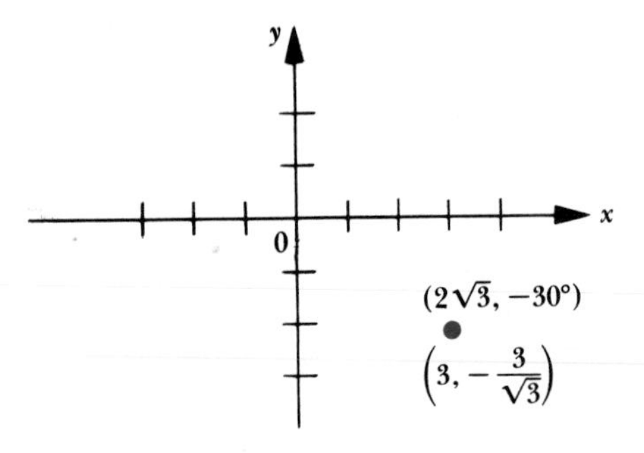

Usually we try to represent a point by that unique set of polar coordinates for which $r > 0$ and $0 \leq \theta \leq 2\pi$ [and we represent the origin as (0, 0), even though $(0, \theta)$ would do for any θ].

4.2 Polar Form of Complex Numbers

A complex number $z = x + yi$ can be expressed as

$$z = r \cos \theta + ir \sin \theta = r(\cos \theta + i \sin \theta)$$

since $x = r \cos \theta$, $y = r \sin \theta$, and $r = |z|$. $r(\cos \theta + i \sin \theta)$ is called the *polar form* or *trigonometric form* of the complex number z (Figure 16).

The number θ in this representation is called an *argument* of the complex number z.

Figure 16

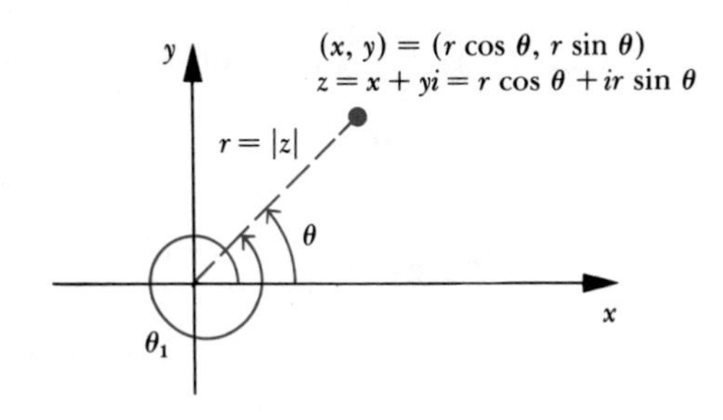

Notice that the modulus r is determined uniquely by the relation $r = \sqrt{x^2 + y^2}$, whereas the argument θ is not unique, since $r(\cos\theta + i\sin\theta) = r(\cos\theta_1 + i\sin\theta_1)$ holds whenever $\theta - \theta_1$ is an integral multiple of 2π. Hence, if θ is an argument of $z = x + iy$, so is $\theta + 2\pi k$ (or $\theta + k \cdot 360°$) for each $k \in I$. For example, the complex number $z = -1 - \sqrt{3}i$ can be expressed in polar form as

$$z = 2\left(\cos\frac{4\pi}{3} + i\sin\frac{4\pi}{3}\right)$$

or as

$$z = 2\left(\cos\frac{10\pi}{3} + i\sin\frac{10\pi}{3}\right) \qquad \text{(Figure 17)}$$

Figure 17

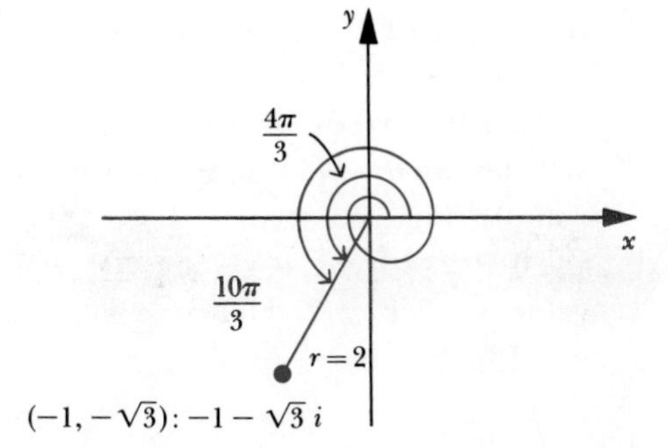

Notice that $(2, 4\pi/3)$ and $(2, 10\pi/3)$ are two possible pairs of polar coordinates of $(-1, -\sqrt{3})$.

EXAMPLES

1 Change each of the following complex numbers from polar form to the rectangular form.

a) $z = 7(\cos 0° + i \sin 0°)$
b) $z = 20[\cos (7\pi/6) + i \sin (7\pi/6)]$
c) $z = 10[\cos (\pi/2) + i \sin (\pi/2)]$
d) $z = 4(\cos 240° + i \sin 240°)$

SOLUTION

a) $z = 7 \cos 0° + i\, 7 \sin 0° = 7 + 0i$
b) $z = 20 \cos (7\pi/6) + i\, 20 \sin (7\pi/6)$
$= 20(-\sqrt{3}/2) + i\, 20(-\frac{1}{2}) = -10\sqrt{3} - 10i$
c) $z = 10 \cos (\pi/2) + i\, 10 \sin (\pi/2) = 10 \cdot 0 + i\, 10(1)$
$= 0 + 10i$
d) $z = 4 \cos 240° + i\, 4 \sin 240° = 4(-\frac{1}{2}) + i\, 4(-\sqrt{3}/2)$
$= -2 - 2\sqrt{3}i$

2 Express each of the following complex numbers in polar form.
a) $z = \sqrt{3} - i$
b) $z = -1 + i\sqrt{3}$
c) $z = 3 + 0i$
d) $z = 2i$

SOLUTION.
a) $z = \sqrt{3} - i = r(\cos \theta + i \sin \theta)$, where $r = \sqrt{(\sqrt{3})^2 + (-1)^2} = 2$ and θ satisfies $\tan \theta = (-1/\sqrt{3})$ with θ in quadrant IV, so one value of θ is $\theta = (11\pi/6)$ (Figure 18). Hence,

$$z = 2\left[\cos \frac{11\pi}{6} + i \sin \frac{11\pi}{6}\right]$$

b) $z = -1 + i\sqrt{3} = r(\cos \theta + i \sin \theta)$, where $r = \sqrt{(-1)^2 + (\sqrt{3})^2} = 2$ and θ satisfies $\tan \theta = \sqrt{3}/(-1)$, with θ, in quadrant II, so that one value of θ is $2\pi/3$. Hence, $z = 2[\cos (2\pi/3) + i \sin (2\pi/3)]$ (Figure 18).
c) $z = 3 + 0i = r(\cos \theta + i \sin \theta)$, where $r = \sqrt{3^2 + 0^2} = 3$ and θ satisfies $\tan \theta = 0/3 = 0$, so one value of θ is 0. Hence, $z = 3(\cos 0 + i \sin 0)$ (Figure 18).
d) $z = 0 + 2i = r(\cos \theta + i \sin \theta)$, where $r = \sqrt{0 + (2)^2} = 2$. One value of θ is $\pi/2$. Hence, $z = 2[\cos (\pi/2) + i \sin (\pi/2)]$ (Figure 18).

Figure 18

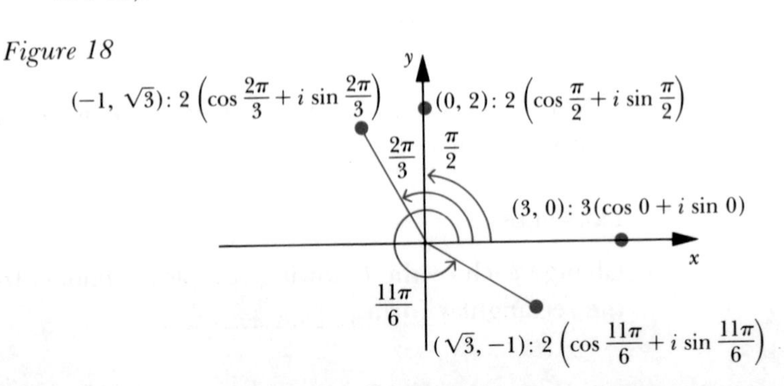

4.3 Multiplication and Division of Complex Numbers in Polar Form

Consider the complex numbers $z = 2\sqrt{3} + 2i$ and $w = -2 + 2i$. The polar representations of z and w are given by

$$z = 4\left(\cos\frac{\pi}{6} + i\sin\frac{\pi}{6}\right) \qquad \text{and} \qquad w = 2\sqrt{2}\left(\cos\frac{3\pi}{4} + i\sin\frac{3\pi}{4}\right)$$

Notice that the moduli of z and w are 4 and $2\sqrt{2}$, respectively, whereas their arguments are, respectively, $\pi/6$ and $3\pi/4$. By multiplication, we have

$$\begin{aligned} zw &= \left[4\left(\cos\frac{\pi}{6} + i\sin\frac{\pi}{6}\right)\right]\left[2\sqrt{2}\left(\cos\frac{3\pi}{4} + i\sin\frac{3\pi}{4}\right)\right] \\ &= 8\sqrt{2}\left[\left(\cos\frac{\pi}{6}\cos\frac{3\pi}{4} - \sin\frac{\pi}{6}\sin\frac{3\pi}{4}\right)\right. \\ &\quad \left. + i\left(\sin\frac{\pi}{6}\cos\frac{3\pi}{4} + \sin\frac{3\pi}{4}\cos\frac{\pi}{6}\right)\right] \end{aligned}$$

Using trigonometric identities (Chapter 5, page 192) we have

$$\begin{aligned} zw &= 8\sqrt{2}\left[\cos\left(\frac{\pi}{6} + \frac{3\pi}{4}\right) + i\sin\left(\frac{\pi}{6} + \frac{3\pi}{4}\right)\right] \\ &= 8\sqrt{2}\left(\cos\frac{11\pi}{12} + i\sin\frac{11\pi}{12}\right) \qquad \text{(Figure 19)} \end{aligned}$$

Figure 19

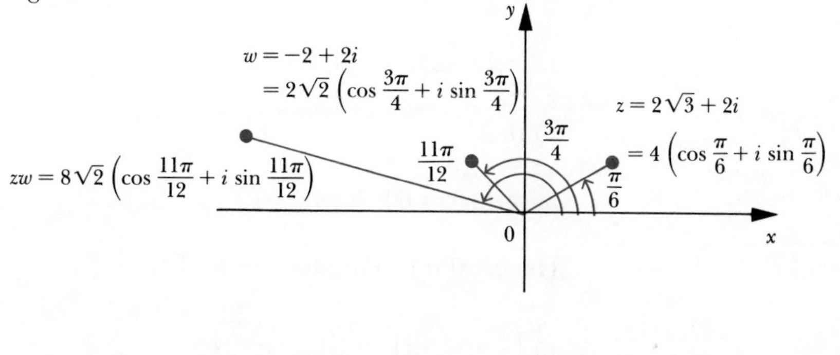

Notice in the example that the modulus of the product zw is $|z||w| = (4)(2\sqrt{2}) = 8\sqrt{2}$, whereas the argument of zw is $\pi/6 + 3\pi/4 = 11\pi/12$. This example can be generalized as follows.

THEOREM 1

Suppose that z and w are complex numbers expressed in polar form as

$$z = r(\cos\alpha + i\sin\alpha) \qquad \text{and} \qquad w = s(\cos\beta + i\sin\beta)$$

Then

$$zw = rs[\cos(\alpha+\beta) + i\sin(\alpha+\beta)]$$

PROOF

$$\begin{aligned} zw &= r(\cos\alpha + i\sin\alpha)\cdot s(\cos\beta + i\sin\beta) \\ &= rs(\cos\alpha + i\sin\alpha)(\cos\beta + i\sin\beta) \\ &= rs[(\cos\alpha\cos\beta - \sin\alpha\sin\beta) + i(\sin\alpha\cos\beta + \sin\beta\cos\alpha)] \end{aligned}$$

so, by the trigonometric identities (see Chapter 5, page 192),

$$zw = rs[\cos(\alpha+\beta) + i\sin(\alpha+\beta)]$$

In other words, the theorem states that the product of two complex numbers is a complex number whose modulus is the product of the moduli and whose argument is the sum of the arguments of the given complex numbers.

EXAMPLES

1 Let $z = 3(\cos 45° + i\sin 45°)$ and $w = 5(\cos 15° + i\sin 15°)$. Determine each of the following expressions in polar and rectangular forms.

a) zw b) z^2

SOLUTION

a) Using Theorem 1, we have

$$zw = (3)(5)[\cos(45° + 15°) + i\sin(45° + 15°)]$$

so that

$$zw = 15(\cos 60° + i\sin 60°)$$

Hence, the rectangular form of zw is

$$15\left(\frac{1}{2} + \frac{\sqrt{3}}{2}i\right) = \frac{15}{2} + \frac{15\sqrt{3}}{2}i$$

b) $$\begin{aligned} z^2 = z\cdot z &= (3)(3)[\cos(45° + 45°) + i\sin(45° + 45°)] \\ &= 9(\cos 90° + i\sin 90°) \end{aligned}$$

Hence, the rectangular form of z^2 is

$$9(0 + i) = 9i$$

2 Express $z = -\sqrt{3} - i$ and $w = -1 - \sqrt{3}i$ in polar form, and then compute zw in polar form and in rectangular form.

SOLUTION. $z = -\sqrt{3} - i = r(\cos\theta + i \sin\theta)$, where

$$r = \sqrt{(-\sqrt{3})^2 + (-1)^2} = 2$$

and θ satisfies $\tan\theta = -1/-\sqrt{3} = 1/\sqrt{3}$ with θ in quadrant III (Figure 20), so that one value of θ is $\theta = 7\pi/6$. Hence,

$$z = 2\left(\cos\frac{7\pi}{6} + i \sin\frac{7\pi}{6}\right)$$

and

$$w = -1 - \sqrt{3}i = r(\cos\theta + i \sin\theta)$$

where $r = \sqrt{(-1)^2 + (-\sqrt{3})^2} = 2$ and θ satisfies $\tan\theta = -\sqrt{3}/(-1) = \sqrt{3}$ with θ in quadrant III (Figure 20), so that one value of θ is $4\pi/3$. Hence,

$$w = 2\left(\cos\frac{4\pi}{3} + i \sin\frac{4\pi}{3}\right)$$

Therefore,

$$zw = (2)(2)\left[\cos\left(\frac{7\pi}{6} + \frac{4\pi}{3}\right) + i \sin\left(\frac{7\pi}{6} + \frac{4\pi}{3}\right)\right]$$

or

$$zw = 4\left(\cos\frac{5\pi}{2} + i \sin\frac{5\pi}{2}\right) = 4\left(\cos\frac{\pi}{2} + i \sin\frac{\pi}{2}\right)$$

The rectangular form of zw is $4(0 + 1i) = 4i$ (Figure 20).

Figure 20

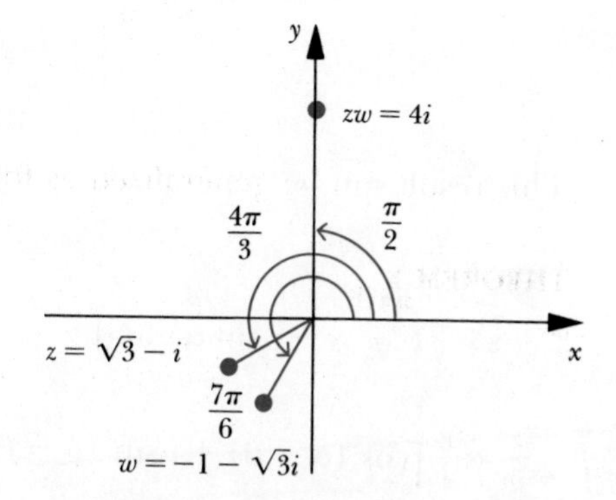

Let us find the quotient z/w of the two complex numbers

$$z = 24(\cos 113° + i \sin 113°)$$

and

$$w = 8(\cos 53° + i \sin 53°).$$

The quotient

$$\frac{z}{w} = \frac{24(\cos 113° + i \sin 113°)}{8(\cos 53° + i \sin 53°)}$$

can be found by multiplying the numerator and denominator of the above expression by $\cos 53° - i \sin 53°$, so that

$$\begin{aligned}
\frac{z}{w} &= \frac{24}{8}\left[\frac{(\cos 113° + i \sin 113°)(\cos 53° - i \sin 53°)}{(\cos 53° + i \sin 53°)(\cos 53° - i \sin 53°)}\right] \\
&= \frac{24}{8}\left[\frac{(\cos 113° \cos 53° + \sin 113° \sin 53°) + i(\sin 113° \cos 53° - \sin 53° \cos 113°)}{\cos^2 53° + \sin^2 53°}\right] \\
&= \frac{24}{8}[\cos(113° - 53°) + i \sin(113° - 53°)] \\
&= 3(\cos 60° + i \sin 60°) = 3\left(\frac{1}{2} + i\frac{\sqrt{3}}{2}\right) = \frac{3}{2} + \frac{i\sqrt{3}}{2} \qquad \text{(Figure 21)}
\end{aligned}$$

Figure 21

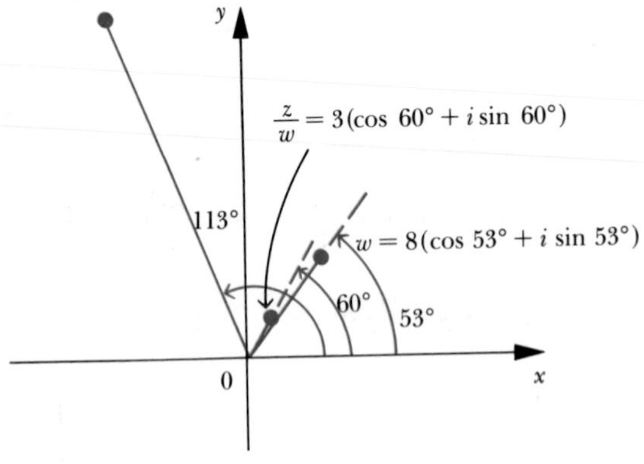

This result can be generalized as follows:

THEOREM 2

Let $z = r(\cos\alpha + i \sin\alpha)$ and $w = s(\cos\beta + i \sin\beta)$; then

$$\frac{z}{w} = \frac{r}{s}[\cos(\alpha - \beta) + i \sin(\alpha - \beta)] \qquad w \neq 0$$

PROOF

$$\frac{z}{w} = \frac{r(\cos \alpha + i \sin \alpha)}{s(\cos \beta + i \sin \beta)}$$

Multiplying the denominator and the numerator by $\cos \beta - i \sin \beta$, we have

$$\begin{aligned}
&= \frac{r}{s}\left[\frac{(\cos \alpha + i \sin \alpha)(\cos \beta - i \sin \beta)}{(\cos \beta + i \sin \beta)(\cos \beta - i \sin \beta)}\right] \\
&= \frac{r}{s}\left[\frac{(\cos \alpha \cos \beta + \sin \alpha \sin \beta) + i(\sin \alpha \cos \beta - \sin \beta \cos \alpha)}{\cos^2 \beta + \sin^2 \beta}\right] \\
&= \frac{r}{s}\left[\cos (\alpha - \beta) + i \sin (\alpha - \beta)\right]
\end{aligned}$$

In other words, the theorem states: The quotient of two complex numbers is a complex number whose modulus is the quotient of the two moduli and whose argument is the difference of the two arguments.

EXAMPLES

1 Let $z = 8[\cos 176° + i \sin 176°]$ and $w = 2[\cos 116° + i \sin 116°]$. Find z/w and express the result in rectangular form.

SOLUTION. Using Theorem 2, we have

$$\begin{aligned}
\frac{z}{w} &= \frac{8}{2}\left[\cos (176° - 116°) + i \sin (176° - 116°)\right] \\
&= 4(\cos 60° + i \sin 60°)
\end{aligned}$$

The rectangular form of z/w is

$$\frac{z}{w} = 4\left(\frac{1}{2} + i\,\frac{\sqrt{3}}{2}\right) = 2 + 2\sqrt{3}i$$

2 Express $z = 1 - i$ and $w = \sqrt{3} + i$ in polar form, and then compute z/w in polar form. Also, write the result in rectangular form.

SOLUTION

$$z = 1 - i = r(\cos \theta + i \sin \theta)$$

where $r = \sqrt{1^2 + (-1)^2} = \sqrt{2}$ and θ satisfies $\tan \theta = -1/1 = -1$ with θ is in quadrant IV, so that one value of θ is $7\pi/4$. Hence,

$$z = \sqrt{2}\left(\cos \frac{7\pi}{4} + i \sin \frac{7\pi}{4}\right)$$

$$w = \sqrt{3} + i = r(\cos\theta + i\sin\theta)$$

where $r = \sqrt{(\sqrt{3})^2 + 1^2} = 2$ and θ satisfies $\tan\theta = 1/\sqrt{3}$, with θ in quadrant I, so that one value of θ is $\pi/6$. Hence,

$$w = 2\left(\cos\frac{\pi}{6} + i\sin\frac{\pi}{6}\right)$$

Then, by Theorem 2,

$$\begin{aligned}\frac{z}{w} &= \frac{\sqrt{2}}{2}\left[\cos\left(\frac{7\pi}{4} - \frac{\pi}{6}\right) + i\sin\left(\frac{7\pi}{4} - \frac{\pi}{6}\right)\right]\\ &= \frac{\sqrt{2}}{2}\left(\cos\frac{19\pi}{12} + i\sin\frac{19\pi}{12}\right)\end{aligned}$$

The rectangular form of z/w is

$$\begin{aligned}\frac{z}{w} &= \frac{\sqrt{2}}{2}[0.2588 + i(-0.9659)]\\ &= 0.1830 - 0.6830i \qquad \text{(approximately)}\end{aligned}$$

PROBLEM SET 2

1 Plot each of the rectangular coordinates and then convert to polar coordinates.
a) $(-1, \sqrt{3})$
b) $(-3, 0)$
c) $(4, 3)$
d) $(-6, 6\sqrt{3})$
e) $(5, 5)$
f) $(0, -2)$
g) $(-3, 3\sqrt{3})$
h) $(2\sqrt{3}, -2)$
i) $(-5, -5)$
j) $(-3, 5)$

2 Plot the polar points of the following pairs and convert the representation to rectangular form.
a) $(6, 30°)$
b) $(10, \pi/3)$
c) $(7, 120°)$
d) $(-4, -\pi/6)$

3 For each of the following complex numbers, find the modulus of z and represent z graphically.
a) $z = -2 + 3i$
b) $z = 5 - 3i$
c) $z = \frac{1}{2} - (\sqrt{3}/2)i$
d) $z = 5i$
e) $z = (1/\sqrt{2}) + (1/\sqrt{2})i$
f) $z = -4 + 5i$
g) $z = -i$
h) $z = (1 - 2i)^2$

4 For each of the following complex numbers, find the modulus and the argument of z. Express z in polar form and represent it graphically.

a) $z = -2 - 2i$
b) $z = -1 + i$
c) $z = -2$
d) $z = -i$
e) $z = 1 + \sqrt{3}i$
f) $z = -\frac{1}{2} + (\sqrt{3}/2)i$
g) $z = -2 - 2\sqrt{3}i$
h) $z = 3i$
i) $z = 1 - \sqrt{3}i$
j) $z = (\sqrt{3}/2) + \frac{1}{2}i$

5 Represent each of the following complex numbers graphically and then express the number in the form of $a + bi$.

a) $z = 3(\cos 30° + i \sin 30°)$
b) $z = 8(\cos 45° + i \sin 45°)$
c) $z = 4(\cos 135° + i \sin 135°)$
d) $z = 3[\cos (-90°) + i \sin (-90°)]$
e) $z = 10(\cos 270° + i \sin 270°)$
f) $z = 6(\cos 180° + i \sin 180°)$
g) $z = 2[\cos (\pi/3) + i \sin (\pi/3)]$
h) $z = 8[\cos (5\pi/6) + i \sin (5\pi/6)]$
i) $z = 10[\cos (4\pi/3) + i \sin (4\pi/3)]$
j) $z = 6[\cos (-210°) + i \sin (-210°)]$

6 Find zw and z/w for each of the following numbers. Express the results in polar form and in rectangular form.

a) $z = 5(\cos 30° + i \sin 30°)$ and $w = 6(\cos 240° + i \sin 240°)$
b) $z = 10(\cos 90° + i \sin 90°)$ and $w = 6(\cos 300° + i \sin 300°)$
c) $z = 4[\cos (\pi/3) + i \sin (\pi/3)]$ and $w = 8(\cos 0 + i \sin 0)$
d) $z = 3\sqrt{2}[\cos (5\pi/4) + i \sin (5\pi/4)]$ and $w = 7\sqrt{2}[\cos (3\pi/2) + i \sin (3\pi/2)]$
e) $z = 5\sqrt{2}(\cos 45° + i \sin 45°)$ and $w = \sqrt{2}(\cos 315° + i \sin 315°)$

7 Write $z = 3 + 3\sqrt{3}i$ and $w = -5\sqrt{3} + 5i$ in polar form, and then compute each of the following values in polar form and in the rectangular form.

a) zw
b) z/w
c) w/z
d) $(zw)^2$

8 Prove that $|z| = \sqrt{z \cdot \bar{z}}$.

5 Powers and Roots of Complex Numbers

Let $z = r(\cos \theta + i \sin \theta)$; we have shown in Example 1b) on page 268 that $z^2 = r^2(\cos 2\theta + i \sin 2\theta)$. Multiplying this result by $r(\cos \theta + i \sin \theta)$ yields

$$\begin{aligned} z^3 = z^2 \cdot z &= r^2 \cdot r[\cos (2\theta + \theta) + i \sin (2\theta + \theta)] \\ &= r^3[\cos 3\theta + i \sin 3\theta] \end{aligned}$$

If the process is repeated once more, we get

$$z^4 = z^3 \cdot z = r^4(\cos 4\theta + i \sin 4\theta)$$

Further repeated multiplication of the complex number in polar form leads to the generalized form

$$z^n = r^n(\cos n\theta + i \sin n\theta)$$

where n is a positive integer. This form is known as *DeMoivre's theorem.* In particular, if $r = 1$, then

$$(\cos \theta + i \sin \theta)^n = \cos n\theta + i \sin n\theta$$

EXAMPLES

1 Use DeMoivre's theorem to determine each of the following values. Express the results in rectangular form.

a) $[\cos (\pi/4) + i \sin (\pi/4)]^{40}$ b) $[4(\cos 6° + i \sin 6°)]^5$

c) $[(\sqrt{3}/2) + \frac{1}{2}i]^9$ d) $(1 + i)^6$

SOLUTION

a) Using DeMoivre's theorem, we have

$$\begin{aligned}\left(\cos \frac{\pi}{4} + i \sin \frac{\pi}{4}\right)^{40} &= \cos \left[40\left(\frac{\pi}{4}\right)\right] + i \sin \left[40\left(\frac{\pi}{4}\right)\right] \\ &= \cos 10\pi + i \sin 10\pi \\ &= 1 + 0i\end{aligned}$$

b)
$$\begin{aligned}[4(\cos 6° + i \sin 6°)]^5 &= 4^5[\cos 5(6°) + i \sin 5(6°)] \\ &= 1024(\cos 30° + i \sin 30°) \\ &= 1024\left(\frac{\sqrt{3}}{2} + \frac{1}{2}i\right) \\ &= 512\sqrt{3} + 512i\end{aligned}$$

c) We first express the number in polar form

$$\left(\frac{\sqrt{3}}{2} + \frac{1}{2}i\right) = r(\cos \theta + i \sin \theta)$$

with

$$r = \sqrt{\left(\frac{\sqrt{3}}{2}\right)^2 + \left(\frac{1}{2}\right)^2} = 1$$

and θ satisfies $\tan \theta = 1/\sqrt{3}$, so that one value of θ is $\pi/6$ since θ is in quadrant I. Thus

$$\left(\frac{\sqrt{3}}{2}+\frac{1}{2}i\right)^9=\left(\cos\frac{\pi}{6}+i\sin\frac{\pi}{6}\right)^9=\cos\left[9\left(\frac{\pi}{6}\right)\right]+i\sin\left[9\left(\frac{\pi}{6}\right)\right]$$
$$=\cos\frac{9\pi}{6}+i\sin\frac{9\pi}{6}=\cos\frac{3\pi}{2}+i\sin\frac{3\pi}{2}=0-i=-i$$

d) $(1+i)=r(\cos\theta+i\sin\theta)$ with $r=\sqrt{1^2+1^2}=\sqrt{2}$ and θ satisfies $\tan\theta=1/1=1$, so that one value of θ is $\pi/4$ since θ is in quadrant I. Thus

$$(1+i)^6=\left[\sqrt{2}\left(\cos\frac{\pi}{4}+i\sin\frac{\pi}{4}\right)\right]^6=8\left(\cos\frac{6\pi}{4}+i\sin\frac{6\pi}{4}\right)$$
$$=8(0-i)=-8i$$

2 Use the fact that $(\cos\theta+i\sin\theta)^2=\cos 2\theta+i\sin 2\theta$ to find identities for $\cos 2\theta$ and $\sin 2\theta$.

SOLUTION. $(\cos\theta+i\sin\theta)^2=\cos 2\theta+i\sin 2\theta$ follows from DeMoivre's theorem, so that after expanding the left-hand side, we get

$$(\cos^2\theta-\sin^2\theta)+(2\sin\theta\cos\theta)i=\cos 2\theta+\sin 2\theta i$$

Therefore, after equating the real and imaginary parts, we get

$$\cos 2\theta=\cos^2\theta-\sin^2\theta$$

and

$$\sin 2\theta=2\sin\theta\cos\theta$$

3 Let

$$z=\frac{(\cos 15°+i\sin 15°)^{10}}{(1+i)^6}$$

Use DeMoivre's theorem to write z in the form $a+bi$, then find $|z|$.

SOLUTION. First, we express $1+i$ in polar form, so that $1+i=\sqrt{2}(\cos 45°+i\sin 45°)$. Then using DeMoivre's theorem, we have

$$z=\frac{(\cos 15°+i\sin 15°)^{10}}{[\sqrt{2}(\cos 45°+i\sin 45°)]^6}=\frac{\cos 150°+i\sin 150°}{8(\cos 270°+i\sin 270°)}$$
$$=\frac{1}{8}[\cos(150°-270°)+i\sin(150°-270°)]$$
$$=\frac{1}{8}[\cos(-120°)+i\sin(-120°)]$$
$$=\frac{1}{8}(\cos 120°-i\sin 120°)=\frac{1}{8}\left(-\frac{1}{2}-\frac{\sqrt{3}}{2}i\right)=-\frac{1}{16}-\frac{\sqrt{3}}{16}i$$

Since $|z|=r$, we have $|z|=\frac{1}{8}$.

5.1 Roots of Complex Numbers

In this section, we will use the polar form of a complex number and DeMoivre's theorem to find the roots of a complex number. For example, to find the square roots of -1, let $z = r(\cos\theta + i\sin\theta)$, so that $z^2 = [r(\cos\theta + i\sin\theta)]^2 = r^2(\cos 2\theta + i\sin 2\theta) = 1$. But the polar form of -1 is $1(\cos\pi + i\sin\pi)$, so that

$$r^2(\cos 2\theta + i\sin 2\theta) = 1(\cos\pi + i\sin\pi)$$

which is true if $r^2 = 1$ and $2\theta = \pi + 2\pi k$, $k = 0, 1$ or $\theta = (\pi/2) + \pi k$, $k = 0, 1$. If $k = 0$, we have one root, namely,

$$\cos(\pi/2) + i\sin(\pi/2) = i,$$

and if $k = 1$, we have the other root which is

$$\cos[(\pi/2) + \pi] + i\sin[(\pi/2) + \pi] = -i.$$

In general, if $z = r(\cos\theta + i\sin\theta)$ is a solution of the equation $z^n = w = R(\cos\phi + i\sin\phi)$, where n is a positive integer, then,

$$z^n = [r(\cos\theta + i\sin\theta)]^n = r^n(\cos n\theta + i\sin n\theta) = R(\cos\phi + i\sin\phi)$$

This equation will be satisfied if $R = r^n$ and $n\theta = \phi + 2\pi k$ (or $\phi + 360°k$) where $k = 0, \pm 1, \pm 2, \ldots$, so that

$$\theta = \frac{\phi}{n} + \frac{2\pi k}{n} \quad \left(\text{or } \theta = \frac{\phi}{n} + \frac{360°k}{n}\right) \quad \text{and} \quad r = \sqrt[n]{R},$$

where $k = 0, 1, 2, \ldots, n-1$ gives n different values of z equally spaced around a circle of radius r (Figure 1 illustrates the case for $n = 5$). Note that the substitution of any other integer for k will produce one of those roots already found.

Figure 1

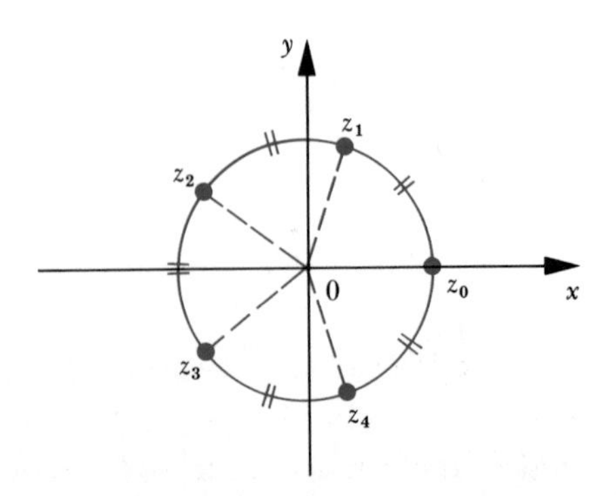

Thus there are n roots of $z^n = R(\cos\phi + i\sin\phi)$ and they are given by

$$z = \sqrt[n]{R}\left[\cos\left(\frac{\phi}{n} + \frac{2\pi k}{n}\right) + i\sin\left(\frac{\phi}{n} + \frac{2\pi k}{n}\right)\right]$$

where $k = 0, 1, 2, \ldots, n-1$.

EXAMPLES

1 Determine the square roots of $-i$ and represent them geometrically.

SOLUTION. As our first step, we determine the polar representation, $R(\cos\phi + i\sin\phi)$, of $-i$. Here $R = \sqrt{0^2 + (-1)^2} = 1$ and $\phi = 3\pi/2$. Thus, we obtain the roots $z = r(\cos\theta + i\sin\theta)$ by using the formulas

$$r = \sqrt{1} = 1 \qquad \text{and} \qquad \theta = \frac{\left(\frac{3\pi}{2}\right)}{2} + \frac{2\pi k}{2} = \frac{3\pi}{4} + \pi k \qquad k = 0, 1$$

so that the roots are

$$\text{for } k = 0\text{: } z_0 = \cos\frac{3\pi}{4} + i\sin\frac{3\pi}{4} = -\frac{\sqrt{2}}{2} + \frac{\sqrt{2}}{2}i$$
$$\text{for } k = 1\text{: } z_1 = \cos\frac{7\pi}{4} + i\sin\frac{7\pi}{4} = \frac{\sqrt{2}}{2} - \frac{\sqrt{2}}{2}i$$

The square roots of $-i$ are equally spaced on the circumference of a circle of radius 1 and differ by an angle of π radians (Figure 2).

Figure 2

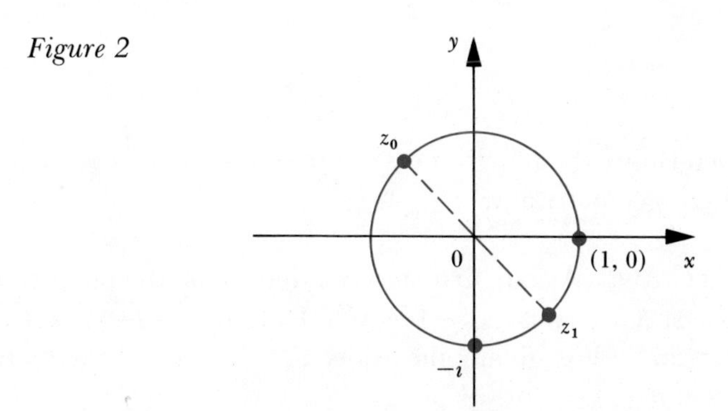

2 Find the cube roots of 27 and represent them geometrically.

SOLUTION. As our first step, we determine the polar representation, $R(\cos\phi + i\sin\phi)$, of 27. Here $R = \sqrt{(27)^2 + 0^2} = 27$ and $\phi = 0$. Finally, we obtain the roots $z = r(\cos\theta + i\sin\theta)$ by using the formulas

$$r = \sqrt[3]{27} \qquad \text{and} \qquad \theta = 0 + \frac{2\pi k}{3} \qquad k = 0, 1, 2$$

so that

$$\text{for } k = 0: z_0 = \sqrt[3]{27}(\cos 0 + i \sin 0) = 3(1 + 0i) = 3$$

$$\text{for } k = 1: z_1 = \sqrt[3]{27}\left(\cos \frac{2\pi}{3} + i \sin \frac{2\pi}{3}\right) = 3\left(-\frac{1}{2} + \frac{\sqrt{3}}{2} i\right)$$

$$= \frac{-3}{2} + i \frac{3\sqrt{3}}{2}$$

$$\text{for } k = 2: z_2 = \sqrt[3]{27}\left(\cos \frac{4\pi}{3} + i \sin \frac{4\pi}{3}\right) = 3\left(-\frac{1}{2} - \frac{\sqrt{3}}{2} i\right)$$

$$= \frac{-3}{2} - \frac{3\sqrt{3}}{2} i$$

The cube roots of 27 lie on the circle with radius $\sqrt[3]{27} = 3$, and the difference in the arguments of any two of them is $2\pi/3$ (Figure 3).

Figure 3

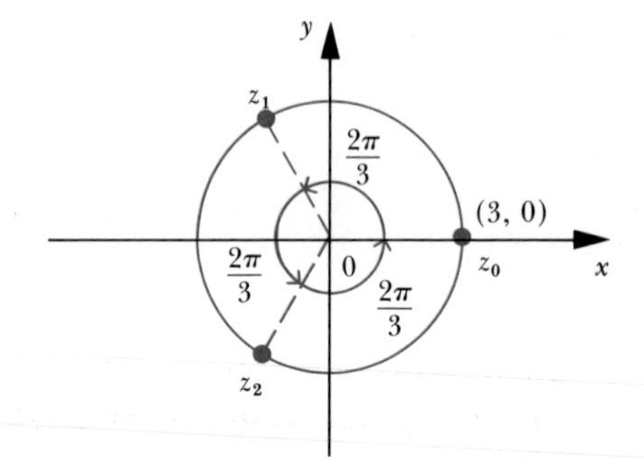

3 Determine the fourth roots of $-1 + \sqrt{3}i$ in polar form and represent them geometrically.

SOLUTION. As our first step, we determine the polar representation, $R(\cos \phi + i \sin \phi)$, of $-1 + \sqrt{3}i$. Here $R = \sqrt{(-1)^2 + (\sqrt{3})^2} = 2$ and $\phi = 2\pi/3$. We obtain the roots $z = r(\cos \theta + i \sin \theta)$ by using the formulas

$$r = \sqrt[4]{2} \qquad \text{and} \qquad \theta = \frac{\pi}{6} + \frac{\pi k}{2} \qquad k = 0, 1, 2, 3$$

so the roots are

$$\text{for } k=0\text{: } z_0 = \sqrt[4]{2}\left(\cos\frac{\pi}{6} + i\sin\frac{\pi}{6}\right)$$

$$\text{for } k=1\text{: } z_1 = \sqrt[4]{2}\left(\cos\frac{2\pi}{3} + i\sin\frac{2\pi}{3}\right)$$

$$\text{for } k=2\text{: } z_2 = \sqrt[4]{2}\left(\cos\frac{7\pi}{6} + i\sin\frac{7\pi}{6}\right)$$

$$\text{for } k=3\text{: } z_3 = \sqrt[4]{2}\left(\cos\frac{5\pi}{3} + i\sin\frac{5\pi}{3}\right)$$

The fourth roots of $-1+\sqrt{3}i$ are equally spaced on the circumference of a circle of radius $\sqrt[4]{2}$ and differ by angles of $\pi/2$ radians (Figure 4).

Figure 4

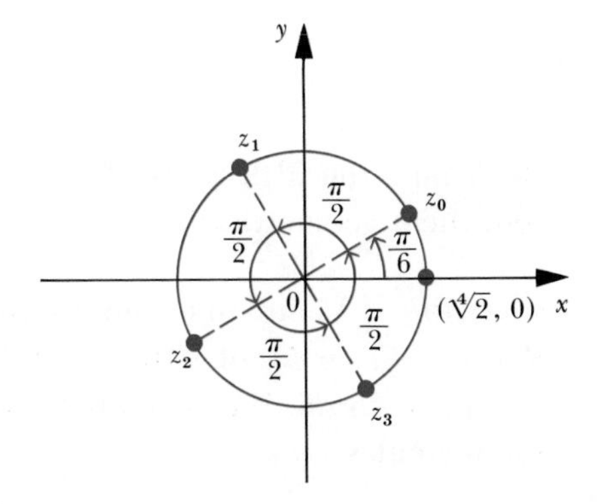

4 Find the fourth roots of $1+i$ in polar form and represent them geometrically.

SOLUTION. As our first step, we determine the polar representation, $R(\cos\phi + i\sin\phi)$, of $1+i$. Here $R=\sqrt{2}$ and $\phi=\pi/4$. Finally, we obtain the roots $z=r(\cos\theta + i\sin\theta)$ by using the formulas

$$r = \sqrt[4]{\sqrt{2}} = \sqrt[8]{2} \qquad \text{and} \qquad \theta = \frac{\pi}{16} + \frac{\pi k}{2} \qquad k=0, 1, 2, 3$$

so that

$$\text{for } k=0\text{: } z_0 = \sqrt[8]{2}\left(\cos\frac{\pi}{16} + i\sin\frac{\pi}{16}\right)$$

$$\text{for } k=1\text{: } z_1 = \sqrt[8]{2}\left(\cos\frac{9\pi}{16} + i\sin\frac{9\pi}{16}\right)$$

$$\text{for } k=2\text{: } z_2 = \sqrt[8]{2}\left(\cos\frac{17\pi}{16} + i\sin\frac{17\pi}{16}\right)$$

$$\text{for } k=3\text{: } z_3 = \sqrt[8]{2}\left(\cos\frac{25\pi}{16} + i\sin\frac{25\pi}{16}\right)$$

The fourth roots of $1 + i$ are equally spaced on the circumference of a circle of radius $\sqrt[8]{2}$ and differ by angles of $\pi/2$ radians (Figure 5).

Figure 5

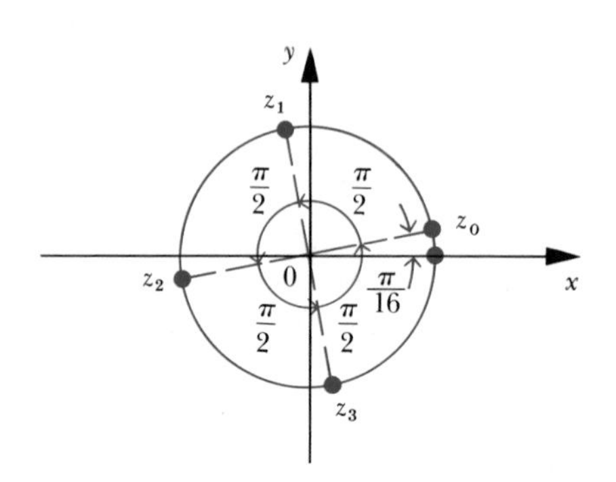

5 Determine the fifth roots of $-16\sqrt{3} + 16i$ in polar form and represent them geometrically.

SOLUTION. As our first step, we determine the polar representation, $R(\cos\phi + i\sin\phi)$, of $-16\sqrt{3} + 16i$. Here $R = \sqrt{(-16\sqrt{3})^2 + (16)^2} = 32$ and $\phi = 5\pi/6$. We obtain the roots of $z = r(\cos\theta + i\sin\theta)$ by using the formulas

$$r = \sqrt[5]{32} = 2 \qquad \text{and} \qquad \theta = \frac{\pi}{6} + \frac{2\pi k}{5} \qquad k = 0, 1, 2, 3, 4$$

so that

$$\text{for } k = 1\text{: } z_0 = 2\left(\cos\frac{\pi}{6} + i\sin\frac{\pi}{6}\right)$$

$$\text{for } k = 1\text{: } z_1 = 2\left(\cos\frac{17\pi}{30} + i\sin\frac{17\pi}{30}\right)$$

$$\text{for } k = 2\text{: } z_2 = 2\left(\cos\frac{29\pi}{30} + i\sin\frac{29\pi}{30}\right)$$

$$\text{for } k = 3\text{: } z_3 = 2\left(\cos\frac{41\pi}{30} + i\sin\frac{41\pi}{30}\right)$$

$$\text{for } k = 4\text{: } z_4 = 2\left(\cos\frac{53\pi}{30} + i\sin\frac{53\pi}{30}\right)$$

The fifth roots of $-16\sqrt{3} + 16i$ are equally spaced on the circumference of a circle of radius 2 and differ by an angle of $2\pi/5$ radians (Figure 6).

Figure 6

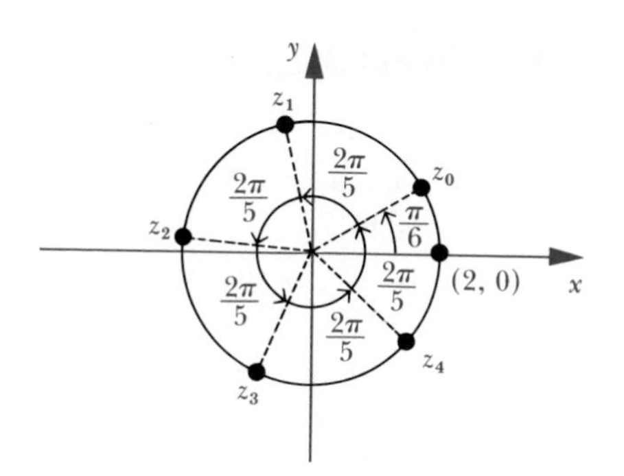

6 The solutions to $z^n = 1$ are called the nth *roots of unity*. Find the fifth roots of unity in polar form and represent them geometrically.

SOLUTION. The number 1 can be written in polar form as $1 = 1(\cos 0 + i \sin 0)$. We obtain the five roots $z = r(\cos\theta + i \sin\theta)$ by using the formulas

$$r = \sqrt[5]{1} \qquad \text{and} \qquad \theta = \frac{0}{5} + \frac{2\pi k}{5} \qquad k = 0, 1, 2, 3, 4$$

so that

$$\text{for } k = 0\text{: } z_0 = 1(\cos 0 + i \sin 0)$$
$$\text{for } k = 1\text{: } z_1 = 1\left(\cos\frac{2\pi}{5} + i \sin\frac{2\pi}{5}\right)$$
$$\text{for } k = 2\text{: } z_2 = 1\left(\cos\frac{4\pi}{5} + i \sin\frac{4\pi}{5}\right)$$
$$\text{for } k = 3\text{: } z_3 = 1\left(\cos\frac{6\pi}{5} + i \sin\frac{6\pi}{5}\right)$$
$$\text{for } k = 4\text{: } z_4 = 1\left(\cos\frac{8\pi}{5} + i \sin\frac{8\pi}{5}\right)$$

We observe that all the fifth roots of unity are on a unit circle and they are equally spaced at angles of $2\pi/5$ radians (Figure 7).

Figure 7

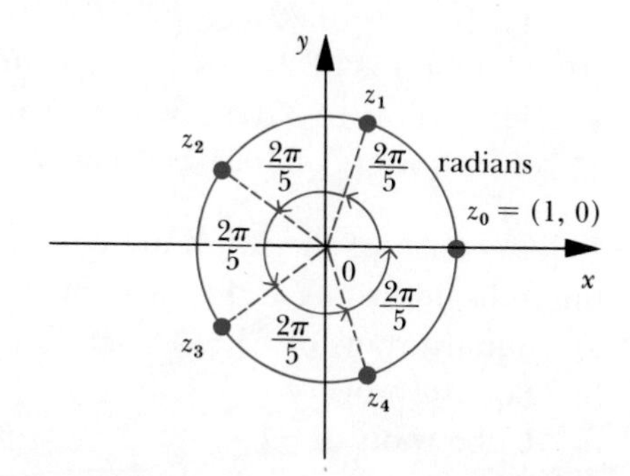

PROBLEM SET 3

1 Use DeMoivre's theorem to compute each of the following powers. Express your answer in the form $a + bi$.

a) $[2(\cos 30° + i \sin 30°)]^{10}$ b) $[2(\cos 75° + i \sin 75°)]^{11}$
c) $[3(\cos 10° + i \sin 10°)]^{6}$ d) $(\cos 55° + i \sin 55°)^{18}$
e) $[3(\cos 330° + i \sin 330°)]^{5}$ f) $[2(\cos 225° + i \sin 225°)]^{8}$
g) $[4(\cos 36° + i \sin 36°)]^{5}$ h) $[4(\cos 36° + i \sin 36°)]^{10}$

2 Express each of the following complex numbers in polar form, then use DeMoivre's theorem to calculate the following indicated power. Express the results in rectangular form.

a) $(2 + 2i)^6$ b) $(1 + \sqrt{3}i)^5$
c) $\left[-\frac{3}{2} - \frac{3\sqrt{3}}{2}i\right]^6$ d) $\left[-\frac{1}{2} - \frac{\sqrt{3}}{2}i\right]^{15}$
e) $(-\sqrt{2} + \sqrt{2}i)^{10}$ f) $(2\sqrt{3} - 2i)^4$
g) $\left[\frac{\sqrt{2}}{2} + \frac{\sqrt{2}}{2}i\right]^{20}$ h) $\left[-\frac{1}{2} + \frac{\sqrt{3}}{2}i\right]^{30}$
i) $\left[-\frac{\sqrt{3}}{2} + \frac{1}{2}i\right]^{100}$ j) $\left(\frac{\sqrt{3}}{2} + \frac{1}{2}i\right)^{60}$
k) $(-1 + i)^{28}$ l) $\left(\frac{1}{2} - \frac{1}{2}i\right)^{12}$
m) $(2 - 2i)^{16}$ n) $(3 + 3\sqrt{3}i)^{15}$

3 Using DeMoivre's theorem to find expressions for $\cos 3\theta$ and $\sin 3\theta$ in terms of $\cos \theta$ and $\sin \theta$.

4 We define $z = \sqrt{w}$ to be a number that satisfies $z^2 = w$. Write each of the following numbers in the form $a + bi$.

a) $\sqrt{i}$ b) $\sqrt{i^5}$
c) $\sqrt{3 - 3i}$ d) $\sqrt{16i}$

5 Find the indicated roots of the complex numbers and represent them graphically.

a) Fourth roots of $81(\cos 240° + i \sin 240°)$
b) Cube roots of $8(\cos 84° + i \sin 84°)$
c) Fifth roots of $32(\cos 200° + i \sin 200°)$
d) Cube roots of $27(\cos 135° + i \sin 135°)$
e) Fifth roots of $32(\cos 300° + i \sin 300°)$
f) Tenth roots of $1024(\cos 300° + i \sin 300°)$

6 Express each of the following complex numbers in polar form. Find the indicated roots and represent them graphically.

a) Square roots of $(3 + 3\sqrt{3}i)$
b) Square roots of i
c) Cube roots of $-i$

d) Cube roots of $-1 + i$
e) Fourth roots of -1
f) Fifth roots of $-16\sqrt{2} - 16\sqrt{2}i$
g) Fifth roots of i

REVIEW PROBLEM SET

1 Perform the following operations and express the result in the form of $a + bi$.

a) $(20 - i) + (3 + i)$
b) $(10 + 3i) + (-2 + 3i)$
c) $(7 - 4i) + (9 + 5i)$
d) $(-1 + \sqrt{2}i) + (-5 - \sqrt{2}i)$
e) $(5 - 7i) - (13 + i)$
f) $(6 + 9i) - (1 - 2i)$
g) $(3 + 3i) - (2 - i)$
h) $i - (-2 + 3i)$
i) $(4 - 5i)(1 - 3i)$
j) $(10 + 3i)(-2 + 3i)$
k) $(-1 - 4i)(2 + 5i)$
l) $(6 + i)(6 + i)$
m) $(3 + i)^2$
n) $(3 - 5i)^2$
o) $i^{26} + 13i$
p) $i^{17} - 3i^{13}$
q) $\dfrac{3}{1 + 2i}$
r) $\dfrac{-2 - 3i}{1 + 5i}$
s) $\dfrac{6 - i}{7 - 6i}$
t) $\dfrac{7 + 5i}{3 - 2i}$

2 Find the real numbers x and y such that each of the following equations is true.

a) $x + yi = 3 - 7i$
b) $5x - 3yi = -10 - 9i$
c) $5 + 3yi = x + 15i$
d) $3x + 17i = 15 + 34yi$
e) $x + 3yi = \dfrac{1 + i}{1 - i}$
f) $5x - 2yi = \dfrac{3 + 2i}{5 + 3i}$

3 a) Suppose that $z = x + yi$. Under what conditions does $z^2 = \bar{z}^2$?
b) If $z = x + yi$ and $z = -\bar{z}$, what is Re (z)?

4 Plot each of the following points in polar coordinates, and then find the rectangular coordinates that corresponds to each of them.

a) $(5, \pi/4)$
b) $(-2, -90°)$
c) $(\sqrt{2}, -135°)$
d) $(3, \pi)$
e) $(3, 270°)$
f) $(2, 4\pi/3)$

5 Find the modulus and an argument of each of the following complex numbers. Express z in polar form.

a) $z = -5$
b) $z = -5\sqrt{3} + 5i$
c) $z = 4\sqrt{3} + 4i$
d) $z = -4 + 4i$
e) $z = 7i$
f) $z = -5 - 5i$
g) $z = -12i$
h) $z = (2 + 2\sqrt{3}i)^4$
i) $z = 6\sqrt{2} + 6\sqrt{2}i$

6 Find zw and z/w for each of the following pairs of complex numbers, give the results in polar form and in rectangular form.

a) $z = 5(\cos 45° + i \sin 45°)$, $w = 3(\cos 15° + i \sin 15°)$
b) $z = 2(\cos 20° + i \sin 20°)$, $w = 5(\cos 50° + i \sin 50°)$
c) $z = 3(\cos 43° + i \sin 43°)$, $w = 18(\cos 17° + i \sin 17°)$
d) $z = 16(\cos 116° + i \sin 116°)$, $w = 2(\cos 176° + i \sin 176°)$
e) $z = 10(\cos 75° + i \sin 75°)$, $w = 4(\cos 165° + i \sin 165°)$

7 Express each of the following powers in polar form and in rectangular form.

a) $(\cos 30° + i \sin 30°)^{10}$
b) $(\cos 0° + i \sin 0°)^{300}$
c) $(\cos 55° + i \sin 55°)^{18}$
d) $\left[\frac{1}{2} + \frac{\sqrt{3}}{2} i\right]^{100}$
e) $\left[\frac{1}{\sqrt{2}} - \frac{1}{\sqrt{2}} i\right]^{40}$
f) $(\sqrt{3} + i)^{30}$
g) $\left(\frac{\sqrt{3}}{2} - \frac{1}{2} i\right)^{-5}$
h) $32(1 + i)^{-6}$

8 Let $z = x + yi$; show that

$$\left|\frac{x + yi}{x - yi}\right|^2 = 1$$

9 Use De Moivre's theorem to find expressions for $\cos 4\theta$ and $\sin 4\theta$ in terms of $\cos \theta$ and $\sin \theta$.

10 Let

$$z = \frac{2(\cos 6° + i \sin 6°)^{10}}{(1 - i)^2}$$

Find Re z, Im z, and $|z|$.

11 Find the indicated roots in each of the following complex numbers.

a) Square roots of i^3
b) Cube roots of $8i$
c) Cube roots of $27(\cos 300° + i \sin 300°)$
d) Cube roots of -64
e) Fourth roots of $8 + 8i$
f) Sixth roots of -64

CHAPTER 8

Vectors in the Plane

CHAPTER 8

Vectors in the Plane

1 Introduction

Vectors can be used to integrate many of the basic notions of algebra, geometry, and trigonometry. They are used in physics to identify such quantities as force, velocity, and acceleration. In this chapter we shall consider vectors in the plane from both the geometric and the algebraic points of view, and then relate them to applications from plane geometry and trigonometry.

2 Geometric Representation of Vectors

A *vector* in the plane can be represented by a directed line segment that is determined by two points, A and B. Such a directed line segment is usually drawn as an arrow (Figure 1); and is denoted by boldface letters as **AB**.

Figure 1

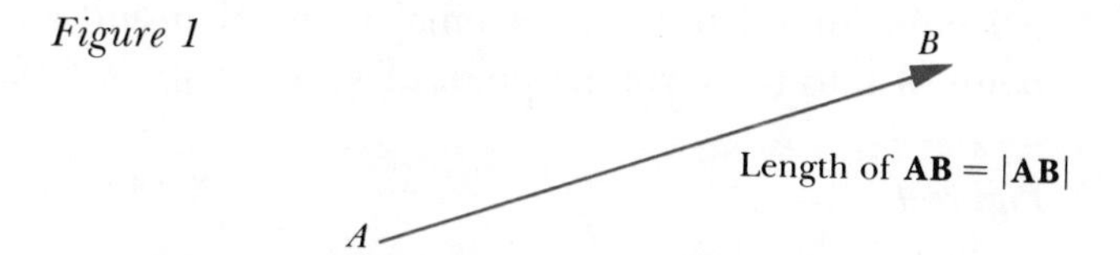

The point A is called the *initial point* of the vector **AB**, and the point B is called the *terminal point* of the vector **AB**. The length of the line segment $\overline{AB}$ is called the *length* or *magnitude* of the vector and is denoted by the symbol $|\mathbf{AB}|$ (Figure 1).

If the initial point and terminal point are the same, the vector is called a *zero vector* and is denoted by **0**. Notice that $|\mathbf{0}| = 0$.

The vector **AB** is determined by the initial point A and the terminal point B, whereas the vector **BA** has the initial point B and the terminal point A. The vector **BA** can also be denoted by $-\mathbf{AB}$ (Figure 2).

Figure 2

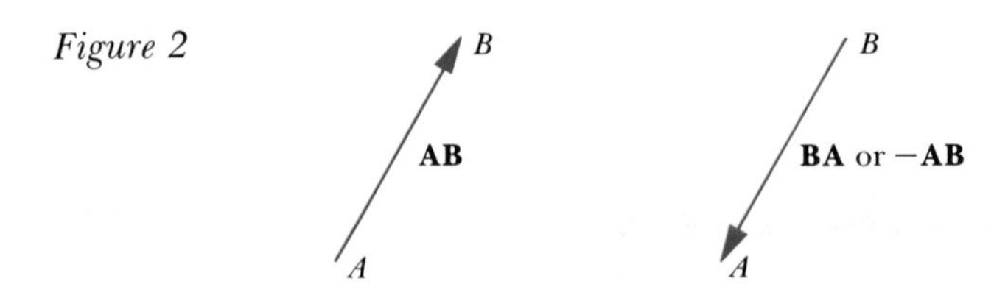

A vector may also be denoted by a single lower case boldface letter. Hence, we speak of a "vector u" and write it as **u**.

Two vectors **u** and **v** are considered to be equal if, and only if, they have the same magnitude or length and the same direction (Figure 3). (Note that we do not require that they have the initial and terminal points.)

Figure 3

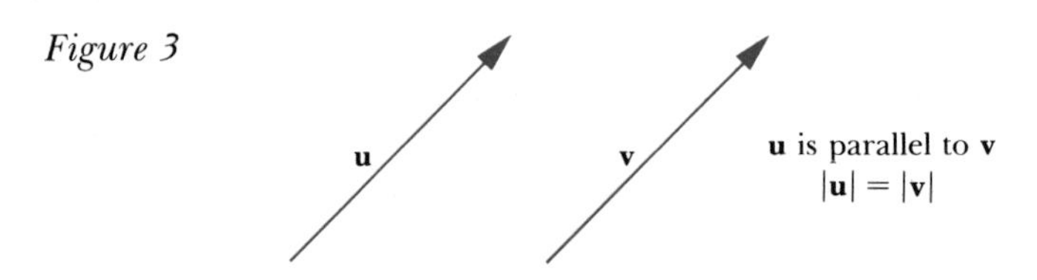

2.1 The Geometry of Vector Addition, Scalar Multiplication, and Vector Subtraction

1 *Vector addition*

Suppose that we have two vectors **u** and **v** (Figure 4). We obtain the "vector sum" $\mathbf{u} + \mathbf{v}$ geometrically by "translating" **v** so that the initial point of **v** is on the terminal point of **u** and joining the initial point of **u** to the terminal point of **v** to form $\mathbf{u} + \mathbf{v}$ (Figure 5).

Figure 4

Figure 5

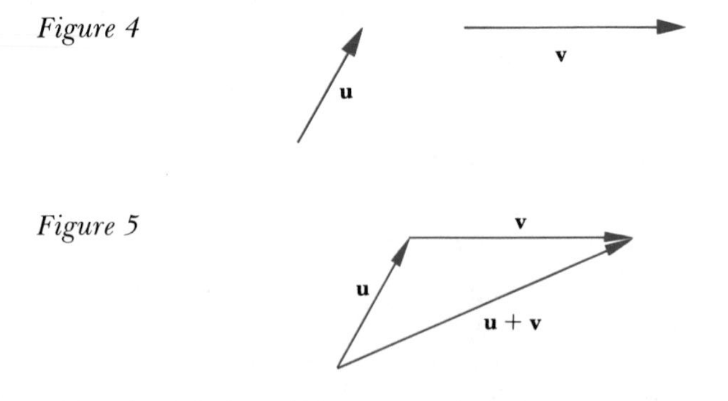

Given two vectors **u** and **v**, using the geometry of vectors, we have $\mathbf{u} + \mathbf{v} = \mathbf{v} + \mathbf{u}$, since $\mathbf{u} + \mathbf{v}$ and $\mathbf{v} + \mathbf{u}$ represent the diagonal **AC** of the parallelogram *ABCD* (Figure 6).

Figure 6

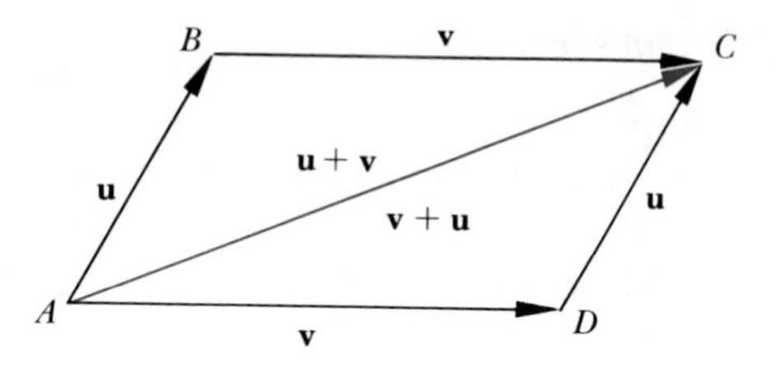

EXAMPLE

Use the geometry of vector addition to show that

$$(\mathbf{u} + \mathbf{v}) + \mathbf{w} = \mathbf{u} + (\mathbf{v} + \mathbf{w})$$

SOLUTION. First, let us assume that $\mathbf{u} = \mathbf{AB}$, $\mathbf{v} = \mathbf{BC}$, and $\mathbf{w} = \mathbf{CD}$. Then

$$\begin{aligned}(\mathbf{u} + \mathbf{v}) + \mathbf{w} &= (\mathbf{AB} + \mathbf{BC}) + \mathbf{CD} \\ &= \mathbf{AC} + \mathbf{CD} \\ &= \mathbf{AD} \qquad \text{(Figure 7)}\end{aligned}$$

On the other hand,

$$\begin{aligned}\mathbf{u} + (\mathbf{v} + \mathbf{w}) &= \mathbf{AB} + (\mathbf{BC} + \mathbf{CD}) \\ &= \mathbf{AB} + \mathbf{BD} \\ &= \mathbf{AD} \qquad \text{(Figure 7)}\end{aligned}$$

Figure 7

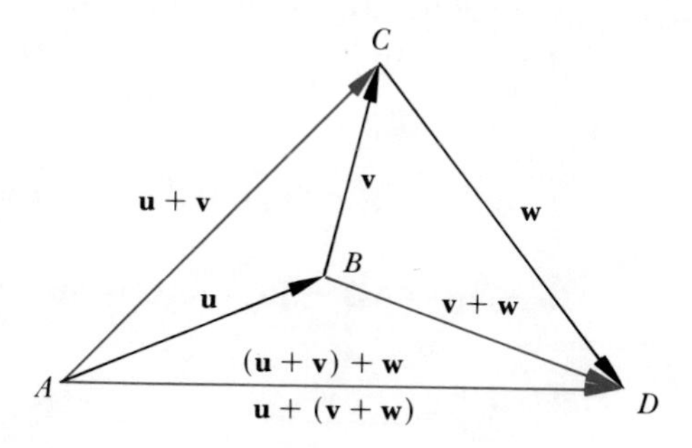

Hence,

$$(\mathbf{u} + \mathbf{v}) + \mathbf{w} = \mathbf{u} + (\mathbf{v} + \mathbf{w})$$

If $\mathbf{u}$ is any vector, then $\mathbf{u} + \mathbf{0} = \mathbf{0} + \mathbf{u} = \mathbf{u}$. Furthermore, the rules of vector addition suggest that $\mathbf{u} + (-\mathbf{u}) = \mathbf{0}$

2 *Scalar multiplication*

Suppose we are given a real number a (a is called a *scalar*) and $\mathbf{u} \neq \mathbf{0}$ is a vector; then $a\mathbf{u}$ is defined to be a vector with magnitude $|a||\mathbf{u}|$, with the same direction as $\mathbf{u}$ if $a > 0$ and with opposite direction as $\mathbf{u}$ if $a < 0$. For example, $3\mathbf{u}$ has the same direction as $\mathbf{u}$ with magnitude three times the magnitude of $\mathbf{u}$, whereas $-2\mathbf{u}$ has a direction opposite to that of $\mathbf{u}$ with magnitude two times the magnitude of $\mathbf{u}$ (Figure 8).

Figure 8

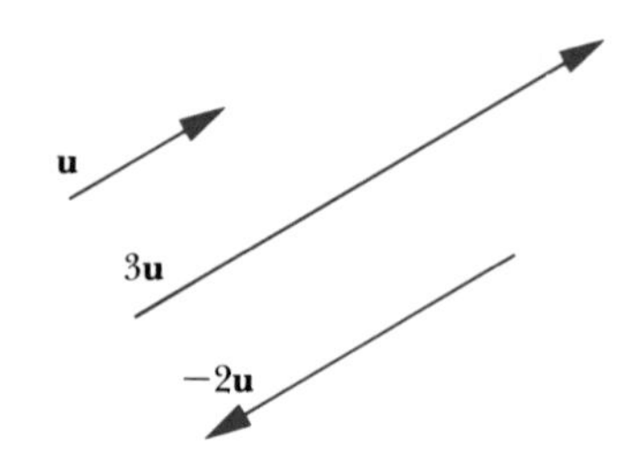

3 *Vector subtraction*

The *vector difference* $\mathbf{u} - \mathbf{v}$ is defined as $\mathbf{u} - \mathbf{v} = \mathbf{u} + (-\mathbf{v})$ (Figure 9).

Figure 9

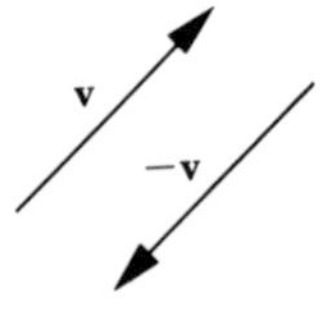

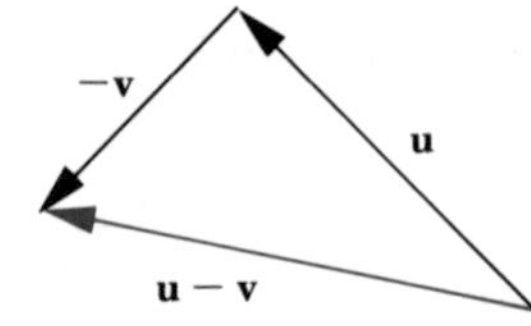

EXAMPLES

Use the geometry of vectors to show that each of the following statements are true.

1 $2\mathbf{u} + 3\mathbf{u} = 5\mathbf{u}$

SOLUTION. Since $2\mathbf{u}$ and $3\mathbf{u}$ have the same direction,

$$2\mathbf{u} + 3\mathbf{u} = \mathbf{AB} + \mathbf{BC} = \mathbf{AC} = 5\mathbf{u} \qquad \text{(Figure 10).}$$

Figure 10

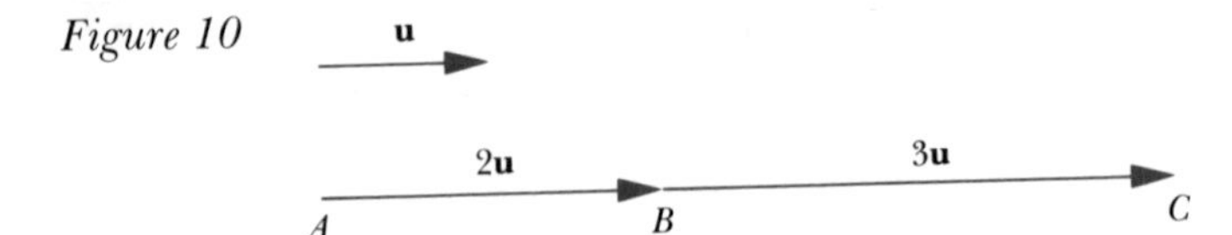

2 $3(\mathbf{u} + \mathbf{v}) = 3\mathbf{u} + 3\mathbf{v}$

SOLUTION. First, assume that $\mathbf{u} = \mathbf{AB}$ and $\mathbf{v} = \mathbf{BC}$ in Figure 11. Then $\triangle ABC$ is similar to $\triangle ADE$, where $\mathbf{AD}$ is 3 times $\mathbf{AB}$ and $\mathbf{DE}$ is 3 times $\mathbf{BC}$. Hence, $\mathbf{AE}$ is 3 times $\mathbf{AC}$, that is,

$$\mathbf{AE} = 3\mathbf{AC} = 3(\mathbf{u} + \mathbf{v})$$

Also,

$$\begin{aligned}\mathbf{AE} &= \mathbf{AD} + \mathbf{DE} \\ &= 3\mathbf{u} + 3\mathbf{v}\end{aligned}$$

so that

$$3(\mathbf{u} + \mathbf{v}) = 3\mathbf{u} + 3\mathbf{v} \qquad \text{(Figure 11)}$$

Figure 11

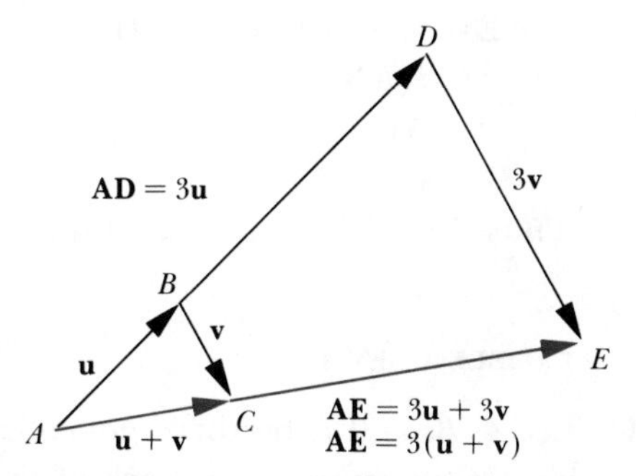

3 The diagonals AC and BD of a parallelogram $ABCD$ bisect each other.

SOLUTION. Assume that F is the midpoint of AC and E is the midpoint of BD (Figure 12).

Figure 12

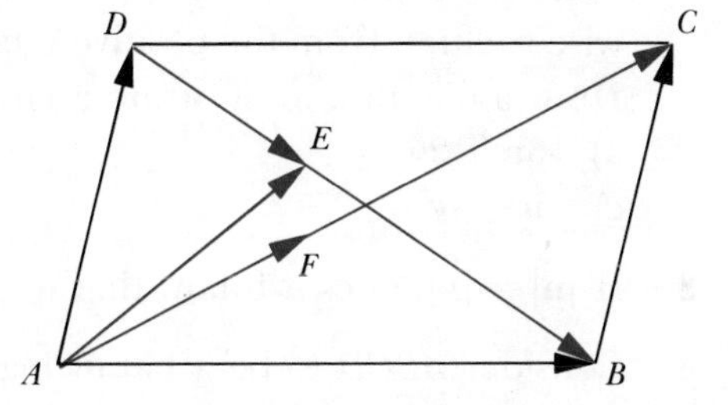

If the two diagonals bisect each other, then E and F must coincide, that is, we wish to show that $\mathbf{AE} = \mathbf{AF}$. However,

$$\mathbf{AF} = \tfrac{1}{2}\mathbf{AC} \qquad \text{and} \qquad \mathbf{AC} = \mathbf{AB} + \mathbf{BC}$$

so that

$$\mathbf{AF} = \tfrac{1}{2}(\mathbf{AB} + \mathbf{BC})$$

Also,

$$\mathbf{AE} = \mathbf{AD} + \mathbf{DE}$$

where

$$\mathbf{DE} = \tfrac{1}{2}\mathbf{DB} \qquad \text{and} \qquad \mathbf{DB} = \mathbf{AB} - \mathbf{AD}$$

so that

$$\mathbf{DE} = \tfrac{1}{2}(\mathbf{AB} - \mathbf{AD})$$

Hence,

$$\begin{aligned}\mathbf{AE} &= \mathbf{AD} + \tfrac{1}{2}(\mathbf{AB} - \mathbf{AD})\\ &= \tfrac{1}{2}(\mathbf{AB} + \mathbf{AD})\\ &= \mathbf{AF}\end{aligned}$$

Thus, E and F coincide and the diagonals bisect each other.

PROBLEM SET 1

1 Let A, B, and C be three noncollinear points in the plane. Find and illustrate geometrically, each of the following expressions.

a) $3\mathbf{AB} + 3\mathbf{BC} + 2\mathbf{CA}$ b) $5\mathbf{AB} - 5\mathbf{CB}$
c) $-2\mathbf{AC} - 2\mathbf{CB}$ d) $4\mathbf{AB} + 4\mathbf{BC} + 3\mathbf{CA}$

2 Let $\mathbf{u}$ be a vector in the positive direction of the x axis and $\mathbf{v}$ be a vector that makes an angle of 45° with vector $\mathbf{u}$ by a counterclockwise rotation from the positive x axis; assume that $|\mathbf{u}| = 4$ and $|\mathbf{v}| = 3$. Draw a diagram to illustrate each of the following vectors.

a) $3\mathbf{u} + 2\mathbf{v}$ b) $-\tfrac{1}{2}\mathbf{u} + 3\mathbf{v}$
c) $\mathbf{u} - \tfrac{3}{2}\mathbf{v}$ d) $-2\mathbf{u} - 5\mathbf{v}$

3 If $|\mathbf{u}| = |\mathbf{v}|$, does it follow that $\mathbf{u} = \mathbf{v}$? Explain.

4 Consider $ABCD$ to be a parallelogram (Figure 13). Find each of the following vectors.

a) $\mathbf{AB} + \mathbf{AD}$ b) $\mathbf{AB} - \mathbf{AD}$
c) A vector equal to $\mathbf{DC}$ d) A vector equal to $\mathbf{BC}$

Figure 13

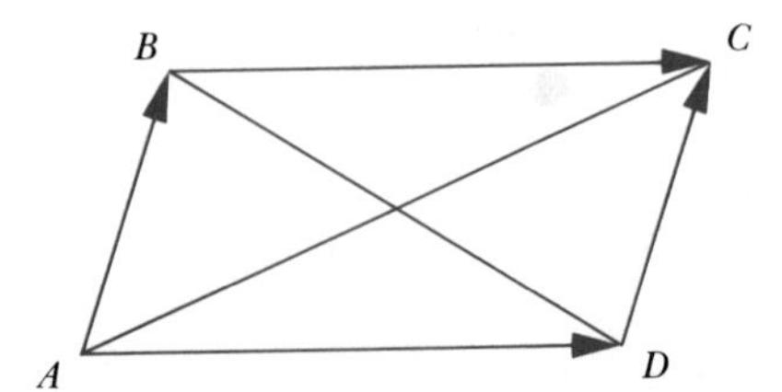

5 Let $\mathbf{u} = \mathbf{AB}$ and $\mathbf{v} = \mathbf{AC}$. Express $\mathbf{AD}$ in terms of $\mathbf{u}$ and $\mathbf{v}$, where D is $\tfrac{3}{8}$ of the way from B to C.

6 Let $ABCD$ be a square (Figure 14). If $\mathbf{AB} = \mathbf{u}$ and $\mathbf{BC} = \mathbf{v}$ are the sides of the square and E is the midpoint of $\mathbf{CD}$, express $\mathbf{AE}$ in terms of $\mathbf{u}$ and $\mathbf{v}$.

Figure 14

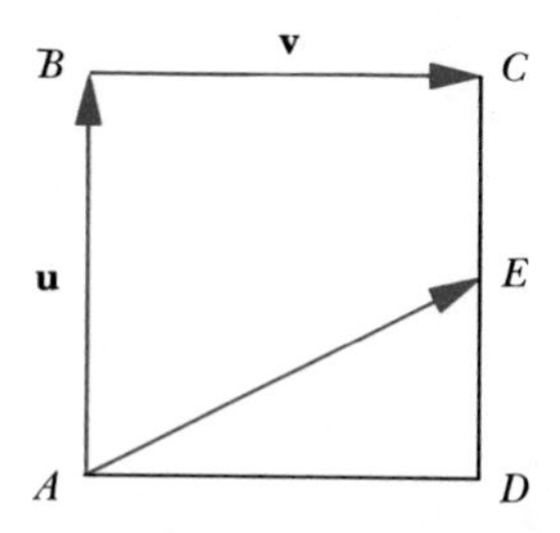

7 Let $\mathbf{u} = \mathbf{AB}$ and $\mathbf{v} = \mathbf{AC}$ be two vectors (Figure 15). If D is the midpoint of $\mathbf{AB}$ and E is the midpoint of $\mathbf{AC}$, express $\mathbf{AE}$, $\mathbf{CD}$, and $\mathbf{DE}$ in terms of $\mathbf{u}$ and $\mathbf{v}$.

Figure 15

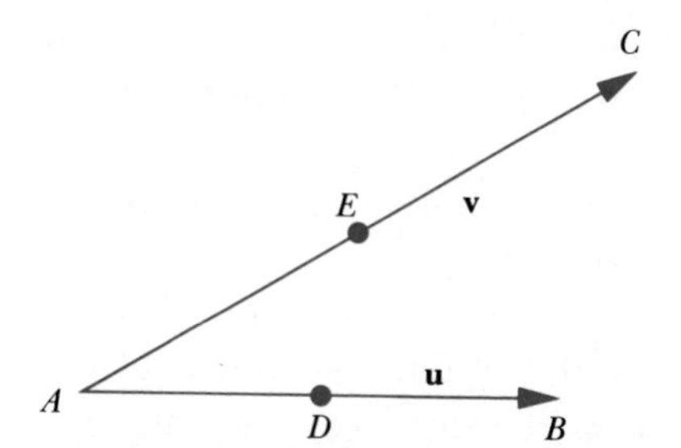

8 Use vectors to prove that the line segment joining the midpoints of two sides of a triangle in the plane has a length equal to one half the length of the third side.

3 Analytic Representation of Vectors

Suppose that a given vector $\mathbf{u}$ is placed in a plane with a Cartesian coordinate system so that its initial point is the origin and its terminal point is the point determined by the pair of numbers (a, b) (Figure 1). $\mathbf{u}$ is called a *radius vector* or *position vector* and is denoted by $\mathbf{u} = \langle a, b \rangle$. a is called the *x component* of $\mathbf{u}$ and b is called the *y component* of $\mathbf{u}$.

Figure 1

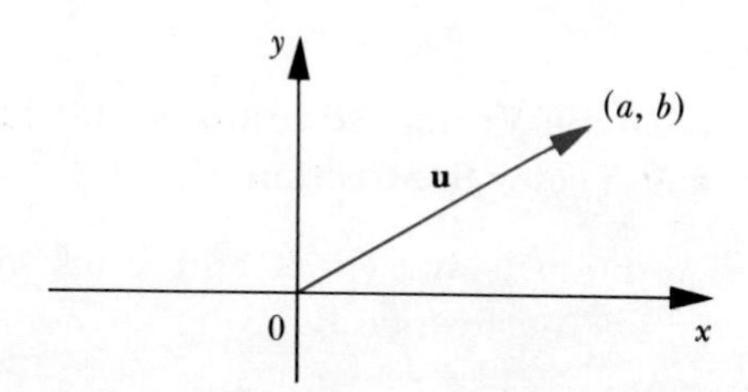

For example, if $\mathbf{u} = \langle 3, -5 \rangle$, then 3 is its x component and -5 is its y component (Figure 2). The zero vector $\mathbf{0} = \langle 0, 0 \rangle$ has x component 0 and y component 0.

Figure 2

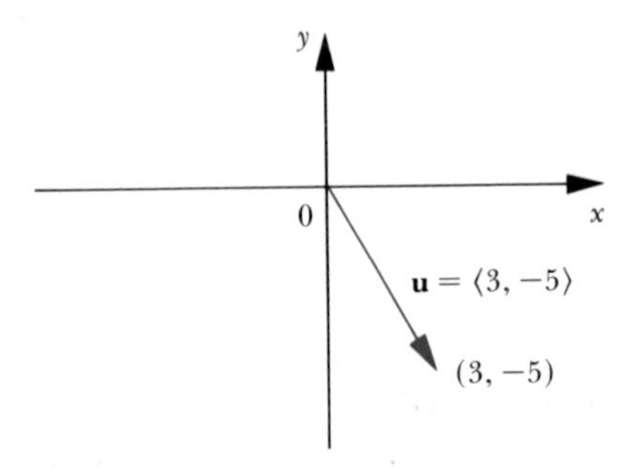

As stated previously, two vectors $\mathbf{u}$ and $\mathbf{v}$ are equal whenever $|\mathbf{u}| = |\mathbf{v}|$ and $\mathbf{u}$ and $\mathbf{v}$ have the same direction. Thus, for position vectors if $\mathbf{u} = \langle a, b \rangle$ and $\mathbf{v} = \langle c, d \rangle$, then $\mathbf{u} = \mathbf{v}$ whenever $a = c$ and $b = d$.

For example, if $\mathbf{u} = \mathbf{v}$, $\mathbf{u} = \langle x, 5 \rangle$ and $\mathbf{v} = \langle 3, y \rangle$, then $x = 3$ and $y = 5$.

The *magnitude* or *length* of a position vector $\mathbf{u}$ can be determined by the distance formula. If $\mathbf{u} = \langle a, b \rangle$, then $|\mathbf{u}|$, the magnitude of $\mathbf{u}$, is the distance from $(0, 0)$ to the point (a, b). Hence, by the distance formula,

$$|\mathbf{u}| = \sqrt{(a - 0)^2 + (b - 0)^2} = \sqrt{a^2 + b^2} \qquad \text{(Figure 3)}$$

Figure 3

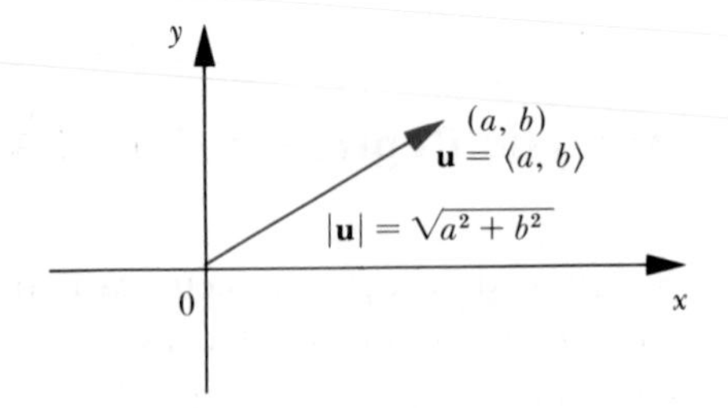

For example, if $\mathbf{u} = \langle -4, 3 \rangle$, then $|\mathbf{u}| = \sqrt{(-4)^2 + 3^2} = 5$, and if $\mathbf{u} = \langle 5, -12 \rangle$, then $|\mathbf{u}| = \sqrt{5^2 + (-12)^2} = 13$.

3.1 Analytic Vector Addition, Scalar Multiplication, and Vector Subtraction

Addition, subtraction, and scalar multiplication of position vectors can be performed by using the components.

1 Vector addition

The sum of two vectors $\mathbf{u} = \langle a, b \rangle$ and $\mathbf{v} = \langle c, d \rangle$ is the vector $\mathbf{u} + \mathbf{v}$ represented by $\mathbf{u} + \mathbf{v} = \langle a + c,\ b + d \rangle$ (Figure 4). For example, if $\mathbf{u} = \langle -3, 4 \rangle$ and $\mathbf{v} = \langle 5, 7 \rangle$, then $\mathbf{u} + \mathbf{v} = \langle -3 + 5, 4 + 7 \rangle = \langle 2, 11 \rangle$.

Figure 4

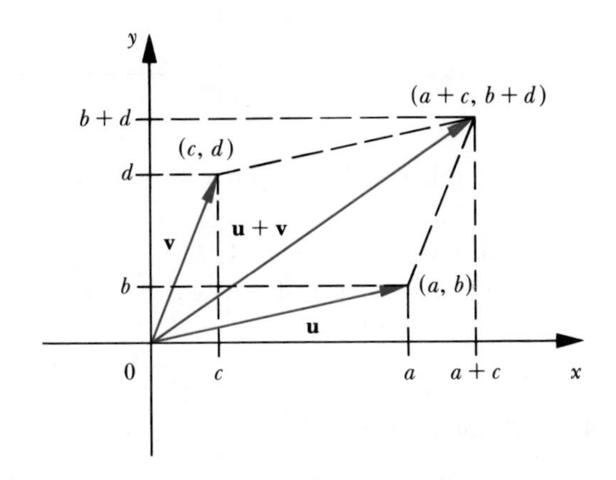

2 Scalar multiplication

If c is a scalar and $\mathbf{u}$ is the vector $\langle a, b \rangle$, then the product $c\mathbf{u}$ is the vector given by $c\mathbf{u} = c\langle a, b \rangle = \langle ca, cb \rangle$ (Figure 5). For example, if $\mathbf{u} = \langle -7, 3 \rangle$, then $4\mathbf{u} = 4\langle -7, 3 \rangle = \langle -28, 12 \rangle$.

Figure 5

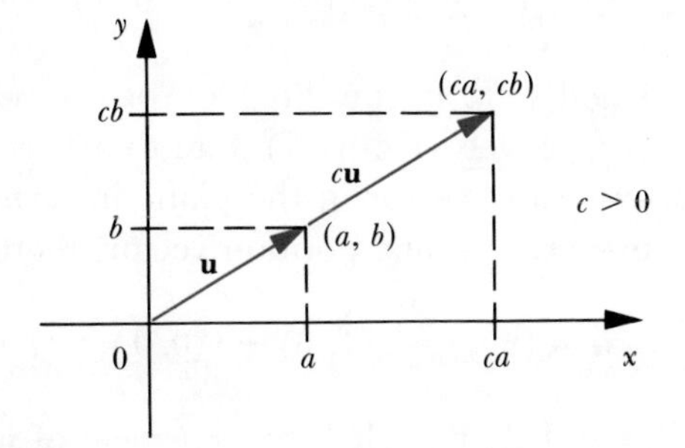

3 Vector subtraction

The difference of two vectors $\mathbf{u} = \langle a, b \rangle$ and $\mathbf{v} = \langle c, d \rangle$ is given by

$$\begin{aligned} \mathbf{u} - \mathbf{v} = \mathbf{u} + (-\mathbf{v}) &= \langle a, b \rangle + (-\langle c, d \rangle) \\ &= \langle a, b \rangle + \langle -c, -d \rangle \\ &= \langle a - c, b - d \rangle \qquad \text{(Figure 6)} \end{aligned}$$

Figure 6

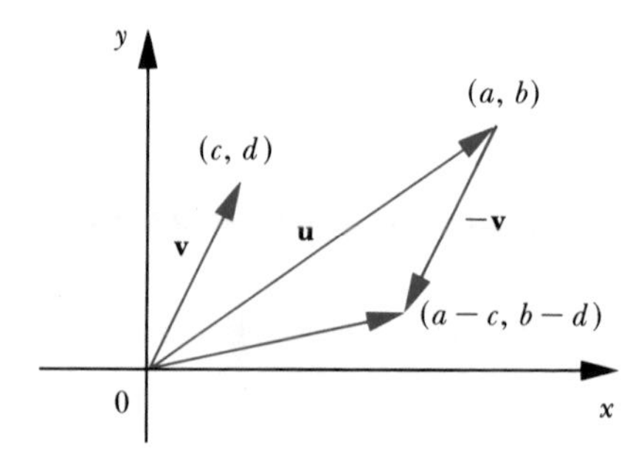

Thus, if $\mathbf{u} = \langle 3, 4\rangle$ and $\mathbf{v} = \langle -1, 5\rangle$, then

$$\begin{aligned}\mathbf{u} - \mathbf{v} &= \langle 3, 4\rangle - \langle -1, 5\rangle \\ &= \langle 3 + 1, 4 - 5\rangle \\ &= \langle 4, -1\rangle\end{aligned}$$

3.2 Basis Vectors

Consider a position vector $\mathbf{u} = \langle x, y\rangle$. We can write $\mathbf{u}$ as

$$\mathbf{u} = \langle x, y\rangle = \langle x, 0\rangle + \langle 0, y\rangle = x\langle 1, 0\rangle + y\langle 0, 1\rangle$$

Since the magnitude of each of the two vectors $\langle 1, 0\rangle$ and $\langle 0, 1\rangle$ is one unit, they are called *unit vectors.* The following symbols are used for these two unit vectors:

$$\mathbf{i} = \langle 1, 0\rangle \qquad \text{and} \qquad \mathbf{j} = \langle 0, 1\rangle$$

$\mathbf{i}$ and $\mathbf{j}$ are in the direction of the positive x axis and positive y axis, respectively (Figure 7). $\mathbf{i}$ and $\mathbf{j}$ are called *basis vectors.* It is possible to write any vector in the plane in terms of these basis vectors, for, if $\mathbf{u} = \langle x, y\rangle$ is any position vector, then

$$\mathbf{u} = \langle x, y\rangle = x\langle 1, 0\rangle + y\langle 0, 1\rangle = x\mathbf{i} + y\mathbf{j}$$

$x\mathbf{i}$ and $y\mathbf{j}$ are called the *projections* of $\mathbf{u}$ along the x axis and y axis, respectively (Figure 7).

Figure 7

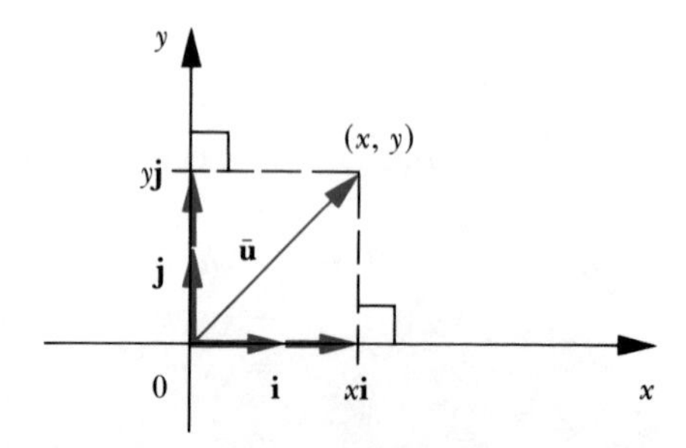

For example, the vector $\mathbf{u} = \langle 3, 7 \rangle$ can be expressed in terms of $\mathbf{i}$ and $\mathbf{j}$ as $\mathbf{u} = 3\mathbf{i} + 7\mathbf{j}$.

If $\mathbf{u} = \langle a, b \rangle$ is not the zero vector, the vector $\mathbf{v}$, having the same direction as $\mathbf{u}$, given by

$$\mathbf{v} = \frac{a}{|\mathbf{u}|}\,\mathbf{i} + \frac{b}{|\mathbf{u}|}\,\mathbf{j} = \frac{1}{|\mathbf{u}|}\,(a\mathbf{i} + b\mathbf{j})$$

is a unit vector since

$$|\mathbf{v}| = \sqrt{\left(\frac{a}{|\mathbf{u}|}\right)^2 + \left(\frac{b}{|\mathbf{u}|}\right)^2} = \sqrt{\frac{a^2 + b^2}{|\mathbf{u}|^2}} = \frac{\sqrt{a^2 + b^2}}{|\mathbf{u}|} = \frac{|\mathbf{u}|}{|\mathbf{u}|} = 1$$

Since $1/|\mathbf{u}| > 0$, $\mathbf{v}$ has the same direction as $\mathbf{u}$. $\mathbf{v}$ is called the *normalized* $\mathbf{u}$ vector.

Let $\mathbf{u}$ be a vector determined by two points P_1 and P_2, where $P_1 = (x_1, y_1)$ and $P_2 = (x_2, y_2)$ are any two points in the plane. Then $\mathbf{u} = \mathbf{u}_2 - \mathbf{u}_1$, where $\mathbf{u}_1$ and $\mathbf{u}_2$ are the position vectors determined by P_1 and P_2, as illustrated in Figure 8.

Figure 8

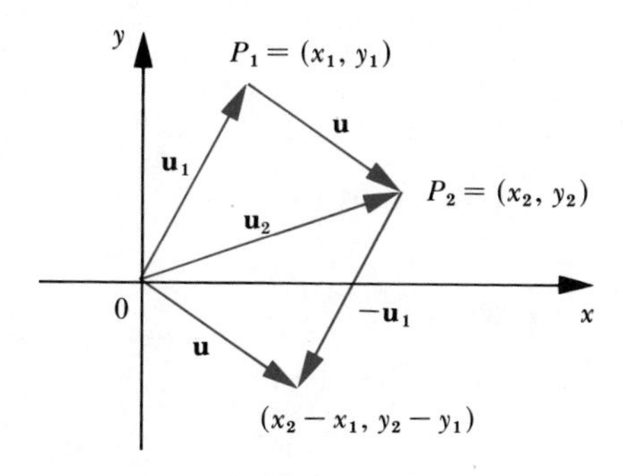

Since $\mathbf{u}_1 = \langle x_1, y_1 \rangle$ and $\mathbf{u}_2 = \langle x_2, y_2 \rangle$

$$\mathbf{u} = \mathbf{u}_2 - \mathbf{u}_1 = \langle x_2, y_2 \rangle - \langle x_1, y_1 \rangle = \langle x_2 - x_1, y_2 - y_1 \rangle$$

so $\mathbf{u} = (x_2 - x_1)\mathbf{i} + (y_2 - y_1)\mathbf{j}$, where (x_1, y_1) is the initial point of $\mathbf{u}$ and (x_2, y_2) is the terminal point of $\mathbf{u}$. The magnitude of $\mathbf{u}$ is given by

$$|\mathbf{u}| = \sqrt{(x_2 - x_1)^2 + (y_2 - y_1)^2} = \sqrt{u_x^{\,2} + u_y^{\,2}}$$

where $u_x = x_2 - x_1$ and $u_y = y_2 - y_1$.

EXAMPLES

1 Let $\mathbf{u} = 3\mathbf{i} + 4\mathbf{j}$ and $\mathbf{v} = 4\mathbf{i} - 5\mathbf{j}$. Find the components of the following vectors.

a) $3\mathbf{u}$ b) $5\mathbf{v}$ c) $3\mathbf{u} - 5\mathbf{v}$

d) $3\mathbf{u} + 5\mathbf{v}$ e) $\mathbf{u} - 3\mathbf{v}$

SOLUTION

a) $3\mathbf{u} = 3(3\mathbf{i} + 4\mathbf{j}) = 9\mathbf{i} + 12\mathbf{j} = \langle 9, 12\rangle$

b) $5\mathbf{v} = 5(4\mathbf{i} - 5\mathbf{j}) = 20\mathbf{i} - 25\mathbf{j} = \langle 20, -25\rangle$

c) $$3\mathbf{u} - 5\mathbf{v} = (9\mathbf{i} + 12\mathbf{j}) - (20\mathbf{i} - 25\mathbf{j}) = (9 - 20)\mathbf{i} + (12 + 25)\mathbf{j}$$
$$= -11\mathbf{i} + 37\mathbf{j} = \langle -11, 37\rangle$$

d) $$3\mathbf{u} + 5\mathbf{v} = (9\mathbf{i} + 12\mathbf{j}) + (20\mathbf{i} - 25\mathbf{j})$$
$$= (9 + 20)\mathbf{i} + (12 - 25)\mathbf{j} = 29\mathbf{i} - 13\mathbf{j} = \langle 29, -13\rangle$$

e) $$\mathbf{u} - 3\mathbf{v} = (3\mathbf{i} + 4\mathbf{j}) - 3(4\mathbf{i} - 5\mathbf{j}) = (3\mathbf{i} + 4\mathbf{j}) - (12\mathbf{i} - 15\mathbf{j})$$
$$= (3 - 12)\mathbf{i} + (4 + 15)\mathbf{j} = -9\mathbf{i} + 19\mathbf{j} = \langle -9, 19\rangle$$

2 Write $\mathbf{u}$ in the form $\mathbf{u} = u_x\mathbf{i} + u_y\mathbf{j}$ and find $|\mathbf{u}|$ if $\mathbf{u}$ is the vector whose initial point is the first point and whose terminal point is second point of the following pairs of points.

a) $(-1, 4)$ and $(3, 5)$

b) $(5, 3)$ and $(-3, -3)$

c) $(2, -1)$ and $(-1, -2)$

d) $(5, 6)$ and $(0, -4)$

SOLUTION

a) $\mathbf{u} = [3 - (-1)]\mathbf{i} + (5 - 4)\mathbf{j} = 4\mathbf{i} + \mathbf{j}$ and $|\mathbf{u}| = \sqrt{4^2 + 1^2} = \sqrt{17}$

b) $\mathbf{u} = (-3 - 5)\mathbf{i} + (-3 - 3)\mathbf{j} = -8\mathbf{i} - 6\mathbf{j}$ and
$|\mathbf{u}| = \sqrt{(-8)^2 + (-6)^2} = 10$

c) $\mathbf{u} = (-1 - 2)\mathbf{i} + [-2 - (-1)]\mathbf{j} = -3\mathbf{i} - \mathbf{j}$ and
$|\mathbf{u}| = \sqrt{(-3)^2 + (-1)^2} = \sqrt{10}$

d) $\mathbf{u} = (0 - 5)\mathbf{i} + (-4 - 6)\mathbf{j} = -5\mathbf{i} - 10\mathbf{j}$ and
$|\mathbf{u}| = \sqrt{(-5)^2 + (-10)^2} = 5\sqrt{5}$

3 Use vectors to find a point $\frac{3}{4}$ of the way from $(-5, -9)$ to $(7, 7)$.

SOLUTION. (See Figure 9.) Let $P = (x, y)$ be a point $\frac{3}{4}$ of the way from $A = (-5, -9)$ to $B = (7, 7)$; then

$$\mathbf{AP} = \langle x + 5, y + 9\rangle = \tfrac{3}{4}\langle 7 + 5, 7 + 9\rangle$$
$$= \tfrac{3}{4}\langle 12, 16\rangle$$
$$= \langle 9, 12\rangle$$

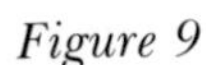

Figure 9

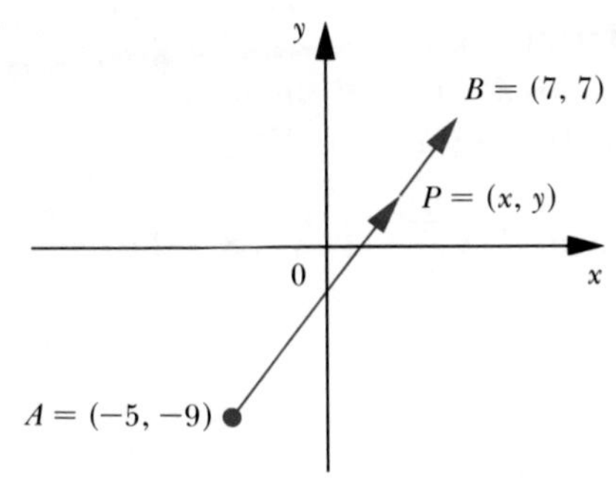

so that

$$x + 5 = 9 \qquad \text{and} \qquad y + 9 = 12$$

or

$$x = 4 \qquad \text{and} \qquad y = 3$$

Hence, the point is (4, 3).

4 Write the vector $\mathbf{u} = 2\mathbf{i} + 2\mathbf{j}$ in the form $\mathbf{u} = r(\cos\theta\mathbf{i} + \sin\theta\mathbf{j})$.

SOLUTION. First, **u** can be sketched in the plane (Figure 10).

Figure 10

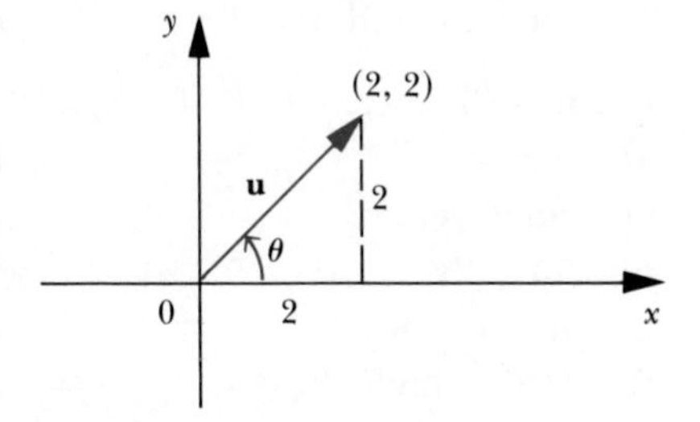

Using right triangle trigonometry, we know that the acute angle formed by **u** and the positive x axis is 45°. Consequently,

$$\cos 45° = \frac{2}{|\mathbf{u}|} \qquad \text{and} \qquad \sin 45° = \frac{2}{|\mathbf{u}|}$$

so that

$$2 = |\mathbf{u}| \cos 45° \qquad \text{and} \qquad 2 = |\mathbf{u}| \sin 45°$$

Thus

$$\begin{aligned}\mathbf{u} &= 2\mathbf{i} + 2\mathbf{j}\\ &= |\mathbf{u}| \cos 45^\circ\, \mathbf{i} + |\mathbf{u}| \sin 45^\circ\, \mathbf{j}\\ &= |\mathbf{u}| \,(\cos 45^\circ\, \mathbf{i} + \sin 45^\circ\, \mathbf{j})\end{aligned}$$

But since $|\mathbf{u}| = \sqrt{2^2 + 2^2} = \sqrt{8} = 2\sqrt{2}$,

$$\mathbf{u} = 2\sqrt{2}\,(\cos 45^\circ\, \mathbf{i} + \sin 45^\circ\, \mathbf{j}).$$

PROBLEM SET 2

1 Find $\mathbf{u} + \mathbf{v}$ and $\mathbf{u} - \mathbf{v}$ and illustrate the results geometrically for each of the following pairs of vectors.

a) $\mathbf{u} = \langle 3, 2\rangle$ and $\mathbf{v} = \langle -5, 1\rangle$
b) $\mathbf{u} = \langle -3, 0\rangle$ and $\mathbf{v} = \langle 5, 6\rangle$
c) $\mathbf{u} = \langle -5, 10\rangle$ and $\mathbf{v} = \langle 3, 5\rangle$
d) $\mathbf{u} = \langle 1, 2\rangle$ and $\mathbf{v} = \langle 4, 3\rangle$
e) $\mathbf{u} = \langle 7, 2\rangle$ and $\mathbf{v} = \langle 1, 6\rangle$
f) $\mathbf{u} = \langle 4, -5\rangle$ and $\mathbf{v} = \langle 0, -3\rangle$
g) $\mathbf{u} = \langle 3, -4\rangle$ and $\mathbf{v} = \langle 6, 0\rangle$
h) $\mathbf{u} = \langle 0, 0\rangle$ and $\mathbf{v} = \langle 1, 8\rangle$

2 Let $\mathbf{u} = \langle 2, -4\rangle$, $\mathbf{v} = \langle -3, 1\rangle$, and $\mathbf{w} = \langle 2, 4\rangle$. Find each of the following vectors and illustrate them geometrically.

a) $\mathbf{u} + \mathbf{v}$ and $\mathbf{v} + \mathbf{u}$
b) $\mathbf{u} + (\mathbf{v} + \mathbf{w})$ and $(\mathbf{u} + \mathbf{v}) + \mathbf{w}$
c) $3\mathbf{u} + 3\mathbf{v}$ and $3(\mathbf{u} + \mathbf{v})$
d) $-3\mathbf{u} + 3\mathbf{v}$ and $-3(\mathbf{u} - \mathbf{v})$

3 Let $\mathbf{u} = \langle 3, -4\rangle$, $\mathbf{v} = \langle -4, 3\rangle$, and $\mathbf{w} = \langle 5, 6\rangle$. Find

a) $\mathbf{u} + \mathbf{v}$
b) $\mathbf{u} + \mathbf{w}$
c) $|\mathbf{u}| + |\mathbf{v}|$
d) $|\mathbf{u}| + |\mathbf{w}|$
e) $3\mathbf{u} + 2\mathbf{v}$
f) $\mathbf{u} + (\mathbf{v} + 2\mathbf{w})$
g) $\mathbf{u} - \mathbf{v} - \mathbf{w}$
h) $|2\mathbf{u} - 3\mathbf{w}|$
i) $|3\mathbf{u}| - |2\mathbf{v}|$
j) $|3\mathbf{u} - 2\mathbf{v}|$

4 Write each of the following vectors in the form of $r(\cos \theta\mathbf{i} + \sin \theta\mathbf{j})$, where $r = |\mathbf{u}|$ and θ is a radian measure of the smallest positive angle that defines the direction of the vector.

a) $\mathbf{u} = \mathbf{i} - \mathbf{j}$
b) $\mathbf{u} = -\mathbf{i} + 3\mathbf{j}$
c) $\mathbf{u} = -4\mathbf{i}$

5 Let $\mathbf{u} = 3\mathbf{i} + 4\mathbf{j}$ and $\mathbf{v} = 5\mathbf{i} - 2\mathbf{j}$. Find a unit vector having the same direction as $\mathbf{u} + \mathbf{v}$.

6 Let $\mathbf{u} = -2\mathbf{i} + \mathbf{j}$, $\mathbf{v} = 2\mathbf{i} - 3\mathbf{j}$, and $\mathbf{w} = 5\mathbf{i} + 2\mathbf{j}$. Find scalar numbers a and b such that $\mathbf{w} = a\mathbf{u} + b\mathbf{v}$.

7 Write $\mathbf{u}$ in the form $\mathbf{u} = u_x\mathbf{i} + u_y\mathbf{j}$, find $|\mathbf{u}|$, and normalize $\mathbf{u}$ if $\mathbf{u}$ is the vector whose initial point is the first point and whose terminal point is the second point given in each of the following parts.

a) (3, 4), (−4, 3)
b) (−3, 0), (2, 4)
c) (3, 2), (3, 5)
d) (7, 2), (1, 6)
e) (3, −4), (6, 0)
f) (0, 0), (4, −5)
g) (4, 1), (2, 3)
h) (5, −2), (3, 4)

8 Let $P = (-3, -5)$ and $Q = (4, 4)$ be two points in the plane.
a) Use vectors to find the point that is $\frac{2}{3}$ of the way from P to Q.
b) Use vectors to find the point 2 units from P along the directed line from P to Q.

4 Inner Product

Consider the problem of finding the angle between two vectors $\mathbf{u} = \langle a, b \rangle$ and $\mathbf{v} = \langle c, d \rangle$. The angle θ between $\mathbf{u}$ and $\mathbf{v}$ is the angle whose sides contain $\mathbf{u}$ and $\mathbf{v}$. This angle is unique, if we specify $0 \leqslant \theta \leqslant 180°$ (Figure 1).

Figure 1

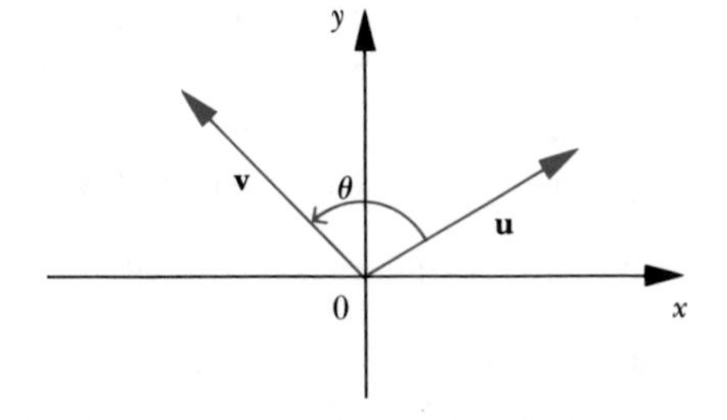

Now, we will define a special product of two vectors $\mathbf{u}$ and $\mathbf{v}$ called the *inner product* or *dot product,* denoted by $\mathbf{u} \cdot \mathbf{v}$, as

$$\mathbf{u} \cdot \mathbf{v} = |\mathbf{u}||\mathbf{v}| \cos \theta$$

where θ is the angle between $\mathbf{u}$ and $\mathbf{v}$.

If either $\mathbf{u}$ or $\mathbf{v}$ is the zero vector, then $\mathbf{u} \cdot \mathbf{v}$ is defined to be the scalar zero. Notice that the inner product is an operation that assigns to each pair of vectors a scalar number. If $\mathbf{v} \neq \mathbf{0}$, then it follows from $\mathbf{u} \cdot \mathbf{v} = |\mathbf{u}||\mathbf{v}| \cos \theta$ that

$$|\mathbf{u}| \cos \theta = \frac{\mathbf{u} \cdot \mathbf{v}}{|\mathbf{v}|}$$

The product $|\mathbf{u}| \cos \theta$ is called the *scalar projection* of $\mathbf{u}$ onto $\mathbf{v}$ (Figure 2).

Figure 2

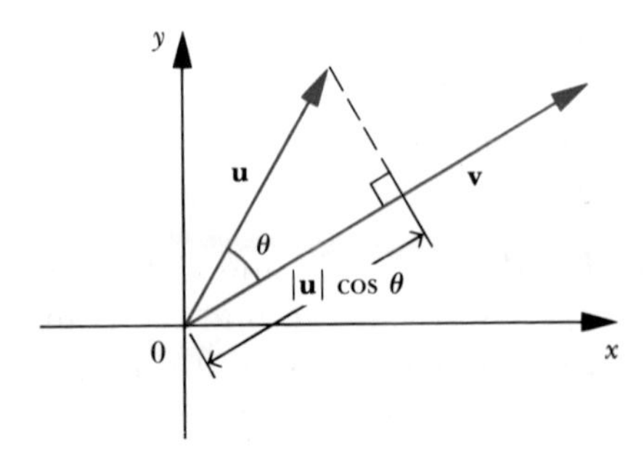

EXAMPLES

1 Find $\mathbf{u} \cdot \mathbf{v}$ if $|\mathbf{u}| = 3$ and $|\mathbf{v}| = 4$ and

a) $\theta = 0°$ b) $\theta = 30°$ c) $\theta = 45°$

d) $\theta = 90°$ e) $\theta = 180°$

SOLUTION

a) $\mathbf{u} \cdot \mathbf{v} = |\mathbf{u}||\mathbf{v}| \cos \theta$

$= 3 \cdot 4 \cos 0° = 12(1) = 12$

b) $\mathbf{u} \cdot \mathbf{v} = 3 \cdot 4 \cos 30° = 12(\sqrt{3}/2) = 6\sqrt{3}$

c) $\mathbf{u} \cdot \mathbf{v} = 3 \cdot 4 \cos 45° = 12(\sqrt{2}/2) = 6\sqrt{2}$

d) $\mathbf{u} \cdot \mathbf{v} = 3 \cdot 4 \cos 90° = 12(0) = 0$

e) $\mathbf{u} \cdot \mathbf{v} = 3 \cdot 4 \cos 180° = 12(-1) = -12$

Find the scalar projection of **u** onto **v** for each part in Example 1.

SOLUTION

a) $\dfrac{\mathbf{u} \cdot \mathbf{v}}{|\mathbf{v}|} = \dfrac{12}{4} = 3$

b) $\dfrac{\mathbf{u} \cdot \mathbf{v}}{|\mathbf{v}|} = \dfrac{6\sqrt{3}}{4} = \dfrac{3\sqrt{3}}{2}$

c) $\dfrac{\mathbf{u} \cdot \mathbf{v}}{|\mathbf{v}|} = \dfrac{6\sqrt{2}}{4} = \dfrac{3\sqrt{2}}{2}$

d) $\dfrac{\mathbf{u} \cdot \mathbf{v}}{|\mathbf{v}|} = \dfrac{0}{4} = 0$

e) $\dfrac{\mathbf{u} \cdot \mathbf{v}}{|\mathbf{v}|} = \dfrac{-12}{4} = -3$

The inner product has many applications. One of these is a test of *orthogonality* or perpendicularity.

THEOREM 1

If **u** and **v** are nonzero vectors, then they are orthogonal (perpendicular) if and only if $\mathbf{u} \cdot \mathbf{v} = 0$.

PROOF. By definition, the inner product of **u** and **v** is

$$\mathbf{u} \cdot \mathbf{v} = |\mathbf{u}||\mathbf{v}| \cos \theta$$

This product of real numbers is zero if and only if one of its factors is zero. Since $\mathbf{u}$ and $\mathbf{v}$ are nonzero vectors, $|\mathbf{u}| \neq 0$ and $|\mathbf{v}| \neq 0$. Therefore, the product is zero if and only if $\cos \theta = 0$. This implies that $\theta = 90°$, which is the case if and only if $\mathbf{u}$ and $\mathbf{v}$ are orthogonal.

The reason that Theorem 1 is useful is that we can prove a very simple way of computing the inner product of two vectors when they are expressed in component form. (In the proof, we will see an application of the Law of Cosines.)

THEOREM 2

If $\mathbf{u} = u_x\mathbf{i} + u_y\mathbf{j}$ and $\mathbf{v} = v_x\mathbf{i} + v_y\mathbf{j}$, then $\mathbf{u} \cdot \mathbf{v} = u_xv_x + u_yv_y$.

PROOF. Suppose that θ is the angle between $\mathbf{u}$ and $\mathbf{v}$ (Figure 3). Using the definition of the inner product, we have

Figure 3

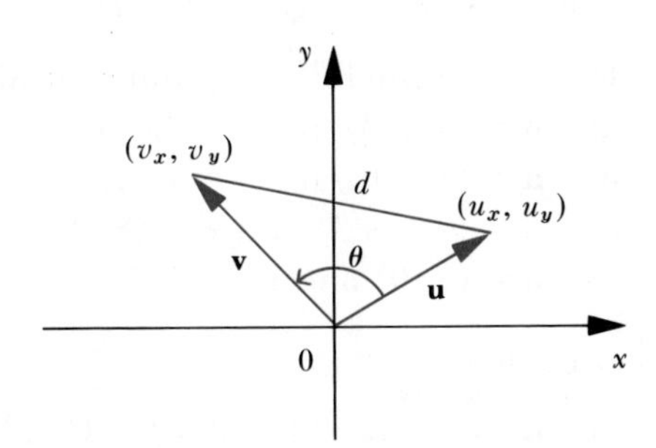

$$\mathbf{u} \cdot \mathbf{v} = |\mathbf{u}||\mathbf{v}| \cos \theta$$

By the Law of Cosines,

$$d^2 = |\mathbf{u}|^2 + |\mathbf{v}|^2 - 2|\mathbf{u}||\mathbf{v}| \cos \theta$$

so that

$$\begin{aligned}
|\mathbf{u}||\mathbf{v}| \cos \theta &= \tfrac{1}{2}(|\mathbf{u}|^2 + |\mathbf{v}|^2 - d^2) \\
&= \tfrac{1}{2}\{u_x^2 + u_y^2 + v_x^2 + v_y^2 - [(v_x - u_x)^2 + (v_y - u_y)^2]\} \\
&= \tfrac{1}{2}[u_x^2 + u_y^2 + v_x^2 + v_y^2 \\
&\quad - (v_x^2 - 2u_xv_x + u_x^2 + v_y^2 - 2u_yv_y + u_y^2)] \\
&= \tfrac{1}{2}[2(u_xv_x + u_yv_y)] \\
&= u_xv_x + u_yv_y
\end{aligned}$$

Since $|\mathbf{u}||\mathbf{v}| \cos \theta = \mathbf{u} \cdot \mathbf{v}$, we get $\mathbf{u} \cdot \mathbf{v} = u_xv_x + u_yv_y$.

EXAMPLES

1 Compute each of the following inner products.

a) $\mathbf{i} \cdot \mathbf{i}$ b) $\mathbf{i} \cdot \mathbf{j}$ c) $\mathbf{j} \cdot \mathbf{j}$

SOLUTION

a) $\mathbf{i} \cdot \mathbf{i} = (1\mathbf{i} + 0\mathbf{j}) \cdot (1\mathbf{i} + 0\mathbf{j}) = 1 \cdot 1 + 0 \cdot 0 = 1$

b) $\mathbf{i} \cdot \mathbf{j} = (1\mathbf{i} + 0\mathbf{j}) \cdot (0\mathbf{i} + 1\mathbf{j}) = 0 \cdot 1 + 0 \cdot 1 = 0$; consequently, $\mathbf{i}$ and $\mathbf{j}$ are orthogonal.

c) $\mathbf{j} \cdot \mathbf{j} = (0\mathbf{i} + 1\mathbf{j}) \cdot (0\mathbf{i} + 1\mathbf{j}) = 0 \cdot 0 + 1 \cdot 1 = 1$

2 Verify that $u_x = \mathbf{u} \cdot \mathbf{i}$ and $u_y = \mathbf{u} \cdot \mathbf{j}$ for $\mathbf{u} = u_x\mathbf{i} + u_y\mathbf{j}$.

SOLUTION

$$\mathbf{u} \cdot \mathbf{i} = (u_x\mathbf{i} + u_y\mathbf{j}) \cdot (\mathbf{i} + 0\mathbf{j}) = u_x + 0 = u_x$$

and

$$\mathbf{u} \cdot \mathbf{j} = (u_x\mathbf{i} + u_y\mathbf{j}) \cdot (0\mathbf{i} + \mathbf{j}) = 0 + u_y = u_y$$

3 Determine the following inner products, $\mathbf{u} \cdot \mathbf{v}$ in each case.

a) $\mathbf{u} = \langle 4, -3 \rangle$ and $\mathbf{v} = \langle 2, 6 \rangle$

b) $\mathbf{u} = \langle -5, 6 \rangle$ and $\mathbf{v} = \langle -1, -1 \rangle$

c) $\mathbf{u} = \langle -3, -3 \rangle$ and $\mathbf{v} = \langle 8, 2 \rangle$

d) $\mathbf{u} = \langle 7, 0 \rangle$ and $\mathbf{v} = \langle 0, 7 \rangle$

SOLUTION

a) $\mathbf{u} \cdot \mathbf{v} = \langle 4, -3 \rangle \cdot \langle 2, 6 \rangle = (4)(2) + (-3)(6) = 8 - 18 = -10$

b) $\mathbf{u} \cdot \mathbf{v} = \langle -5, 6 \rangle \cdot \langle -1, -1 \rangle = (-5)(-1) + (6)(-1) = 5 - 6 = -1$

c) $\mathbf{u} \cdot \mathbf{v} = \langle -3, -3 \rangle \cdot \langle 8, 2 \rangle = (-3)(8) + (-3)(2) = -24 - 6 = -30$

d) $\mathbf{u} \cdot \mathbf{v} = \langle 7, 0 \rangle \cdot \langle 0, 7 \rangle = (7)(0) + (0)(7) = 0 + 0 = 0$

4 Find the angle between the vectors $\mathbf{u} = 4\mathbf{i} + 3\mathbf{j}$ and $\mathbf{v} = -2\mathbf{i} + 2\mathbf{j}$ to the nearest degree.

SOLUTION

$$\mathbf{u} \cdot \mathbf{v} = |\mathbf{u}||\mathbf{v}| \cos \theta$$

so that

$$\cos \theta = \frac{\mathbf{u} \cdot \mathbf{v}}{|\mathbf{u}||\mathbf{v}|} = \frac{\langle 4, 3 \rangle \cdot \langle -2, 2 \rangle}{\sqrt{4^2 + 3^2}\sqrt{(-2)^2 + 2^2}}$$
$$= \frac{(4)(-2) + (3)(2)}{(5)(2\sqrt{2})}$$
$$= -\frac{1}{5\sqrt{2}} = -0.1414$$

Hence, $\theta = 98°$ approximately (Figure 4).

Figure 4

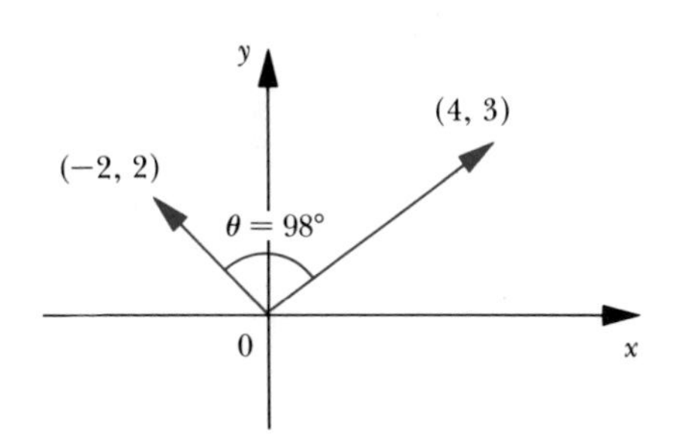

5 Let $\mathbf{u} = 8\mathbf{i} - 6\mathbf{j}$ and $\mathbf{v} = 3\mathbf{i} + 4\mathbf{j}$. Show that $\mathbf{u}$ and $\mathbf{v}$ are orthogonal.

SOLUTION

$$\begin{aligned}\mathbf{u} \cdot \mathbf{v} &= (8\mathbf{i} - 6\mathbf{j}) \cdot (3\mathbf{i} + 4\mathbf{j}) \\ &= (8)(3) + (-6)(4) \\ &= 24 - 24 \\ &= 0\end{aligned}$$

Hence, $\mathbf{u}$ and $\mathbf{v}$ are orthogonal by Theorem 1.

6 Let $\mathbf{u} = \langle -4, 2\rangle$, $\mathbf{v} = \langle 4, -3\rangle$, and $\mathbf{w} = \langle -2, 1\rangle$. Verify each of the following equations.

a) $\mathbf{u} \cdot \mathbf{u} = |\mathbf{u}|^2$

b) $\mathbf{u} \cdot (\mathbf{v} + \mathbf{w}) = \mathbf{u} \cdot \mathbf{v} + \mathbf{u} \cdot \mathbf{w}$

c) $(5\mathbf{u}) \cdot \mathbf{v} = 5(\mathbf{u} \cdot \mathbf{v})$

d) $\mathbf{u} \cdot (3\mathbf{u} + 2\mathbf{w}) = 3(\mathbf{u} \cdot \mathbf{u}) + 2(\mathbf{u} \cdot \mathbf{w})$

SOLUTION

a) Since

$$\mathbf{u} \cdot \mathbf{u} = \langle -4, 2\rangle \cdot \langle -4, 2\rangle = (-4)(-4) + (2)(2) = 16 + 4 = 20$$

and

$$|\mathbf{u}|^2 = (\sqrt{(-4)^2 + 2^2})^2 = 16 + 4 = 20$$

we have

$$\mathbf{u} \cdot \mathbf{u} = |\mathbf{u}|^2$$

b)
$$\begin{aligned}\mathbf{u} \cdot (\mathbf{v} + \mathbf{w}) &= \langle -4, 2\rangle \cdot (\langle 4, -3\rangle + \langle -2, 1\rangle)\\ &= \langle -4, 2\rangle \cdot \langle 2, -2\rangle\\ &= -8 - 4 = -12\end{aligned}$$

$$\begin{aligned}\mathbf{u} \cdot \mathbf{v} + \mathbf{u} \cdot \mathbf{w} &= \langle -4, 2\rangle \cdot \langle 4, -3\rangle + \langle -4, 2\rangle \cdot \langle -2, 1\rangle\\ &= (-16 - 6) + (8 + 2)\\ &= -22 + 10 = -12\end{aligned}$$

Hence,

$$\mathbf{u} \cdot (\mathbf{v} + \mathbf{w}) = \mathbf{u} \cdot \mathbf{v} + \mathbf{u} \cdot \mathbf{w}$$

c)
$$\begin{aligned}(5\mathbf{u}) \cdot \mathbf{v} &= (5\langle -4, 2\rangle) \cdot \langle 4, -3\rangle\\ &= \langle -20, 10\rangle \cdot \langle 4, -3\rangle\\ &= -80 - 30\\ &= -110\end{aligned}$$

$$\begin{aligned}5(\mathbf{u} \cdot \mathbf{v}) &= 5(\langle -4, 2\rangle \cdot \langle 4, -3\rangle)\\ &= 5(-16 - 6)\\ &= 5(-22)\\ &= -110\end{aligned}$$

Hence,

$$(5\mathbf{u}) \cdot \mathbf{v} = 5(\mathbf{u} \cdot \mathbf{v})$$

d)
$$\begin{aligned}\mathbf{u} \cdot (3\mathbf{u} + 2\mathbf{w}) &= \langle -4, 2\rangle \cdot (3\langle -4, 2\rangle + 2\langle -2, 1\rangle)\\ &= \langle -4, 2\rangle \cdot (\langle -12, 6\rangle + \langle -4, 2\rangle)\\ &= \langle -4, 2\rangle \cdot \langle -16, 8\rangle\\ &= 64 + 16\\ &= 80\end{aligned}$$

$$\begin{aligned}3(\mathbf{u} \cdot \mathbf{u}) + 2(\mathbf{u} \cdot \mathbf{w}) &= 3(\langle -4, 2\rangle \cdot \langle -4, 2\rangle)\\ &\quad + 2(\langle -4, 2\rangle \cdot \langle -2, 1\rangle)\\ &= 3(16 + 4) + 2(8 + 2)\\ &= 3(20) + 2(10)\\ &= 60 + 20\\ &= 80\end{aligned}$$

Hence,

$$\mathbf{u} \cdot (3\mathbf{u} + 2\mathbf{w}) = 3(\mathbf{u} \cdot \mathbf{u}) + 2(\mathbf{u} \cdot \mathbf{w})$$

PROBLEM SET 3

1 Find $\mathbf{u} \cdot \mathbf{v}$ if $|\mathbf{u}| = 4$ and $|\mathbf{v}| = 5$ and

a) $\theta = 60°$ b) $\theta = 135°$
c) $\theta = 150°$ d) $\theta = 120°$

2 Find the scalar projections of **u** onto **v** in Problem 1.

3 Find $\mathbf{u} \cdot \mathbf{v}$ and the angle between **u** and **v** if

a) $\mathbf{u} = \langle 1, 1\rangle$ and $\mathbf{v} = \langle 1, -1\rangle$ b) $\mathbf{u} = \langle 2, 3\rangle$ and $\mathbf{v} = \langle 4, -5\rangle$
c) $\mathbf{u} = \langle 7, 0\rangle$ and $\mathbf{v} = \langle 0, 8\rangle$ d) $\mathbf{u} = \langle 1, 2\rangle$ and $\mathbf{v} = \langle 2, -3\rangle$
e) $\mathbf{u} = \langle 4, 1\rangle$ and $\mathbf{v} = \langle 1, -2\rangle$ f) $\mathbf{u} = \langle -4, 2\rangle$ and $\mathbf{v} = \langle 5, 8\rangle$

4 Find the scalar projections of **u** onto **v** in Problem 3.

5 In each of the following, find x so that **u** and **v** are orthogonal if $\mathbf{u} = 3\mathbf{i} + 4\mathbf{j}$ and

a) $\mathbf{v} = x\mathbf{i} + 4\mathbf{j}$ b) $\mathbf{v} = x\mathbf{i} - 4\mathbf{j}$ c) $\mathbf{v} = 4\mathbf{i} + x\mathbf{j}$

6 Use vectors to show that $\triangle ABC$ is a right triangle and identify the right angle if $A = (-2, 1)$, $B = (4, -4)$, and $C = (-1, -10)$.

7 Suppose that **u** and **v** are orthogonal vectors in the plane. Express the following values in terms of $|\mathbf{u}|$ and $|\mathbf{v}|$.

a) $|\mathbf{u} + \mathbf{v}|$ b) $|\mathbf{u} - \mathbf{v}|$ c) $|3\mathbf{u} - 4\mathbf{v}|$
d) $|-3\mathbf{u} + 4\mathbf{v}|$ e) $\left|\dfrac{\mathbf{u}}{|\mathbf{u}|}\right|$

8 Let **u** and **v** be two vectors in the plane. Show that

a) $|\mathbf{u} + \mathbf{v}|^2 = |\mathbf{u}|^2 + 2\mathbf{u} \cdot \mathbf{v} + |\mathbf{v}|^2$
b) $|\mathbf{u} - \mathbf{v}|^2 = |\mathbf{u}|^2 - 2\mathbf{u} \cdot \mathbf{v} + |\mathbf{v}|^2$
c) $|\mathbf{u} + \mathbf{v}|^2 + |\mathbf{u} - \mathbf{v}|^2 = 2|\mathbf{u}|^2 + 2|\mathbf{v}|^2$

9 Let **u** and **v** be unit position vectors and θ the measure of the angle between them, $0 \leqslant \theta \leqslant \pi$. Show that $\frac{1}{2}|\mathbf{v} - \mathbf{u}| = \sin(\theta/2)$.

10 Let $\mathbf{u} = -1\mathbf{i} + 3\mathbf{j}$, $\mathbf{v} = 3\mathbf{i} + 5\mathbf{j}$ and $\mathbf{w} = -2\mathbf{i} - 3\mathbf{j}$. Use **u**, **v**, and **w** to verify each of the following assertions.

a) $\mathbf{v} \cdot \mathbf{v} = |\mathbf{v}|^2$
b) $\mathbf{w} \cdot \mathbf{w} = |\mathbf{w}|^2$
c) $\mathbf{u} \cdot (\mathbf{v} + \mathbf{w}) = \mathbf{u} \cdot \mathbf{v} + \mathbf{u} \cdot \mathbf{w}$
d) $(2\mathbf{u}) \cdot (3\mathbf{v}) = 6(\mathbf{u} \cdot \mathbf{v})$
e) $\mathbf{u} \cdot (4\mathbf{v} + 2\mathbf{w}) = 2\mathbf{u} \cdot 2\mathbf{v} + 2\mathbf{u} \cdot \mathbf{w}$

11 Show that $\mathbf{u} \cdot \mathbf{u} = |\mathbf{u}|^2$ for any vector **u**.

12 Show that $\mathbf{u} \cdot (\mathbf{v} + \mathbf{w}) = \mathbf{u} \cdot \mathbf{v} + \mathbf{u} \cdot \mathbf{w}$ for all vectors **u**, **v**, and **w**.

5 Vector Applications

In this section we will consider examples of applications of vectors to solving problems in geometry and trigonometry.

5.1 Inner Product Applications

EXAMPLES

1 Prove that the diagonals of a rectangle are equal.

PROOF. Let $\mathbf{AB} = \mathbf{u}$ and $\mathbf{AD} = \mathbf{v}$; for the rectangle *ADCB*. Then the two diagonals represent the vectors $\mathbf{u} + \mathbf{v}$ and $\mathbf{u} - \mathbf{v}$ (Figure 1).

Figure 1

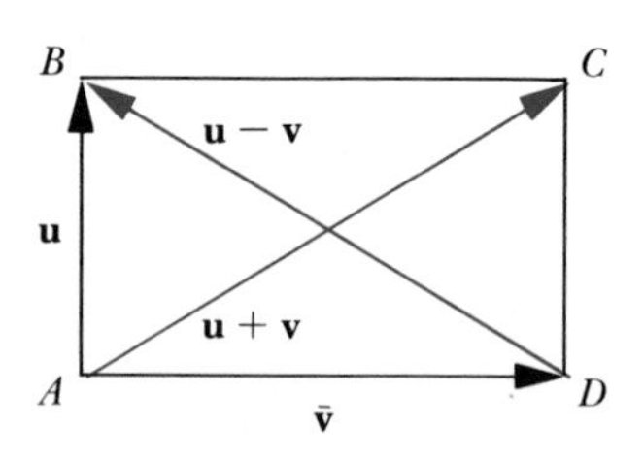

The equality of the diagonals means that $|\mathbf{u} + \mathbf{v}| = |\mathbf{u} - \mathbf{v}|$. This is equivalent to

$$|\mathbf{u} + \mathbf{v}|^2 = |\mathbf{u} - \mathbf{v}|^2$$

or

$$(\mathbf{u} + \mathbf{v}) \cdot (\mathbf{u} + \mathbf{v}) = (\mathbf{u} - \mathbf{v}) \cdot (\mathbf{u} - \mathbf{v})$$

or

$$\mathbf{u} \cdot \mathbf{u} + \mathbf{u} \cdot \mathbf{v} + \mathbf{v} \cdot \mathbf{u} + \mathbf{v} \cdot \mathbf{v} = \mathbf{u} \cdot \mathbf{u} - \mathbf{u} \cdot \mathbf{v} - \mathbf{v} \cdot \mathbf{u} + \mathbf{v} \cdot \mathbf{v}$$
$$|\mathbf{u}|^2 + 2\mathbf{u} \cdot \mathbf{v} + |\mathbf{v}|^2 = |\mathbf{u}|^2 - 2\mathbf{u} \cdot \mathbf{v} + |\mathbf{v}|^2$$

or

$$2\mathbf{u} \cdot \mathbf{v} = -2\mathbf{u} \cdot \mathbf{v}$$

or

$$4\mathbf{u} \cdot \mathbf{v} = 0$$

or

$$\mathbf{u} \cdot \mathbf{v} = 0$$

But since $\mathbf{u}$ and $\mathbf{v}$ are perpendicular, $\mathbf{u} \cdot \mathbf{v} = 0$, so that if we reverse the steps we get $|\mathbf{u} + \mathbf{v}| = |\mathbf{u} - \mathbf{v}|$.

2 Prove that the diagonals of a rhombus are perpendicular to each other.

PROOF. Let $\mathbf{AB} = \mathbf{u}$ and $\mathbf{AD} = \mathbf{v}$; for rhombus $ADCB$. Then the two diagonals represent the vectors $\mathbf{u} + \mathbf{v}$ and $\mathbf{u} - \mathbf{v}$ (Figure 2).

Figure 2

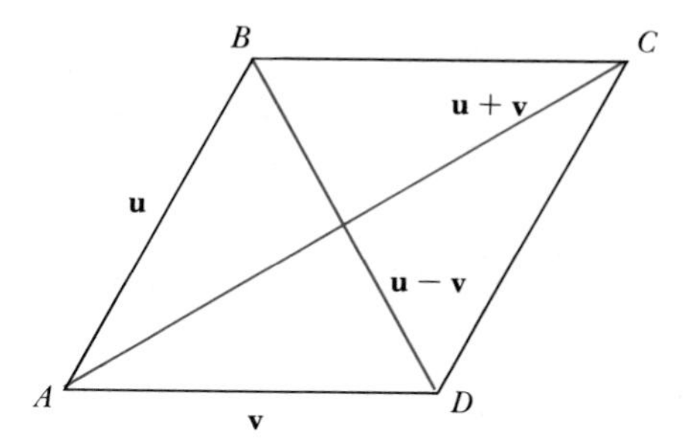

In order to show that the diagonals are perpendicular, we have to show that

$$(\mathbf{u} + \mathbf{v}) \cdot (\mathbf{u} - \mathbf{v}) = 0$$

Since

$$\begin{aligned}(\mathbf{u} + \mathbf{v}) \cdot (\mathbf{u} - \mathbf{v}) &= \mathbf{u} \cdot \mathbf{u} - \mathbf{u} \cdot \mathbf{v} + \mathbf{v} \cdot \mathbf{u} - \mathbf{v} \cdot \mathbf{v} \\ &= |\mathbf{u}|^2 - |\mathbf{v}|^2\end{aligned}$$

we must show that

$$|\mathbf{u}|^2 - |\mathbf{v}|^2 = 0$$

But $|\mathbf{u}| = |\mathbf{v}|$, since $ADCB$ is a rhombus. Hence,

$$|\mathbf{u}|^2 = |\mathbf{v}|^2 \qquad \text{or} \qquad |\mathbf{u}|^2 - |\mathbf{v}|^2 = 0$$

so that

$$(\mathbf{u} + \mathbf{v}) \cdot (\mathbf{u} - \mathbf{v}) = 0$$

Thus the diagonals are perpendicular.

3 Let $\mathbf{u}$ and $\mathbf{v}$ be unit radius vectors that make angles α and β with the positive x axis, respectively. Use the inner products to prove the identity $\cos(\alpha - \beta) = \cos\alpha \cos\beta + \sin\alpha \sin\beta$.

PROOF. (See Figure 3.) Since

Figure 3

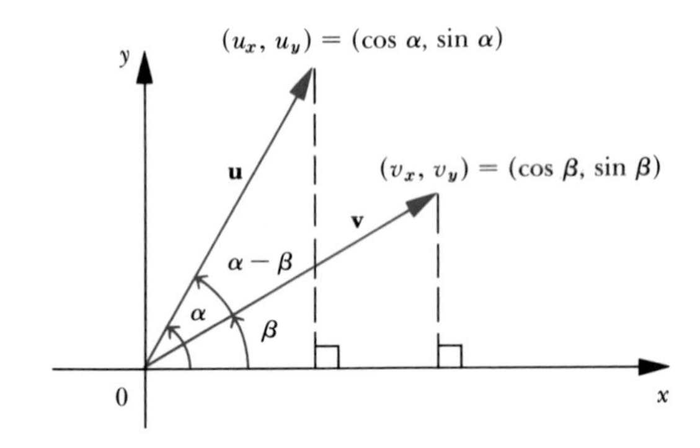

$$\cos\alpha = \frac{u_x}{|\mathbf{u}|} = \frac{u_x}{1} \qquad \text{and} \qquad \sin\alpha = \frac{u_y}{|\mathbf{u}|} = \frac{u_y}{1}$$

u can be written as

$$\mathbf{u} = u_x\mathbf{i} + u_y\mathbf{j} = \cos\alpha\mathbf{i} + \sin\alpha\mathbf{j} \qquad \text{(Figure 3)}$$

Similarly, **v** can be written as

$$\mathbf{v} = v_x\mathbf{i} + v_y\mathbf{j} = \cos\beta\mathbf{i} + \sin\beta\mathbf{j} \qquad \text{(Figure 3)}$$

Hence,

$$\mathbf{u}\cdot\mathbf{v} = \cos\alpha\cos\beta + \sin\alpha\sin\beta$$

Also

$$\begin{aligned}\mathbf{u}\cdot\mathbf{v} &= |\mathbf{u}||\mathbf{v}|\cos(\alpha-\beta)\\ &= \cos(\alpha-\beta)\end{aligned}$$

since

$$|\mathbf{u}| = |\mathbf{v}| = 1$$

Therefore,

$$\cos(\alpha-\beta) = \cos\alpha\cos\beta + \sin\alpha\sin\beta$$

5.2 Rotation

A *rotation* of the plane about the origin is a function that maps each position vector into a unique position vector. Such a rotation is completely defined by an angle called the *angle of rotation.* For example, if the angle of rotation is 90°, then f maps position vectors into other position vectors by a "90° counterclockwise turn" about the origin.

In this case $f(\mathbf{u}) = \mathbf{v}$, where $\mathbf{u}$ and $\mathbf{v}$ determine a 90° angle, and $|\mathbf{u}| = |\mathbf{v}|$ (Figure 4).

Figure 4

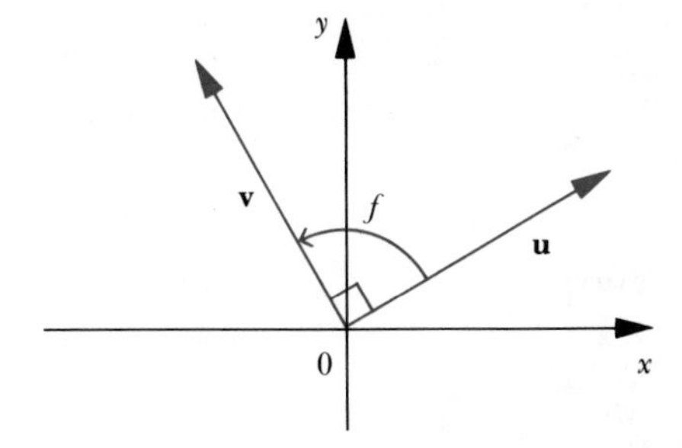

Thus, if f is a 90° rotation, then $f(\mathbf{i}) = \mathbf{j}$, $f(\mathbf{j}) = -\mathbf{i}$, $f(-\mathbf{i}) = -\mathbf{j}$, and $f(-\mathbf{j}) = \mathbf{i}$ (Figure 5).

Figure 5

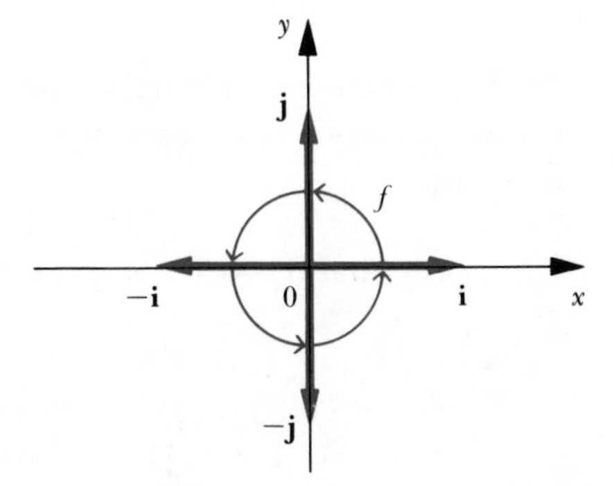

We now consider two important properties of rotations. Let f be a rotation through an angle θ such that $f(\mathbf{u}) = \mathbf{v}$; then the following properties hold.

1 Let $\mathbf{u}$ be a position vector; the rotation of $a\mathbf{u}$ by f (Figure 6*b*) is equivalent to the rotation of $\mathbf{u}$ by f with the result multiplied by a (Figure 6*a*); that is, $f(a\mathbf{u}) = af(\mathbf{u})$.

Figure 6

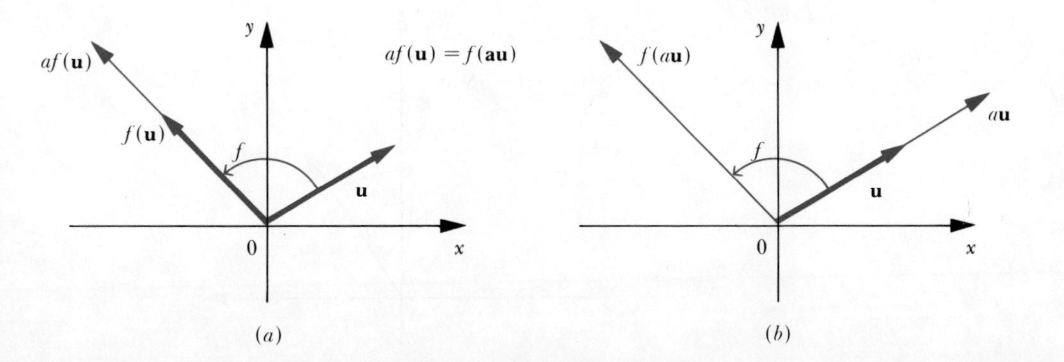

2 If f is a rotation, **u** and **v** are vectors; then $f(\mathbf{u}+\mathbf{v}) = f(\mathbf{u}) + f(\mathbf{v})$ (Figure 7).

Figure 7

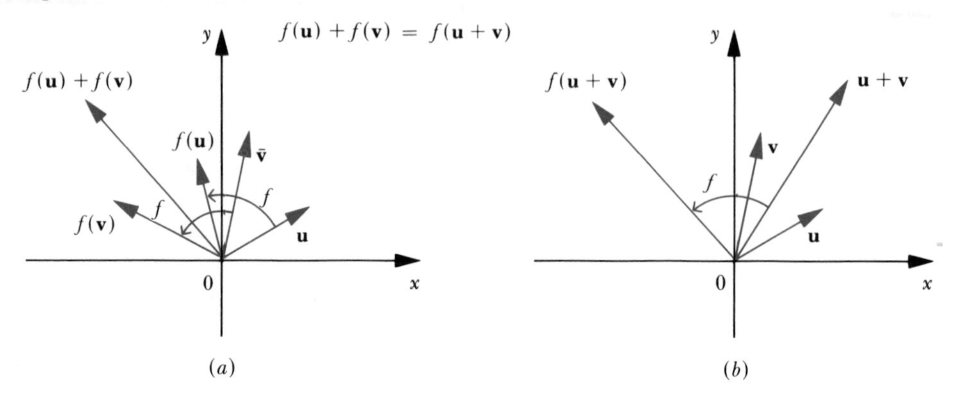

If f is a function satisfying properties 1 and 2, then f is called a *linear transformation* or *linear operation.* Thus, the rotation is a linear transformation.

Suppose that $\mathbf{u} = u_x\mathbf{i} + u_y\mathbf{j}$; then

$$\begin{aligned} f(\mathbf{u}) &= f(u_x\mathbf{i} + u_y\mathbf{j}) \\ &= f(u_x\mathbf{i}) + f(u_y\mathbf{j}) \qquad \text{Property 2} \\ &= u_x f(\mathbf{i}) + u_y f(\mathbf{j}) \qquad \text{Property 1} \end{aligned}$$

In other words, the image of any vector **u** under a rotation f can be determined once we know the images of the basis vectors **i** and **j** under f. For example, if f is a rotation such that $f(\mathbf{i}) = \frac{1}{2}\mathbf{i} + (\sqrt{3}/2)\mathbf{j}$ and $f(\mathbf{j}) = -\frac{1}{2}\mathbf{i} + (\sqrt{3}/2)\mathbf{j}$, then

$$\begin{aligned} f(4\mathbf{i}+6\mathbf{j}) = f(4\mathbf{i}) + f(6\mathbf{j}) &= 4f(\mathbf{i}) + 6(f(\mathbf{j})) \\ &= 4\left(\frac{1}{2}\mathbf{i} + \frac{\sqrt{3}}{2}\mathbf{j}\right) + 6\left(-\frac{1}{2}\mathbf{i} + \frac{\sqrt{3}}{2}\mathbf{j}\right) \\ &= -\mathbf{i} + 5\sqrt{3}\,\mathbf{j} \qquad \text{(Figure 8)} \end{aligned}$$

Figure 8

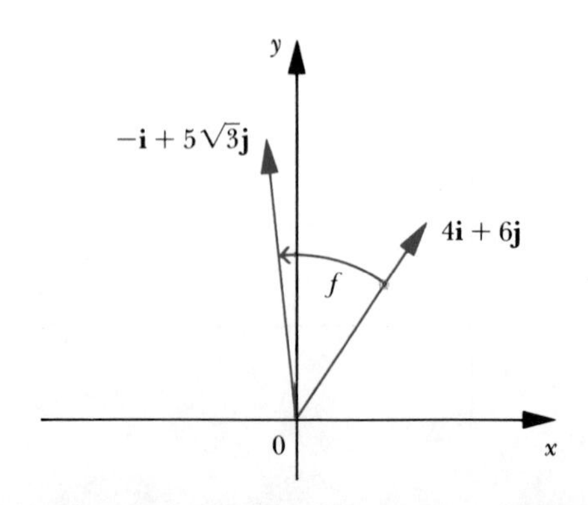

Suppose that g is a rotation function through an angle α and f is a rotation function through an angle β; then, applying the mapping f and g consecutively in any order, yields the rotation function through angle $\alpha + \beta$. Thus, if $f(\mathbf{i}) = x\mathbf{i} + y\mathbf{j}$, then

$$g(f(\mathbf{i})) = g(x\mathbf{i} + y\mathbf{j}) = g(x\mathbf{i}) + g(y\mathbf{j}) = xg(\mathbf{i}) + yg(\mathbf{j})$$

EXAMPLE

Use rotations of vectors to prove the identities

$$\cos(t + s) = \cos t \cos s - \sin t \sin s$$

and

$$\sin(t + s) = \sin t \cos s + \cos t \sin s$$

where s and t are real numbers.

PROOF. Let f be a rotation function which maps the vector $\mathbf{i}$ into the vector $x\mathbf{i} + y\mathbf{j}$ through an angle of radian measure t (Figure 9*a*). Using the definition of the circular functions, we have $x = \cos t$ and $y = \sin t$. Thus,

$$f(\mathbf{i}) = x\mathbf{i} + y\mathbf{j} = (\cos t)\mathbf{i} + (\sin t)\mathbf{j}$$

Figure 9

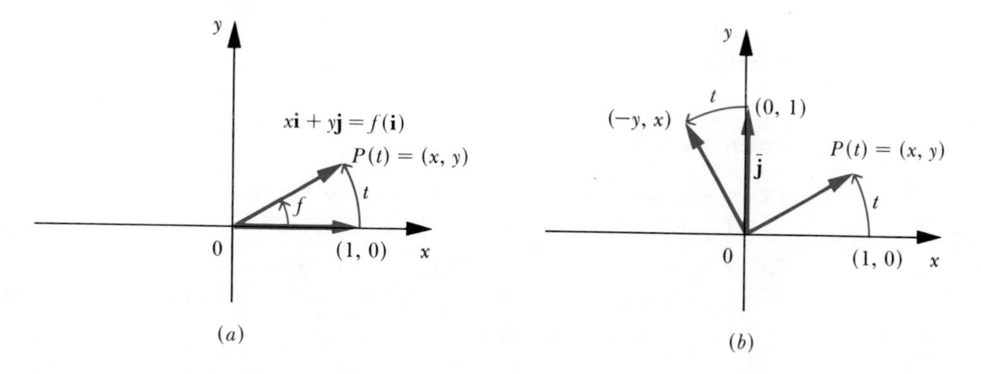

By symmetry (Figure 9*b*), we have

$$f(\mathbf{j}) = -y\mathbf{i} + x\mathbf{j} = (-\sin t)\mathbf{i} + (\cos t)\mathbf{j}$$

Moreover, if g is a rotation function through s radians, then

$$g(\mathbf{i}) = (\cos s)\mathbf{i} + (\sin s)\mathbf{j}$$

and

$$g(\mathbf{j}) = (-\sin s)\mathbf{i} + (\cos s)\mathbf{j}$$

The rotation of f (angle t), followed by g (angle s), is the rotation through angle $t + s$. Hence,

$$\begin{aligned}(1)\quad g[f(\mathbf{i})] &= g[x\mathbf{i} + y\mathbf{i}]\\ &= xg(\mathbf{i}) + yg(\mathbf{j})\\ &= x[(\cos s)\mathbf{i} + (\sin s)\mathbf{j}] + y[(-\sin s)\mathbf{i} + (\cos s)\mathbf{j}]\\ &= (\cos t)[(\cos s)\mathbf{i} + (\sin s)\mathbf{j}]\\ &\quad + (\sin t)[(-\sin s)\mathbf{i} + (\cos s)\mathbf{j}]\\ &= (\cos t \cos s - \sin t \sin s)\mathbf{i} + (\cos t \sin s + \sin t \cos s)\mathbf{j}\end{aligned}$$

Furthermore, the rotation of f followed by g can also be considered as a single rotation through $(t + s)$ radians, so that

$$(2)\quad g[f(\mathbf{i})] = \cos (t + s)\mathbf{i} + \sin (t + s)\mathbf{j}$$

Since two vectors are equal if and only if their components are equal, we have from (1) and (2) that

$$\cos (t + s) = \cos t \cos s - \sin t \sin s$$

and

$$\sin (t + s) = \cos t \sin s + \sin t \cos s$$

PROBLEM SET 4

1 Let f be a rotation function specified by an angle of 60°.
a) Find $f(\mathbf{i})$ by the use of trigonometry.
b) Use symmetry to find $f(\mathbf{j})$ from part (a).
c) Use Properties 1 and 2 in Section 5.2 to find $f(3\mathbf{i})$, $f(-5\mathbf{j})$, $f(3\mathbf{i} - 5\mathbf{j})$, and $f(6\mathbf{i} + 8\mathbf{j})$.

2 Do Problem 1 if f is a rotation function specified by an angle of 30°.

3 Let f be a rotation function through a positive angle less than 360° such that $f(\mathbf{i}) = \frac{1}{2}\mathbf{i} + (\sqrt{3}/2)\mathbf{j}$.
a) What is the angle of rotation of f?
b) What is $f(\mathbf{j})$?
c) If g is a rotation through 30°, describe g followed by f.

4 Assume that f is a rotation function such that $f(\mathbf{i}) = (\sqrt{3}/2)\mathbf{i} + \frac{1}{2}\mathbf{j}$ and $f(\mathbf{j}) = -\frac{1}{2}\mathbf{i} + (\sqrt{3}/2)\mathbf{j}$. Use $\mathbf{u} = 3\mathbf{i} + 2\mathbf{j}$ and $\mathbf{v} = -2\mathbf{i} + 5\mathbf{j}$ to illustrate each of the following statements.

a) $f(\mathbf{u} + \mathbf{v}) = f(\mathbf{u}) + f(\mathbf{v})$
b) $f(5\mathbf{u}) = 5f(\mathbf{u})$
c) $f(6\mathbf{u} + 3\mathbf{v}) = 6f(\mathbf{u}) + 3f(\mathbf{v})$

5 Use vectors to derive the formula for the distance between the points (x_1, y_1) and (x_2, y_2).

6 Use vectors to prove that the altitudes of a triangle are concurrent.

7 If $\mathbf{u} = \langle 3, -6 \rangle$ and $\mathbf{v} = \langle 4, 2 \rangle$ are the vectors representing the sides of a right triangle, prove that the midpoint of the hypotenuse is equidistant from the vertices.

8 Use vectors to prove that the median to the base of an isosceles triangle is perpendicular to the base.

REVIEW PROBLEM SET

1 Let A, B, and C be three points in the plane. Find

a) $2\mathbf{AB} + 2\mathbf{BC} + \mathbf{CA}$
b) $3\mathbf{AB} - 3\mathbf{CB}$
c) $-\mathbf{AC} - \mathbf{CB}$
d) $6\mathbf{AB} + 4\mathbf{CA} + 6\mathbf{BC}$

2 Determine $\mathbf{u}$ in each of the following cases if θ is the angle that $\mathbf{u}$ makes with the positive x axis.

a) $|\mathbf{u}| = 10$ and $\theta = 150°$
b) $|\mathbf{u}| = 6$ and $\theta = 60°$
c) $|\mathbf{u}| = 8$ and $\theta = 135°$

3 Let $\mathbf{u} = \langle 3, -4 \rangle$, $\mathbf{v} = \langle -2, 4 \rangle$, and $\mathbf{w} = \langle 7, 8 \rangle$. Determine each of the following vectors.

a) $\mathbf{u} + \mathbf{v}$
b) $\mathbf{u} - \mathbf{v}$
c) $2\mathbf{u} + \mathbf{v}$
d) $\mathbf{u} + (\mathbf{v} + \mathbf{w})$
e) $2\mathbf{u} - 3\mathbf{w}$
f) $\mathbf{v} - \mathbf{u} - \mathbf{w}$

4 Write $\mathbf{u}$ in the form $\mathbf{u} = u_x\mathbf{i} + u_y\mathbf{j}$ and find $|\mathbf{u}|$ if $\mathbf{u}$ is a vector whose initial point is the first point and whose terminal point is the second point.

a) $(-6, 8)$, $(4, 3)$
b) $(-7, 6)$, $(3, -1)$
c) $(6, -1)$, $(5, 3)$
d) $(7, 1)$, $(-3, 5)$
e) $(1, -3)$, $(5, -1)$
f) $(1, 8)$, $(2, 10)$

5 Let $\mathbf{u} = \langle 1, -1 \rangle$, $\mathbf{v} = \langle 4, 3 \rangle$, and $\mathbf{w} = \langle 6, -4 \rangle$. Find each of the following expressions.

a) $|\mathbf{u} + \mathbf{v}|$
b) $|\mathbf{u} - \mathbf{w}|$

c) $|\mathbf{v} + \mathbf{w}|$
d) $|\mathbf{u}| + |\mathbf{v}|$
e) $|\mathbf{u} + \mathbf{w}|$
f) $|\mathbf{v}| + |\mathbf{w}|$
g) $|\mathbf{u} - \mathbf{v}|^2$
h) $\mathbf{u}/|\mathbf{u}|$
i) $|3\mathbf{u} + \mathbf{v} - \mathbf{w}|^2$
j) $|\mathbf{u}|^2 + 2|\mathbf{v}|^2 + 3|\mathbf{w}|^2$

6 Let $\mathbf{u} = \langle -3, 2\rangle$, $\mathbf{v} = \langle 4, 3\rangle$, and $\mathbf{w} = \langle 5, 1\rangle$. Determine $\mathbf{z}$ in each of the following cases.
a) $\mathbf{u} + \mathbf{v} = \mathbf{w} - 2\mathbf{z}$
b) $3(\mathbf{u} + \mathbf{v}) = 5\mathbf{w} + \mathbf{z}$
c) $5\mathbf{z} - \mathbf{v} = 3\mathbf{z} - 2\mathbf{w}$
d) $\mathbf{z} + 3(\mathbf{z} + \mathbf{u}) = \mathbf{z} - \mathbf{w} + 3\mathbf{v}$

7 Let $\mathbf{u} = \langle -3, 4\rangle$, $\mathbf{v} = \langle 5, 4\rangle$, and $\mathbf{w} = \langle 1, 7\rangle$. Find
a) $\mathbf{u} \cdot \mathbf{v}$
b) $3\mathbf{u} \cdot 4\mathbf{v}$
c) $3\mathbf{u} \cdot (\mathbf{v} + \mathbf{w})$
d) $3\mathbf{v} \cdot (5\mathbf{u} - \mathbf{v})$
e) $(\mathbf{u} - \mathbf{v}) \cdot (\mathbf{u} + \mathbf{v})$
f) $(\mathbf{u} - \mathbf{w}) \cdot (\mathbf{w} + \mathbf{u} + \mathbf{v})$
g) $(\mathbf{u} + \mathbf{v} + \mathbf{w}) \cdot (\mathbf{u} + \mathbf{v} - \mathbf{w})$
h) $\mathbf{v} \cdot (3\mathbf{u} - 2\mathbf{v} + 7\mathbf{w})$

8 Let $\mathbf{u}$ and $\mathbf{v}$ be vectors in the plane such that $|\mathbf{u}| = a$ and $|\mathbf{v}| = b$. Show that
a) $(a\mathbf{v} + b\mathbf{u}) \cdot (a\mathbf{v} - b\mathbf{u}) = 0$
b) $(\mathbf{u} + \mathbf{v}) \cdot (\mathbf{u} - \mathbf{v}) = a^2 - b^2$

9 Find the angle between $\mathbf{u}$ and $\mathbf{v}$ if $|\mathbf{u}| = 2$ and $|\mathbf{v}| = 5$ and $\mathbf{u} \cdot \mathbf{v}$ is
a) 0
b) 1
c) 4
d) -2

10 Find the scalar number k so that $\mathbf{u} = 3\mathbf{i} + 4\mathbf{j}$ is orthogonal to each of the following vectors.
a) $\mathbf{v} = k\mathbf{i} + 3\mathbf{j}$
b) $\mathbf{v} = 5k\mathbf{i} - 2\mathbf{j}$
c) $\mathbf{v} = 4\mathbf{i} - 3k\mathbf{j}$

11 Find a unit vector orthogonal to each of the following vectors.
a) $\langle -3, 4\rangle$
b) $4\mathbf{i} - 3\mathbf{j}$
c) $2\mathbf{i} - 7\mathbf{j}$

12 Write $f(\mathbf{i})$ in the form $x\mathbf{i} + y\mathbf{j}$ if f corresponds to a rotation through each of the following angles.
a) $5\pi/6$
b) $-\pi$
c) $-3\pi/4$
d) $3\pi/2$

13 Let $\mathbf{u} = \mathbf{OP}$, where $P = (\sqrt{2}/2, \sqrt{2}/2)$.
a) Write $\mathbf{u}$ in the form $x\mathbf{i} + y\mathbf{j}$.
b) If $\mathbf{u} = f(\mathbf{i})$, where f is a rotation, find the angle of rotation.
c) Find $f(\mathbf{j})$.
d) Use the results of parts (b) and (c) to determine $f(3\mathbf{i} - 5\mathbf{j})$.

14 Show that the line that joins one vertex of a parallelogram to the midpoint of an opposite side divides the diagonal in the ratio 2 to 1.

APPENDIX

APPENDIX A

Logarithms

1 Logarithms

Let us begin by considering the example $2^3 = 8$. In this example, 2 is the base, 3 is the exponent, and 8 is the result of "raising 2 to the power 3." Suppose that it were desirable to describe this example by highlighting the role of 3, the exponent in $2^3 = 8$. We could say that "3 is the exponent to which the base 2 must be raised in order to get 8."

Similarly, for $5^{-2} = \frac{1}{25}$, we could say that "-2 is the exponent to which the base 5 must be raised in order to get $\frac{1}{25}$"; for $100^{3/2} = 1{,}000$, "$\frac{3}{2}$ is the exponent to which the base 100 must be raised in order to get 1,000."

In general, if $b^y = x$, for $b > 0$, $b \neq 1$, "y is the exponent to which the base b must be raised in order to get x." In this context we refer to y as "the logarithm of x to the base b."

These statements are written more briefly as follows:

$\log_2 8 = 3$	is equivalent to	$2^3 = 8$
$\log_5 \frac{1}{25} = -2$	is equivalent to	$5^{-2} = \frac{1}{25}$
$\log_{100} 1000 = \frac{3}{2}$	is equivalent to	$100^{3/2} = 1000$

More formally, we have the following definition.

Definition

If $b > 0$, $b \neq 1$, then $y = \log_b x$ (reads "y equals the *logarithm* of x to the base b") is equivalent to $b^y = x$. b is called the *base* of the logarithm.

EXAMPLES

1 Write each of the following exponential statements as an equivalent logarithmic statement.

a) $3^2 = 9$ b) $10^0 = 1$ c) $\sqrt[3]{27} = 3$

SOLUTION. By the definition of logarithms we know that $b^y = x$ is equivalent to $y = \log_b x$; hence,

a) $3^2 = 9$ is equivalent to $2 = \log_3 9$
b) $10^0 = 1$ is equivalent to $0 = \log_{10} 1$
c) $\sqrt[3]{27} = 27^{1/3} = 3$ is equivalent to $\frac{1}{3} = \log_{27} 3$

2 Write each of the following logarithmic statements as an equivalent exponential statement.

a) $\log_{10} 10 = 1$ b) $\log_{1/2} 4 = -2$
c) $\log_{1/2} \frac{1}{4} = 2$

SOLUTION. By the definition of logarithms we know that $y = \log_b x$ is equivalent to $b^y = x$; hence,

a) $\log_{10} 10 = 1$ is equivalent to $10^1 = 10$
b) $\log_{1/2} 4 = -2$ is equivalent to $(\frac{1}{2})^{-2} = 4$ (Why?)
c) $\log_{1/2} \frac{1}{4} = 2$ is equivalent to $(\frac{1}{2})^2 = \frac{1}{4}$

3 Use the definition of logarithms to solve each of the following equations.

a) $\log_7 \frac{1}{49} = x$ b) $\log_x 9 = 2$
c) $\log_8 x = -\frac{2}{3}$

SOLUTION

a) $\log_7 \frac{1}{49} = x$ is equivalent to $7^x = \frac{1}{49} = 7^{-2}$, so that $x = -2$.
b) $\log_x 9 = 2$ is equivalent to $x^2 = 9$. Although the solution of this latter equation is $x = \pm 3$, we can only use $x = 3$ because, according to the definition of logarithms, the base of a logarithm must be a positive number.
c) $\log_8 x = -\frac{2}{3}$ is equivalent to $8^{-2/3} = x$, so that $x = \frac{1}{4}$. (Why?)

1.1 Properties of Logarithms

Since $y = \log_b x$ is equivalent to $b^y = x$, the properties of logarithms can be derived from the properties of exponents. Suppose that M, N, and b, $b \neq 1$, are positive real numbers and that r is any real number. Then,

i $b^{\log_b x} = x$

ii $\log_b (MN) = \log_b M + \log_b N$

iii $\log_b N^r = r \log_b N$

iv $\log_b (M/N) = \log_b M - \log_b N$

In order to solve logarithmic equations; we will make use of the following property: $\log_b M = \log_b N$ if and only if $M = N$.

EXAMPLES

1 Evaluate $\log_b b$, $\log_b 1$, and $\log_b b^p$.

SOLUTION. If $\log_b b = t$, then $b^t = b$, so $t = 1$. Hence, $\log_b b = 1$. Also, $\log_b 1 = t$ implies $b^t = 1$, so $t = 0$; that is, $\log_b 1 = 0$. Finally, $\log_b b^p = p \log_b b = p \cdot 1 = p$.

2 Let $\log_b 2 = 0.35$, $\log_b 3 = 0.55$, and $\log_b 5 = 0.82$. Use the properties of logarithms to find each of the following values.

a) $\log_b \frac{2}{3}$ b) $\dfrac{\log_b 2}{\log_b 3}$

c) $\log_b 2^3$ d) $(\log_b 2)^3$

e) $\log_b 24$ f) $\log_b \sqrt{\frac{2}{3}}$

g) $\log_b \dfrac{60}{b}$ h) $\log_b 0.6$

SOLUTIONS

a) $\log_b \frac{2}{3} = \log_b 2 - \log_b 3 = 0.35 - 0.55 = -0.20$

b) $\dfrac{\log_b 2}{\log_b 3} = \dfrac{0.35}{0.55} = \dfrac{7}{11} = 0.64$

c) $\log_b 2^3 = 3 \log_b 2 = 3(0.35) = 1.05$

d) $(\log_b 2)^3 = (0.35)^3 = 0.043$

c) $\log_b 24 = \log_b (2^3 \cdot 3) = \log_b 2^3 + \log_b 3$
$= 3 \log_b 2 + \log_b 3 = 3(0.35) + 0.55 = 1.60$

1.2 Common Logarithms

The two logarithmic bases used most often for purposes of computation are base 10 and base e, where e, an irrational number, is approximately equal to 2.718. Logarithms with base 10 are called *common logarithms.* By convention we usually do not write the base 10 when using logarithmic notation, and *log x* is the abbreviated way of writing $\log_{10} x$. Logarithms with base e are called *natural logarithms,* and the natural logarithm, $\log_e x$, is usually written *ln x*.

Logarithms—base 10

For certain values of x it is easy to determine $\log x$. For example, $\log 10 = 1$ and $\log 100 = \log 10^2 = 2$. (Why?) In general, $\log 10^n = n$. (Why?)

For values of x not expressible as powers of 10, other methods are required to determine the value of $\log x$. Let us consider two examples to see how far we could get in computing $\log x$.

Suppose that $x = 5{,}340$. Using scientific notation we can represent x as $5.34 \cdot 10^3$, so that

$$\log 5{,}340 = \log (5.34 \cdot 10^3)$$
$$= \log 5.34 + \log 10^3$$
$$= \log 5.34 + 3 \qquad \text{(Why?)}$$

Hence, determining log 5,340 has been reduced to finding log 5.34.

Suppose that $x = 0.000234$. Then

$$\log 0.000234 = \log (2.34 \cdot 10^{-4})$$
$$= \log 2.34 + \log 10^{-4} \qquad \text{(Why?)}$$
$$= \log 2.34. + (-4)$$

Here the problem is reduced to finding log 2.34.

We shall now generalize the procedure suggested by the two examples. For any positive number x that can be represented in scientific notation as $x = s \cdot 10^n$, where $1 \leqslant s < 10$ and n is an integer, we have $\log x = \log (s \cdot 10^n) = \log s + \log 10^n = \log s + n$, so $\log x = \log s + n$. This latter form is called the *standard form* of $\log x$, where $\log s$ is called the *mantissa* of $\log x$ and n is called the *characteristic* of $\log x$. Notice that since $y = \log x$ is an increasing function and $1 \leqslant s < 10$, it follows that $\log 1 \leqslant \log s < \log 10$; that is, $0 \leqslant \log s < 1$. In other words, the mantissa is always a number between 0 and 1, possibly equal to 0.

Hence, the task of determining the value of $\log x$ is reduced to determining $\log s$, where s is always between 1 and 10. However, the approximate values of $\log s$ can be determined from the *common log tables* (Table III in Appendix B).

EXAMPLES

In each of the following problems express the number in scientific notation and then determine the common logarithm. Indicate the characteristic and the mantissa.

1 53,900

SOLUTION

$$53{,}900 = (5.39)(10^4)$$

so that

$$\log 53{,}900 = \log 5.39 + 4$$

From Table III we find that $\log 5.39 = 0.7316$, so that

$$\log 53{,}900 = 0.7316 + 4$$
$$= 4.7316$$

Hence, the mantissa is 0.7316 and the characteristic is 4. Notice that

$$10^{4.7316} = 53{,}900$$

2 0.0035

SOLUTION

$$0.0035 = (3.5)(10^{-3})$$

so that

$$\log 0.0035 = \log 3.5 - 3$$

From Table III we find that $\log 3.5 = 0.5441$, so that

$$\log 0.0035 = 0.5441 - 3$$
$$= -2.4559$$

Hence, the mantissa is 0.5441 and the characteristic is -3. Notice that

$$10^{-2.4559} = 0.0035$$

Now we shall "reverse" the process of the preceding examples. That is, given a number r, determine the value of x such that $\log x = r$. This number is called the *antilogarithm* of r and is sometimes written as $x =$ *antilog* r.

As was the case for finding values of the logarithm, it is easy to determine x for some values of r, but not for others. For example, if $\log x = -2$, $x = 0.01$ (why?); if $\log x = 5$, $x = 100{,}000$ (why?); however, if $\log x = 4.4969$, the value of x is not so easy to determine.

The antilog of 4.4969, or the solution of the equation $\log x = 4.4969$, can be determined by reversing the process of determining the logarithm. First, we write $\log x = 4.4969$ in standard form, that is, as the sum of a number between 0 and 1 and an integer:

$$\log x = 4.4969$$
$$= 0.4969 + 4$$

Second, we use Table III to find a value s such that $\log s = 0.4969$. Here $s = 3.14$. Hence,

$$\begin{aligned}\log x &= 4.4969 \\ &= 0.4969 + 4 \\ &= \log 3.14 + 4 = \log 3.14 + \log 10^4 \\ &= \log 3.14(10^4) = \log 31{,}400\end{aligned}$$

so that

$$x = 31{,}400$$

EXAMPLES

1 Solve $\log x = 2.7210$.

SOLUTION

$$\begin{aligned}\log x &= 0.7210 + 2 \\ &= \log s + 2\end{aligned}$$

Using Table III we find that $\log 5.26 = 0.7210$, so that

$$\begin{aligned}\log x &= \log 5.26 + 2 = \log 5.26 + \log 10^2 \\ &= \log 5.26(10^2) = \log 526\end{aligned}$$

so that

$$x = 526$$

2 Solve $\log x = 0.5105 + (-3)$.

SOLUTION

$$\begin{aligned}\log x &= 0.5105 + (-3) \\ &= \log s + (-3)\end{aligned}$$

Using Table III we find that $\log 3.24 = 0.5105$, so that

$$\begin{aligned}\log x &= \log 3.24 + (-3) = \log 3.24 + \log 10^{-3} \\ &= \log 3.24(10^{-3}) = \log 0.00324\end{aligned}$$

so that

$$x = 0.00324$$

Interpolation

The logarithms and antilogarithms which we computed earlier were special in the sense that we were able to find the necessary numbers in Table III. However, this will not always be the case. Suppose, for example, we wanted to find log 1.234 or antilog 0.2217. We would not be able to find log 1.234 or the mantissa 0.2217 in Table III. This problem can be resolved by using an approximation method called *linear interpolation.*

For example, to determine log 1.234 by using linear interpolation, we proceed as follows. From Table III

$$\log 1.24 = 0.0934 \qquad \text{and} \qquad \log 1.23 = 0.0899$$

Note that 1.234 lies between 1.23 and 1.24. Let us examine that portion of the graph of $y = \log x$ where $1.23 < x < 1.24$ (Figure 1).

Figure 1

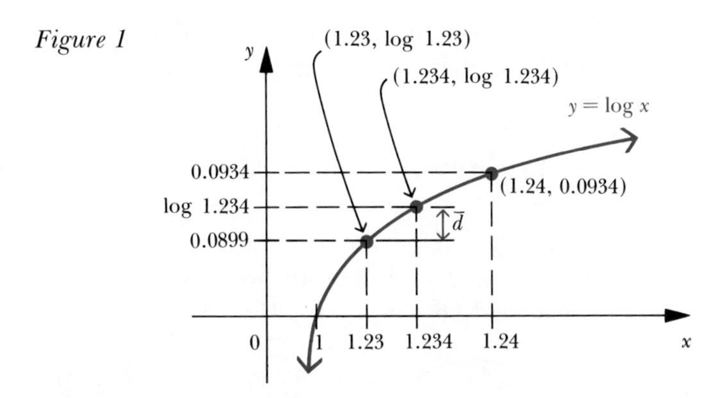

Now, log 1.234 is the length of the ordinate associated with the abscissa 1.234, so that $\log 1.234 = 0.0899 + \bar{d}$, where $\bar{d}$ is the "distance" between log 1.234 and 0.0899. $\bar{d}$ is the number which we will approximate. First, we "replace" the arc of the log curve with a line segment (Figure 2). Next, we assume that $\bar{d}$ is approximately the same as d in Figure 2. Finally, d can be determined by using the proportionality of the sides of the similar right triangles which have been formed. Thus,

Figure 2

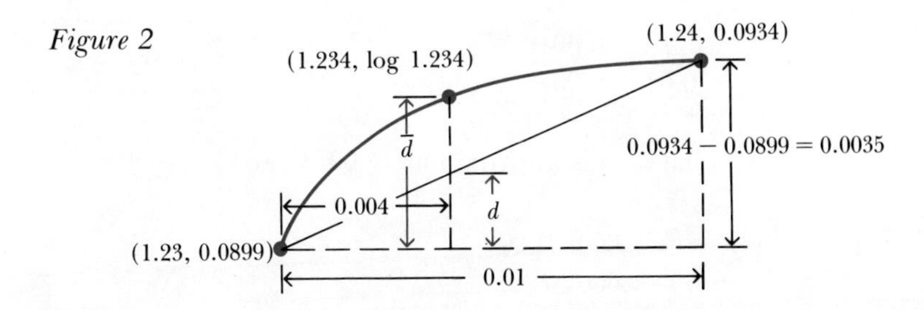

$$\frac{d}{0.0035} = \frac{0.004}{0.01}$$

Hence,

$$d = 0.0014$$

Since $\log 1.234 = \log 1.23 + \bar{d}$, and we approximate $\bar{d}$ by d, we have

$$\log 1.234 = \log 1.23 + d$$

or

$$\begin{aligned}\log 1.234 &= 0.0899 + 0.0014 \\ &= 0.0913\end{aligned}$$

The same process can be used to find antilogs. For example, we can find x so that $\log x = 0.2217$, by linear interpolation. From Table III we find

$$\log 1.67 = 0.2227 \qquad \text{and} \qquad \log 1.66 = 0.2201$$

Note that the given number 0.2217 lies between the two numbers 0.2201 and 0.2227. As before, we examine the graph (Figure 3) of $y = \log x$ for $1.66 < x < 1.67$. Hence,

Figure 3

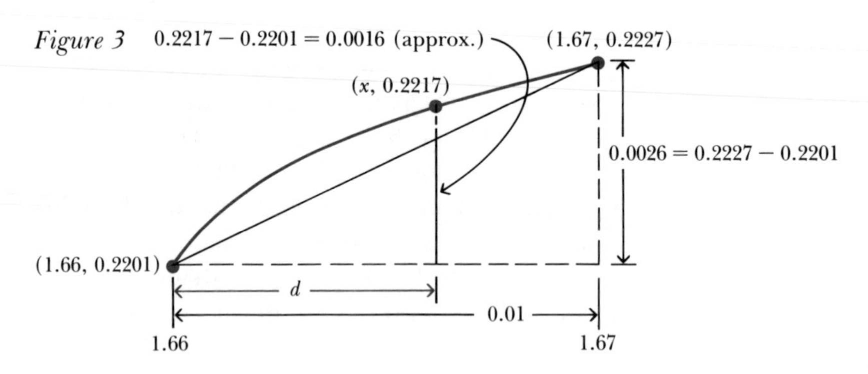

$$\frac{d}{0.01} = \frac{0.0016}{0.0026}$$

That is, d is approximately 0.006, so that

$$\begin{aligned}x &= 1.66 + 0.006 \\ &= 1.666\end{aligned}$$

Essentially, then, linear interpolation is an approximation method which replaces an arc of a curve with a straight-line segment. The accuracy of this method of approximation depends on the "straightness" of the curve between end points.

Now we will present the two examples given above in a manner which simplifies the mechanics involved in linear interpolation.

EXAMPLES

1 Find log 1.234.

SOLUTION

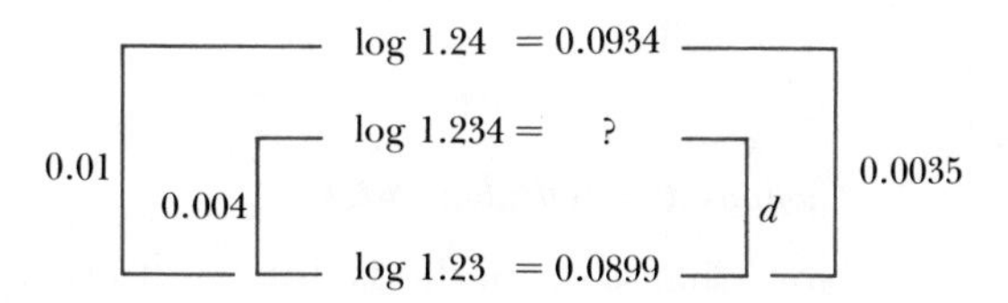

Now,

$$\frac{0.004}{0.01} = \frac{d}{0.0035}$$

so that

$$d = 0.0014$$

Thus,

$$\begin{aligned} \log 1.234 &= 0.0899 + d \\ &= 0.0899 + 0.0014 \\ &= 0.0913 \end{aligned}$$

2 Solve $\log x = 0.2217$.

SOLUTION

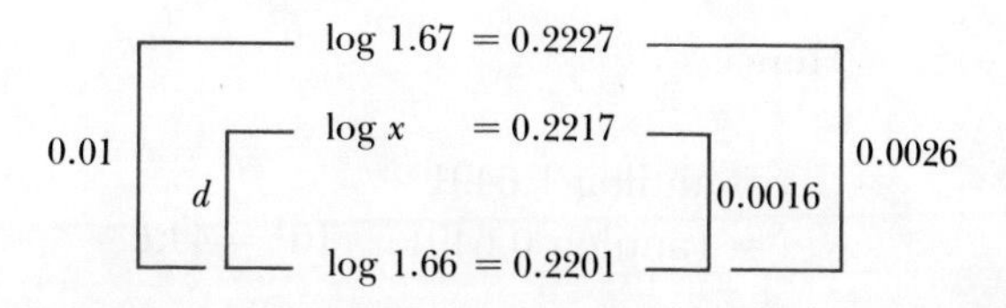

Now,

$$\frac{d}{0.01} = \frac{0.0016}{0.0026}$$

so

$$d = 0.006$$

Hence,

$$\begin{aligned} x &= 1.66 + d \\ &= 1.66 + 0.006 \\ &= 1.666 \end{aligned}$$

Computation with logarithms

Now that the methods for finding common logarithms and antilogarithms have been introduced, we shall investigate the application of the properties of logarithms for computational purposes. These properties will be restated here for reference.

i $\log (MN) = \log M + \log N$

ii $\log (M/N) = \log M - \log N$

iii $\log N^r = r \log N$

EXAMPLES

1 Use logarithms to find (53.7)(0.83).

SOLUTION

Let $x = (53.7)(0.83)$; then

$$\begin{aligned} \log x &= \log [(53.7)(0.83)] \\ &= \log 53.7 + \log 0.83 \qquad \text{[Property (i)]} \\ &= (\log 5.37 + 1) + [\log 8.3 + (-1)] \\ &= 0.7300 + 1 + 0.9191 + (-1) \\ &= 1.6491 \end{aligned}$$

Hence,

$$\begin{aligned} x &= \text{antilog } 1.6491 \\ &= (\text{antilog } 0.6491) \cdot 10^1 = 44.6 \end{aligned}$$

2 Use logarithms to find 0.837/0.00238.

SOLUTION

Let $x = 0.837/0.00238$; then

$$\begin{aligned}\log x &= \log \frac{0.837}{0.00238} \\ &= [\log 8.37 + (-1)] - [\log 2.38 + (-3)] \\ &= (0.9227 - 1) - (0.3766 - 3) \\ &= 2.5461\end{aligned}$$

Hence,

$$\begin{aligned}x &= \text{antilog } 2.5461 \\ &= (\text{antilog } 0.5461) \cdot 10^2 \\ &= (3.517)10^2 = 351.7\end{aligned}$$

3 Use logarithms to find

$$\frac{(289) \cdot (3.47)}{0.0987}$$

SOLUTION. Let

$$x = \frac{(289)(3.47)}{0.0987}$$

Then

$$\log x = \log 289 + \log 3.47 - \log 0.0987$$

From here on, in order to simplify the use of logarithms for computations, we consider the following scheme:

$$\begin{aligned}\log 289 &= 2.4609 \\ +\log 3.47 &= 0.5403 \\ \hline &= 3.0012\end{aligned}$$

$$\begin{aligned}-\log 0.0987 &= -(0.9943 - 2) \\ \hline \log x &= 2.0069 + 2 = 4.0069\end{aligned}$$

Hence,

$$x = 10{,}160$$

PROBLEM SET

1 Write each of the following logarithmic statements as an equivalent exponential statement.

a) $\log_9 81 = 2$
b) $\log_{10} 0.0001 = -4$
c) $\log_{1/3} 9 = -2$
d) $\log_{10} \frac{1}{10} = -1$
e) $\log_{\sqrt{16}} 2 = \frac{1}{2}$
f) $\log_{36} 216 = \frac{3}{2}$
g) $\log_x 2 = 4$
h) $\log_t \frac{1}{z} = \frac{1}{w}$
i) $\log_x 1 = 0$
j) $\log_9 \frac{1}{3} = -\frac{1}{2}$

2 Determine the value of x in each of the following equations.

a) $6^{\log_6 5} + 7^{\log_7 6} = 3^{\log_3 x}$
b) $\log_7 7^4 = x$
c) $9^{\log_x 7} = 7$
d) $\log_3 3^x = 4$
e) $\log_x 3^4 = 4$

3 Prove that $\log_b (xyz) = \log_b x + \log_b y + \log_b z$, where b, x, y, and z are positive, with $b \neq 1$.

4 Use $\log_{10} 2 = 0.3010$ and $\log_{10} 3 = 0.4771$ to find

a) $\log_{10} 4$
b) $\log_{10} 5$
c) $\log_{10} 18$
d) $\log_{10} 60$
e) $\log_{10} \frac{3}{2}$
f) $\log_{10} 0.5$
g) $\log_{10} \sqrt[5]{2}$
h) $\log_{10} \frac{1}{3}$
i) $\log_{10} 3{,}000$
j) $\log_{10} 8$

5 Solve each of the following equations.

a) $\log_{10} (x + 1) = 1$
b) $\log_5 (2x - 7) = 0$
c) $\log_4 (x + \frac{1}{2}) = 3$
d) $\log_7 (2x - 3) = 2$
e) $\log_3 (3x + 1) = 2$
f) $\log_3 (x^2 - 2x) = 1$
g) $\log_{10} (x^2 + 21x) = 2$
h) $\log_3 (x + 1)(x + 2) = 1$

6 Solve each of the following equations.

a) $\log_2 (x^2 - 9) - \log_2 (x + 3) = 2$
b) $\log_{10} (x + 1) - \log_{10} x = 1$
c) $\log_3 (x + 24)x = 4$
d) $\dfrac{\log_{10} (7x - 12)}{\log_{10} x} = 2$

7 Express each of the following numbers in scientific notation and then compute the common logarithm. (Interpolate if necessary.) Indicate the mantissa and characteristic.

a) 0.015
b) 1547
c) 0.531
d) 795.6
e) 33.33
f) 17

8 Solve for the antilogarithm. (Interpolate if necessary.)

a) $\log x = 3.1452$
b) $\log x = -1.505$
c) $\log x = 0.15$
d) $\log x = 2.4969$
e) $\log x = -9.5031$

9 Use common logarithms to compute each of the following values.

a) $(45.6)(0.357)$
b) $(0.00356)(0.786)$
c) $\dfrac{83.4}{20.7}$
d) $\dfrac{0.901}{1.03}$
e) $\sqrt[3]{99}$
f) $\sqrt[7]{0.035}$
g) $\dfrac{(3.87)^2(1.326)}{\sqrt{4.379}}$
h) $\dfrac{\sqrt[3]{0.957}}{\sqrt[5]{32.46}}$
i) $\sqrt{\sqrt[3]{69.83}}$
j) $\sqrt{\sqrt{\sqrt{7}}}$
k) $\sqrt{\dfrac{(3.887)^3(47.32)^2}{(52.37-73.4)^2}}$
l) $\dfrac{(45.07)(0.5689)(2.346)}{(8.379)(100.7)(0.0034)}$

APPENDIX B

Field Axioms for Real Numbers

The set of real numbers R with the operations of addition and multiplication satisfy the following axioms.

Axiom 1. The Closure Laws

If a and b are real numbers, then
i) $a + b$ is a real number and
ii) $a \cdot b$ is a real number.

Axiom 2. The Commutative Laws

If a and b are real numbers, then
i) $a + b = b + a$ and
ii) $a \cdot b = b \cdot a$.

Axiom 3. The Associative Laws

If a, b, and c are real numbers, then
i) $(a + b) + c = a + (b + c)$ and
ii) $(a \cdot b) \cdot c = a \cdot (b \cdot c)$.

Axiom 4. The Identity Elements

i) There exists a real number zero, denoted by 0, such that for any real number a

$$a + 0 = 0 + a = a$$

ii) There exists a real number one, denoted by 1, where zero is different from one, such that for any real number a

$$a \cdot 1 = 1 \cdot a = a$$

Axiom 5. The Inverse Elements

i) For each real number a, there exists a real number, the *additive inverse* of a, denoted by $-a$ such that

$$a + (-a) = (-a) + a = 0$$

ii) For each real number a where $a \neq 0$, there exists a real number, the *multiplicative inverse or reciprocal*, denoted by $1/a$ or a^{-1} such that

$$a \cdot \frac{1}{a} = \frac{1}{a} \cdot a = 1$$

Axiom 6. The Distributive Laws

If a, b and c are real numbers, then
i) $a(b + c) = ab + ac$ and
ii) $(b + c)a = ba + ca$

Any set of elements containing at least two elements with two operations that satisfy these six axioms is called a *field.*

APPENDIX C

Trigonometric and Circular Identities

1. $\sin^2 t + \cos^2 t = 1$
2. $\sin(-t) = -\sin t$
3. $\cos(-t) = \cos t$
4. $\tan t = \dfrac{\sin t}{\cos t}$
5. $\cot t = \dfrac{\cos t}{\sin t}$
6. $\sec t = \dfrac{1}{\cos t}$
7. $\csc t = \dfrac{1}{\sin t}$
8. $\tan t \cot t = 1$
9. $\sec^2 t = 1 + \tan^2 t$
10. $\csc^2 t = 1 + \cot^2 t$
11. $\cos(t + s) = \cos t \cos s - \sin t \sin s$
12. $\cos(t - s) = \cos t \cos s + \sin t \sin s$
13. $\cos\left(\dfrac{\pi}{2} - t\right) = \sin t$
14. $\sin\left(\dfrac{\pi}{2} - t\right) = \cos t$
15. $\sin(t + s) = \sin t \cos s + \cos t \sin s$
16. $\sin(t - s) = \sin t \cos s - \cos t \sin s$
17. $\tan(t + s) = \dfrac{\tan t + \tan s}{1 - \tan t \tan s}$
18. $\tan(t - s) = \dfrac{\tan t - \tan s}{1 + \tan t \tan s}$
19. $\cos 2t = \cos^2 t - \sin^2 t$
20. $\cos 2t = 2\cos^2 t - 1$
21. $\cos 2t = 1 - 2\sin^2 t$
22. $\sin 2t = 2 \sin t \cos t$
23. $\tan 2t = \dfrac{2 \tan t}{1 - \tan^2 t}$
24. $\cos^2 t = \dfrac{1 + \cos 2t}{2}$
25. $\sin^2 t = \dfrac{1 - \cos 2t}{2}$

APPENDIX D

Tables

TABLE I **VALUES OF TRIGONOMETRIC FUNCTIONS**

Degrees	Radians	Sin	Csc	Tan	Cot	Sec	Cos		
0° 0′	.0000	.0000	——	.0000	——	1.000	1.0000	1.5708	90° 0′
10′	029	.0029	343.8	029	343.8	000	000	679	50′
20′	058	.0058	171.9	058	171.9	000	000	650	40′
30′	.0087	.0087	114.6	.0087	114.6	1.000	1.0000	1.5621	30′
40′	116	.0116	85.95	116	85.94	000	0.9999	592	20′
50′	145	.0145	68.76	145	68.75	000	999	563	10′
1° 0′	.0175	.0175	57.30	.0175	57.29	1.000	.9998	1.5533	89° 0′
10′	204	204	49.11	204	49.10	000	998	504	50′
20′	233	233	42.98	233	42.96	000	997	475	40′
30′	.0262	.0262	38.20	.0262	38.19	1.000	.9997	1.5446	30′
40′	291	291	34.38	291	34.37	000	996	417	20′
50′	320	320	31.26	320	31.24	001	995	388	10′
2° 0′	.0349	.0349	28.65	.0349	28.64	1.001	.9994	1.5359	88° 0′
10′	378	378	26.45	378	26.43	001	993	330	50′
20′	407	407	24.56	407	24.54	001	992	301	40′
30′	.0436	.0436	22.93	.0437	22.90	1.001	.9990	1.5272	30′
40′	465	465	21.49	466	21.47	001	989	243	20′
50′	495	494	20.23	495	20.21	001	988	213	10′
3° 0′	.0524	.0523	19.11	.0524	19.08	1.001	.9986	1.5184	87° 0′
10′	553	552	18.10	553	18.07	002	985	155	50′
20′	582	581	17.20	582	17.17	002	983	126	40′
30′	.0611	.0610	16.38	.0612	16.35	1.002	.9981	1.5097	30′
40′	640	640	15.64	641	15.60	002	980	068	20′
50′	669	669	14.96	670	14.92	002	978	039	10′
4° 0′	.0698	.0698	14.34	.0699	14.30	1.002	.9976	1.5010	86° 0′
10′	727	727	13.76	729	13.73	003	974	1.4981	50′
20′	756	756	13.23	758	13.20	003	971	952	40′
30′	.0785	.0785	12.75	.0787	12.71	1.003	.9969	1.4923	30′
40′	814	814	12.29	816	12.25	003	967	893	20′
50′	844	843	11.87	846	11.83	004	964	864	10′
5° 0′	.0873	.0872	11.47	.0875	11.43	1.004	.9962	1.4835	85° 0′
10′	902	901	11.10	904	11.06	004	959	806	50′
20′	931	929	10.76	934	10.71	004	957	777	40′
30′	.0960	.0958	10.43	.0963	10.39	1.005	.9954	1.4748	30′
40′	989	.0987	10.13	.0992	10.08	005	951	719	20′
50′	.1018	.1016	9.839	.1022	9.788	005	948	690	10′
6° 0′	.1047	.1045	9.567	.1051	9.514	1.006	.9945	1.4661	84° 0′
		Cos	Sec	Cot	Tan	Csc	Sin	Radians	Degrees

TABLE I **VALUES OF TRIGONOMETRIC FUNCTIONS** – Cont.

Degrees	Radians	Sin	Csc	Tan	Cot	Sec	Cos		
6° 0′	.1047	.1045	9.567	.1051	9.514	1.006	.9945	1.4661	84° 0′
10′	076	074	9.309	080	9.255	006	942	632	50′
20′	105	103	9.065	110	9.010	006	939	603	40′
30′	.1134	.1132	8.834	.1139	8.777	1.006	.9936	1.4573	30′
40′	164	161	8.614	169	8.556	007	932	544	20′
50′	193	190	8.405	198	8.345	007	929	515	10′
7° 0′	.1222	.1219	8.206	.1228	8.144	1.008	.9925	1.4486	83° 0′
10′	251	248	8.016	257	7.953	008	922	457	50′
20′	280	276	7.834	287	7.770	008	918	428	40′
30′	.1309	.1305	7.661	.1317	7.596	1.009	.9914	1.4399	30′
40′	338	334	7.496	346	7.429	009	911	370	20′
50′	367	363	7.337	376	7.269	009	907	341	10′
8° 0′	.1396	.1392	7.185	.1405	7.115	1.010	.9903	1.4312	82° 0′
10′	425	421	7.040	435	6.968	010	899	283	50′
20′	454	449	6.900	465	6.827	011	894	254	40′
30′	.1484	.1478	6.765	.1495	6.691	1.011	.8980	1.4224	30′
40′	513	507	6.636	524	6.561	012	886	195	20′
50′	542	536	6.512	554	6.435	012	881	166	10′
9° 0′	.1571	.1564	6.392	.1584	6.314	1.012	.9877	1.4137	81° 0′
10′	600	593	277	614	197	013	872	108	50′
20′	629	622	166	644	6.084	013	868	079	40′
30′	.1658	.1650	6.059	.1673	5.976	1.014	.9863	1.4050	30′
40′	687	679	5.955	703	871	014	858	1.4021	20′
50′	716	708	855	733	769	015	853	1.3992	10′
10° 0′	.1745	.1736	5.759	.1763	5.671	1.015	.9848	1.3963	80° 0′
10′	774	765	665	793	576	016	843	934	50′
20′	804	794	575	823	485	016	838	904	40′
30′	.1833	.1822	5.487	.1853	5.396	1.017	.9833	1.3875	30′
40′	862	851	403	883	309	018	827	846	20′
50′	891	880	320	914	226	018	822	817	10′
11° 0′	.1920	.1908	5.241	.1944	5.145	1.019	.9816	1.3788	79° 0′
10′	949	937	164	.1974	5.066	019	811	759	50′
20′	978	965	089	.2004	4.989	020	805	730	40′
30′	.2007	.1994	5.016	.2035	4.915	1.020	.9799	1.3701	30′
40′	036	.2022	4.945	065	843	021	793	672	20′
50′	065	051	876	095	773	022	787	643	10′
12° 0′	.2094	.2079	4.810	.2126	4.705	1.022	.9781	1.3614	78° 0′
10′	123	108	745	156	638	023	775	584	50′
20′	153	136	682	186	574	024	769	555	40′
30′	.2182	.2164	4.620	.2217	4.511	1.024	.9763	1.3526	30′
40′	211	193	560	247	449	025	757	497	20′
50′	240	221	502	278	390	026	750	468	10′
13° 0′	.2269	.2250	4.445	.2309	4.331	1.026	.9744	1.3439	77° 0′
10′	298	278	390	339	275	027	737	410	50′
20′	327	306	336	370	219	028	730	381	40′
30′	.2356	.2334	4.284	.2401	4.165	1.028	.9724	1.3352	30′
40′	385	363	232	432	113	029	717	323	20′
50′	414	391	182	462	061	030	710	294	10′
14° 0′	.2443	.2419	4.134	.2493	4.011	1.031	.9703	1.3265	76° 0′
		Cos	Sec	Cot	Tan	Csc	Sin	Radians	Degrees

TABLE I **VALUES OF TRIGONOMETRIC FUNCTIONS** — Cont.

Degrees	Radians	Sin	Csc	Tan	Cot	Sec	Cos		
14° 0′	.2443	.2419	4.134	.2493	4.011	1.031	.9703	1.3265	76° 0′
10′	473	447	086	524	3.962	031	696	235	50′
20′	502	476	4.039	555	914	032	698	206	40′
30′	.2531	.2504	3.994	.2586	3.867	1.033	.9681	1.3177	30′
40′	560	532	950	617	821	034	674	148	20′
50′	589	560	906	648	776	034	667	119	10′
15° 0′	.2618	.2588	3.864	.2679	3.732	1.035	.9659	1.3090	75° 0′
10′	647	616	822	711	689	036	652	061	50′
20′	676	644	782	742	647	037	644	032	40′
30′	.2705	.2672	3.742	.2773	3.606	1.038	.9636	1.3003	30′
40′	734	700	703	805	566	039	628	1.2974	20′
50′	763	728	665	836	526	039	621	945	10′
16° 0′	.2793	.2756	3.628	.2867	3.487	1.040	.9613	1.2915	74° 0′
10′	822	784	592	899	450	041	605	886	50′
20′	851	812	556	931	412	042	596	857	40′
30′	.2880	.2840	3.521	.2962	3.376	1.043	.9588	1.2828	30′
40′	909	868	487	.2994	340	044	580	799	20′
50′	938	896	453	3026	305	045	572	770	10′
17° 0′	.2967	.2924	3.420	.3057	3.271	1.046	.9563	1.2741	73° 0′
10′	996	952	388	089	237	047	555	712	50′
20′	.3025	.2979	357	121	204	048	546	683	40′
30′	.3054	.3007	3.326	.3153	3.172	1.048	.9537	1.2654	30′
40′	083	035	295	185	140	049	528	625	20′
50′	113	062	265	217	108	050	520	595	10′
18° 0′	.3142	.3090	3.236	.3249	3.078	1.051	.9511	1.2566	72° 0′
10′	171	118	207	281	047	052	502	537	50′
20′	200	145	179	314	3.018	053	492	508	40′
30′	.3229	.3173	3.152	.3346	2.989	1.054	.9483	1.2479	30′
40′	258	201	124	378	960	056	474	450	20′
50′	287	228	098	411	932	057	465	421	10′
19° 0′	.3316	.3256	3.072	.3443	2.904	1.058	.9455	1.2392	71° 0′
10′	345	283	046	476	877	059	446	363	50′
20′	374	311	3.021	508	850	060	436	334	40′
30′	.3403	.3338	2.996	.3541	2.824	1.061	.9426	1.2305	30′
40′	432	365	971	574	798	062	417	275	20′
50′	462	393	947	607	773	063	407	246	10′
20° 0′	.3491	.3420	2.924	.3640	2.747	1.064	.9397	1.2217	70° 0′
10′	520	448	901	673	723	065	387	188	50′
20′	549	475	878	706	699	066	377	159	40′
30′	.3578	.3502	2.855	.3739	2.675	1.068	.9367	1.2130	30′
40′	607	529	833	772	651	069	356	101	20′
50′	636	557	812	805	628	070	346	072	10′
21° 0′	.3665	.3584	2.790	.3839	2.605	1.071	.9336	1.2043	69° 0′
10′	694	611	769	872	583	072	325	1.2014	50′
20′	723	638	749	906	560	074	315	1.1985	40′
30′	.3752	.3665	2.729	.3939	2.539	1.075	.9304	1.1956	30′
40′	782	692	709	.3973	517	076	293	926	20′
50′	811	719	689	.4006	496	077	283	897	10′
22° 0′	.3840	.3746	2.669	.4040	2.475	1.079	.9272	1.1868	68° 0′
		Cos	Sec	Cot	Tan	Csc	Sin	Radians	Degrees

TABLE I **VALUES OF TRIGONOMETRIC FUNCTIONS** — Cont.

Degrees	Radians	Sin	Csc	Tan	Cot	Sec	Cos		
22° 0′	.3840	.3746	2.669	.4040	2.475	1.079	.9272	1.1868	68° 0′
10′	869	773	650	074	455	080	261	839	50′
20′	898	800	632	108	434	081	250	810	40′
30′	.3927	.3827	2.613	.4142	2.414	1.082	.9239	1.1781	30′
40′	956	854	595	176	394	084	228	752	20′
50′	985	881	577	210	375	085	216	723	10′
23° 0′	.4014	.3907	2.559	.4245	2.356	1.086	.9205	1.1694	67° 0′
10′	043	934	542	279	337	088	194	665	50′
20′	072	961	525	314	318	089	182	636	40′
30′	.4102	.3987	2.508	.4348	2.300	1.090	.9171	1.1606	30′
40′	131	.4014	491	383	282	092	159	577	20′
50′	160	041	475	417	264	093	147	548	10′
24° 0′	.4189	.4067	2.459	.4452	2.246	1.095	.9135	1.1519	66° 0′
10′	218	094	443	487	229	096	124	490	50′
20′	247	120	427	522	211	097	112	461	40′
30′	.4276	.4147	2.411	.4557	2.194	1.099	.9100	1.1432	30′
40′	305	173	396	592	177	100	088	403	20′
50′	334	200	381	628	161	102	075	374	10′
25° 0′	.4363	.4226	2.366	.4663	2.145	1.103	.9063	1.1345	65° 0′
10′	392	253	352	699	128	105	051	316	50′
20′	422	279	337	734	112	106	038	286	40′
30′	.4451	.4305	2.323	.4770	2.097	1.108	.9026	1.1257	30′
40′	480	331	309	806	081	109	013	228	20′
50′	509	358	295	841	066	111	.9001	199	10′
26° 0′	.4538	.4384	2.281	.4877	2.050	1.113	.8988	1.1170	64° 0′
10′	567	410	268	913	035	114	975	141	50′
20′	596	436	254	950	020	116	962	112	40′
30′	.4625	.4462	2.241	.4986	2.006	1.117	.8949	1.1083	30′
40′	654	488	228	.5022	1.991	119	936	054	20′
50′	683	514	215	059	977	121	923	1.1025	10′
27° 0′	.4712	.4540	2.203	.5095	1.963	1.122	.8910	1.0996	63° 0′
10′	741	566	190	132	949	124	897	966	50′
20′	771	592	178	169	935	126	884	937	40′
30′	.4800	.4617	2.166	.5206	1.921	1.127	.8870	1.0908	30′
40′	829	643	154	243	907	129	857	879	20′
50′	858	669	142	280	894	131	843	850	10′
28° 0′	.4887	.4695	2.130	.5317	1.881	1.133	.8829	1.0821	62° 0′
10′	916	720	118	354	868	134	.816	792	50′
20′	945	746	107	392	855	136	802	763	40′
30′	.4974	.4772	2.096	.5430	1.842	1.138	.8788	1.0734	30′
40′	.5003	797	085	467	829	140	774	705	20′
50′	032	823	074	505	816	142	760	676	10′
29° 0′	.5061	.4848	2.063	.5543	1.804	1.143	.8746	1.0647	61° 0′
10′	091	874	052	581	792	145	732	617	50′
20′	120	899	041	619	780	147	718	588	40′
30′	.5149	.4924	2.031	.5658	1.767	1.149	.8704	1.0559	30′
40′	178	950	020	696	756	151	689	530	20′
50′	207	.4975	010	735	744	153	675	501	10′
30° 0′	.5236	.5000	2.000	.5774	1.732	1.155	.8660	1.0472	60° 0′
		Cos	Sec	Cot	Tan	Csc	Sin	Radians	Degrees

TABLE I **VALUES OF TRIGONOMETRIC FUNCTIONS** — Cont.

Degrees	Radians	Sin	Csc	Tan	Cot	Sec	Cos		
30° 0′	.5236	.5000	2.000	.5774	1.732	1.155	.8660	1.0472	60° 0′
10′	265	025	1.990	812	720	157	646	443	50′
20′	294	050	980	851	709	159	631	414	40′
30′	.5323	.5075	1.970	.5890	1.698	1.161	.8616	1.0385	30′
40′	352	100	961	930	686	163	601	356	20′
50′	381	125	951	.5969	675	165	587	327	10′
31° 0′	.5411	.5150	1.942	.6009	1.664	1.167	.8572	1.0297	59° 0′
10′	440	175	932	048	653	169	557	268	50′
20′	469	200	923	088	643	171	542	239	40′
30′	.5498	.5225	1.914	.6128	1.632	1.173	.8526	1.0210	30′
40′	527	250	905	168	621	175	511	181	20′
50′	556	275	896	208	611	177	496	152	10′
32° 0′	.5585	.5299	1.887	.6249	1.600	1.179	.8480	1.0123	58° 0′
10′	614	324	878	289	590	181	465	094	50′
20′	643	348	870	330	580	184	450	065	40′
30′	.5672	.5373	1.861	.6371	1.570	1.186	.8434	1.0036	30′
40′	701	398	853	412	560	188	418	1.0007	20′
50′	730	422	844	453	550	190	403	.9977	10′
33° 0′	.5760	.5446	1.836	.6494	1.540	1.192	.8387	.9948	57° 0′
10′	789	471	828	536	530	195	371	919	50′
20′	818	495	820	577	520	197	355	890	40′
30′	.5847	.5519	1.812	.6619	1.511	1.199	.8339	.9861	30′
40′	876	544	804	661	501	202	323	832	20′
50′	905	568	796	703	1.492	204	307	803	10′
34° 0′	.5934	.5592	1.788	.6745	1.483	1.206	.8290	.9774	56° 0′
10′	963	616	781	787	473	209	274	745	50′
20′	992	640	773	830	464	211	258	716	40′
30′	.6021	.5664	1.766	.6873	1.455	1.213	.8241	.9687	30′
40′	050	688	758	916	446	216	225	657	20′
50′	080	712	751	.6959	437	218	208	628	10′
35° 0′	.6109	.5736	1.743	.7002	1.428	1.221	.8192	.9599	55° 0′
10′	138	760	736	046	419	223	175	570	50′
20′	167	783	729	089	411	226	158	541	40′
30′	.6196	.5807	1.722	.7133	1.402	1.228	.8141	.9512	30′
40′	225	831	715	177	393	231	124	483	20′
50′	254	854	708	221	385	233	107	454	10′
36° 0′	.6283	.5878	1.701	.7265	1.376	1.236	.8090	.9425	54° 0′
10′	312	901	695	310	368	239	073	396	50′
20′	341	925	688	355	360	241	056	367	40′
30′	.6370	.5948	1.681	.7400	1.351	1.244	.8039	338	30′
40′	400	972	675	445	343	247	021	308	20′
50′	429	.5995	668	490	335	249	.8004	279	10′
37° 0′	.6458	.6018	1.662	.7536	1.327	1.252	.7986	.9250	53° 0′
10′	487	041	655	581	319	255	969	221	50′
20′	516	065	649	627	311	258	951	192	40′
30′	.6545	.6088	1.643	.7673	1.303	1.260	.7934	.9163	30′
40′	574	111	636	720	295	263	916	134	20′
50′	603	134	630	766	288	266	898	105	10′
38° 0′	.6632	.6157	1.624	.7813	1.280	1.269	.7880	.9076	52° 0′
		Cos	Sec	Cot	Tan	Csc	Sin	Radians	Degrees

TABLE I **VALUES OF TRIGONOMETRIC FUNCTIONS** — Cont.

Degrees	Radians	Sin	Csc	Tan	Cot	Sec	Cos		
38° 0′	.6632	.6157	1.624	.7813	1.280	1.269	.7880	.9076	52° 0′
10′	661	180	618	860	272	272	862	047	50′
20′	690	202	612	907	265	275	844	.9018	40′
30′	.6720	.6225	1.606	.7954	1.257	1.278	.7826	.8988	30′
40′	749	248	601	.8002	250	281	808	959	20′
50′	778	271	595	050	242	284	790	930	10′
39° 0′	.6807	.6293	1.589	.8098	1.235	1.287	.7771	.8901	51° 0′
10′	836	316	583	146	228	290	753	872	50′
20′	865	338	578	195	220	293	735	843	40′
30′	.6894	.6361	1.572	.8243	1.213	1.296	.7716	.8814	30′
40′	923	383	567	292	206	299	698	785	20′
50′	952	406	561	342	199	302	679	756	10′
40° 0′	.6981	.6428	1.556	.8391	1.192	1.305	.7660	.8727	50° 0′
10′	.7010	450	550	441	185	309	642	698	50′
20′	039	472	545	491	178	312	623	668	40′
30′	.7069	.6494	1.540	.8541	1.171	1.315	.7604	.8639	30′
40′	098	517	535	591	164	318	585	610	20′
50′	127	539	529	642	157	322	566	581	10′
41° 0′	.7156	.6561	1.524	.8693	1.150	1.325	.7547	.8552	49° 0′
10′	185	583	519	744	144	328	528	523	50′
20′	214	604	514	796	137	332	509	494	40′
30′	.7243	.6626	1.509	.8847	1.130	1.335	.7490	.8465	30′
40′	272	648	504	899	124	339	470	436	20′
50′	301	670	499	.8952	117	342	451	407	10′
42° 0′	.7330	.6691	1.494	.9004	1.111	1.346	.7431	.8378	48° 0′
10′	359	713	490	057	104	349	412	348	50′
20′	389	734	485	110	098	353	392	319	40′
30′	.7418	.6756	1.480	.9163	1.091	1.356	.7373	.8290	30′
40′	447	777	476	217	085	360	353	261	20′
50′	476	799	471	271	079	364	333	232	10′
43° 0′	.7505	.6820	1.466	.9325	1.072	1.367	.7314	.8203	47° 0′
10′	534	841	462	380	066	371	294	174	50′
20′	563	862	457	435	060	375	274	145	40′
30′	.7592	.6884	1.453	.9490	1.054	1.379	.7254	.8116	30′
40′	621	905	448	545	048	382	234	087	20′
50′	650	926	444	601	042	386	214	058	10′
44° 0′	.7679	.6947	1.440	.9657	1.036	1.390	.7193	.8029	46° 0′
10′	709	967	435	713	030	394	173	.7999	50′
20′	738	.6988	431	770	024	398	153	970	40′
30′	.7767	.7009	1.427	.9827	1.018	1.402	.7133	.7941	30′
40′	796	030	423	884	012	406	112	912	20′
50′	825	050	418	.9942	006	410	092	883	10′
45° 0′	.7854	.7071	1.414	1.000	1.000	1.414	.7071	.7854	45° 0′
		Cos	Sec	Cot	Tan	Csc	Sin	Radians	Degrees

TABLE II **VALUES OF CIRCULAR FUNCTIONS**

t	$\sin t$	$\cos t$	$\tan t$	$\cot t$	$\sec t$	$\csc t$
.00	.0000	1.0000	.0000	—	1.000	—
.01	.0100	1.0000	.0100	99.997	1.000	100.00
.02	.0200	.9998	.0200	49.993	1.000	50.00
.03	.0300	.9996	.0300	33.323	1.000	33.34
.04	.0400	.9992	.0400	24.987	1.001	25.01
.05	.0500	.9988	.0500	19.983	1.001	20.01
.06	.0600	.9982	.0601	16.647	1.002	16.68
.07	.0699	.9976	.0701	14.262	1.002	14.30
.08	.0799	.9968	.0802	12.473	1.003	12.51
.09	.0899	.9960	.0902	11.081	1.004	11.13
.10	.0998	.9950	.1003	9.967	1.005	10.02
.11	.1098	.9940	.1104	9.054	1.006	9.109
.12	.1197	.9928	.1206	8.293	1.007	8.353
.13	.1296	.9916	.1307	7.649	1.009	7.714
.14	.1395	.9902	.1409	7.096	1.010	7.166
.15	.1494	.9888	.1511	6.617	1.011	6.692
.16	.1593	.9872	.1614	6.197	1.013	6.277
.17	.1692	.9856	.1717	5.826	1.015	5.911
.18	.1790	.9838	.1820	5.495	1.016	5.586
.19	.1889	.9820	.1923	5.200	1.018	5.295
.20	.1987	.9801	.2027	4.933	1.020	5.033
.21	.2085	.9780	.2131	4.692	1.022	4.797
.22	.2182	.9759	.2236	4.472	1.025	4.582
.23	.2280	.9737	.2341	4.271	1.027	4.386
.24	.2377	.9713	.2447	4.086	1.030	4.207
.25	.2474	.9689	.2553	3.916	1.032	4.042
.26	.2571	.9664	.2660	3.759	1.035	3.890
.27	.2667	.9638	.2768	3.613	1.038	3.749
.28	.2764	.9611	.2876	3.478	1.041	3.619
.29	.2860	.9582	.2984	3.351	1.044	3.497
.30	.2955	.9553	.3093	3.233	1.047	3.384
.31	.3051	.9523	.3203	3.122	1.050	3.278
.32	.3146	.9492	.3314	3.018	1.053	3.179
.33	.3240	.9460	.3425	2.920	1.057	3.086
.34	.3335	.9428	.3537	2.827	1.061	2.999
.35	.3429	.9394	.3650	2.740	1.065	2.916
.36	.3523	.9359	.3764	2.657	1.068	2.839
.37	.3616	.9323	.3879	2.578	1.073	2.765
.38	.3709	.9287	.3994	2.504	1.077	2.696
.39	.3802	.9249	.4111	2.433	1.081	2.630
.40	.3894	.9211	.4228	2.365	1.086	2.568
.41	.3986	.9171	.4346	2.301	1.090	2.509
.42	.4078	.9131	.4466	2.239	1.095	2.452
.43	.4169	.9090	.4586	2.180	1.100	2.399
.44	.4259	.9048	.4708	2.124	1.105	2.348
.45	.4350	.9004	.4831	2.070	1.111	2.299
.46	.4439	.8961	.4954	2.018	1.116	2.253
.47	.4529	.8916	.5080	1.969	1.122	2.208
.48	.4618	.8870	.5206	1.921	1.127	2.166
.49	.4706	.8823	.5334	1.875	1.133	2.125

TABLE II **VALUES OF CIRCULAR FUNCTIONS** — Cont.

t	$\sin t$	$\cos t$	$\tan t$	$\cot t$	$\sec t$	$\csc t$
.50	.4794	.8776	.5463	1.830	1.139	2.086
.51	.4882	.8727	.5594	1.788	1.146	2.048
.52	.4969	.8678	.5726	1.747	1.152	2.013
$\frac{\pi}{6}$	.5000	.8660	.5774	1.732	1.155	2.000
.53	.5055	.8628	.5859	1.707	1.159	1.978
.54	.5141	.8577	.5994	1.668	1.166	1.945
.55	.5227	.8525	.6131	1.631	1.173	1.913
.56	.5312	.8473	.6269	1.595	1.180	1.883
.57	.5396	.8419	.6410	1.560	1.188	1.853
.58	.5480	.8365	.6552	1.526	1.196	1.825
.59	.5564	.8309	.6696	1.494	1.203	1.797
.60	.5646	.8253	.6841	1.462	1.212	1.771
.61	.5729	.8196	.6989	1.431	1.220	1.746
.62	.5810	.8139	.7139	1.401	1.229	1.721
.63	.5891	.8080	.7291	1.372	1.238	1.697
.64	.5972	.8021	.7445	1.343	1.247	1.674
.65	.6052	.7961	.7602	1.315	1.256	1.652
.66	.6131	.7900	.7761	1.288	1.266	1.631
.67	.6210	.7838	.7923	1.262	1.276	1.610
.68	.6288	.7776	.8087	1.237	1.286	1.590
.69	.6365	.7712	.8253	1.212	1.297	1.571
.70	.6442	.7648	.8423	1.187	1.307	1.552
.71	.6518	.7584	.8595	1.163	1.319	1.534
.72	.6594	.7518	.8771	1.140	1.330	1.517
.73	.6669	.7452	.8949	1.117	1.342	1.500
.74	.6743	.7385	.9131	1.095	1.354	1.483
.75	.6816	.7317	.9316	1.073	1.367	1.467
.76	.6889	.7248	.9505	1.052	1.380	1.452
.77	.6961	.7179	.9697	1.031	1.393	1.437
.78	.7033	.7109	.9893	1.011	1.407	1.422
$\frac{\pi}{4}$	.7071	.7071	1.000	1.000	1.414	1.414
.79	.7104	.7038	1.009	.9908	1.421	1.408
.80	.7174	.6967	1.030	.9712	1.435	1.394
.81	.7243	.6895	1.050	.9520	1.450	1.381
.82	.7311	.6822	1.072	.9331	1.466	1.368
.83	.7379	.6749	1.093	.9146	1.482	1.355
.84	.7446	.6675	1.116	.8964	1.498	1.343
.85	.7513	.6600	1.138	.8785	1.515	1.331
.86	.7578	.6524	1.162	.8609	1.533	1.320
.87	.7643	.6448	1.185	.8437	1.551	1.308
.88	.7707	.6372	1.210	.8267	1.569	1.297
.89	.7771	.6294	1.235	.8100	1.589	1.287
.90	.7833	.6216	1.260	.7936	1.609	1.277
.91	.7895	.6137	1.286	.7774	1.629	1.267
.92	.7956	.6058	1.313	.7615	1.651	1.257
.93	.8016	.5978	1.341	.7458	1.673	1.247
.94	.8076	.5898	1.369	.7303	1.696	1.238

TABLE II **VALUES OF CIRCULAR FUNCTIONS** — Cont.

t	$\sin t$	$\cos t$	$\tan t$	$\cot t$	$\sec t$	$\csc t$
.95	.8134	.5817	1.398	.7151	1.719	1.229
.96	.8192	.5735	1.428	.7001	1.744	1.221
.97	.8249	.5653	1.459	.6853	1.769	1.212
.98	.8305	.5570	1.491	.6707	1.795	1.204
.99	.8360	.5487	1.524	.6563	1.823	1.196
1.00	.8415	.5403	1.557	.6421	1.851	1.188
1.01	.8468	.5319	1.592	.6281	1.880	1.181
1.02	.8521	.5234	1.628	.6142	1.911	1.174
1.03	.8573	.5148	1.665	.6005	1.942	1.166
1.04	.8624	.5062	1.704	.5870	1.975	1.160
$\frac{\pi}{3}$	.8660	.5000	1.732	.5774	2.000	1.155
1.05	.8674	.4976	1.743	.5736	2.010	1.153
1.06	.8724	.4889	1.784	.5604	2.046	1.146
1.07	.8772	.4801	1.827	.5473	2.083	1.140
1.08	.8820	.4713	1.871	.5344	2.122	1.134
1.09	.8866	.4625	1.917	.5216	2.162	1.128
1.10	.8912	.4536	1.965	.5090	2.205	1.122
1.11	.8957	.4447	2.014	.4964	2.249	1.116
1.12	.9001	.4357	2.066	.4840	2.295	1.111
1.13	.9044	.4267	2.120	.4718	2.344	1.106
1.14	.9086	.4176	2.176	.4596	2.395	1.101
1.15	.9128	.4085	2.234	.4475	2.448	1.096
1.16	.9168	.3993	2.296	.4356	2.504	1.091
1.17	.9208	.3902	2.360	.4237	2.563	1.086
1.18	.9246	.3809	2.427	.4120	2.625	1.082
1.19	.9284	.3717	2.498	.4003	2.691	1.077
1.20	.9320	.3624	2.572	.3888	2.760	1.073
1.21	.9356	.3530	2.650	.3773	2.833	1.069
1.22	.9391	.3436	2.733	.3659	2.910	1.065
1.23	.9425	.3342	2.820	.3546	2.992	1.061
1.24	.9458	.3248	2.912	.3434	3.079	1.057
1.25	.9490	.3153	3.010	.3323	3.171	1.054
1.26	.9521	.3058	3.113	.3212	3.270	1.050
1.27	.9551	.2963	3.224	.3102	3.375	1.047
1.28	.9580	.2867	3.341	.2993	3.488	1.044
1.29	.9608	.2771	3.467	.2884	3.609	1.041
1.30	.9636	.2675	3.602	.2776	3.738	1.038
1.31	.9662	.2579	3.747	.2669	3.878	1.035
1.32	.9687	.2482	3.903	.2562	4.029	1.032
1.33	.9711	.2385	4.072	.2456	4.193	1.030
1.34	.9735	.2288	4.256	.2350	4.372	1.027
1.35	.9757	.2190	4.455	.2245	4.566	1.025
1.36	.9779	.2092	4.673	.2140	4.779	1.023
1.37	.9799	.1994	4.913	.2035	5.014	1.021
1.38	.9819	.1896	5.177	.1931	5.273	1.018
1.39	.9837	.1798	5.471	.1828	5.561	1.017

TABLE II **VALUES OF CIRCULAR FUNCTIONS** — Cont.

t	$\sin t$	$\cos t$	$\tan t$	$\cot t$	$\sec t$	$\csc t$
1.40	.9854	.1700	5.798	.1725	5.883	1.015
1.41	.9871	.1601	6.165	.1622	6.246	1.013
1.42	.9887	.1502	6.581	.1519	6.657	1.011
1.43	.9901	.1403	7.055	.1417	7.126	1.010
1.44	.9915	.1304	7.602	.1315	7.667	1.009
1.45	.9927	.1205	8.238	.1214	8.299	1.007
1.46	.9939	.1106	8.989	.1113	9.044	1.006
1.47	.9949	.1006	9.887	.1011	9.938	1.005
1.48	.9959	.0907	10.983	.0910	11.029	1.004
1.49	.9967	.0807	12.350	.0810	12.390	1.003
1.50	.9975	.0707	14.101	.0709	14.137	1.003
1.51	.9982	.0608	16.428	.0609	16.458	1.002
1.52	.9987	.0508	19.670	.0508	19.695	1.001
1.53	.9992	.0408	24.498	.0408	24.519	1.001
1.54	.9995	.0308	32.461	.0308	32.476	1.000
1.55	.9998	.0208	48.078	.0208	48.089	1.000
1.56	.9999	.0108	92.620	.0108	92.626	1.000
1.57	1.0000	.0008	1255.8	.0008	1255.8	1.000
$\frac{\pi}{2}$	1.0000	.0000	—	.0000	—	1.000

TABLE III **COMMON LOGARITHMS**

n	0	1	2	3	4	5	6	7	8	9
10	0000	0043	0086	0128	0170	0212	0253	0294	0334	0374
11	0414	0453	0492	0531	0569	0607	0645	0682	0719	0755
12	0792	0828	0864	0899	0934	0969	1004	1038	1072	1106
13	1139	1173	1206	1239	1271	1303	1335	1367	1399	1430
14	1461	1492	1523	1553	1584	1614	1644	1673	1703	1732
15	1761	1790	1818	1847	1875	1903	1931	1959	1987	2014
16	2041	2068	2095	2122	2148	2175	2201	2227	2253	2279
17	2304	2330	2355	2380	2405	2430	2455	2480	2504	2529
18	2553	2577	2601	2625	2648	2672	2695	2718	2742	2765
19	2788	2810	2833	2856	2878	2900	2923	2945	2967	2989
20	3010	3032	3054	3075	3096	3118	3139	3160	3181	3201
21	3222	3243	3263	3284	3304	3324	3345	3365	3385	3404
22	3424	3444	3464	3483	3502	3522	3541	3560	3579	3598
23	3617	3636	3655	3674	3692	3711	3729	3747	3766	3784
24	3802	3820	3838	3856	3874	3892	3909	3927	3945	3962
25	3979	3997	4014	4031	4048	4065	4082	4099	4116	4133
26	4150	4166	4183	4200	4216	4232	4249	4265	4281	4298
27	4314	4330	4346	4362	4378	4393	4409	4425	4440	4456
28	4472	4487	4502	4518	4533	4548	4564	4579	4594	4609
29	4624	4639	4654	4669	4683	4698	4713	4728	4742	4757
30	4771	4786	4800	4814	4829	4843	4857	4871	4886	4900
31	4914	4928	4942	4955	4969	4983	4997	5011	5024	5038
32	5051	5065	5079	5092	5105	5119	5132	5145	5159	5172
33	5185	5198	5211	5224	5237	5250	5263	5276	5289	5302
34	5315	5328	5340	5353	5366	5378	5391	5403	5416	5428
35	5441	5453	5465	5478	5490	5502	5514	5527	5539	5551
36	5563	5575	5587	5599	5611	5623	5635	5647	5658	5670
37	5682	5694	5705	5717	5729	5740	5752	5763	5775	5786
38	5798	5809	5821	5832	5843	5855	5866	5877	5888	5899
39	5911	5922	5933	5944	5955	5966	5977	5988	5999	6010
40	6021	6031	6042	6053	6064	6075	6085	6096	6107	6117
41	6128	6138	6149	6160	6170	6180	6191	6201	6212	6222
42	6232	6243	6253	6263	6274	6284	6294	6304	6314	6325
43	6335	6345	6355	6365	6375	6385	6395	6405	6415	6425
44	6435	6444	6454	6464	6474	6484	6493	6503	6513	6522
45	6532	6542	6551	6561	6571	6580	6590	6599	6609	6618
46	6628	6637	6646	6656	6665	6675	6684	6693	6702	6712
47	6721	6730	6739	6749	6758	6767	6776	6785	6794	6803
48	6812	6821	6830	6839	6848	6857	6866	6875	6884	6893
49	6902	6911	6920	6928	6937	6946	6955	6964	6972	6981
50	6990	6998	7007	7016	7024	7033	7042	7050	7059	7067
51	7076	7084	7093	7101	7110	7118	7126	7135	7143	7152
52	7160	7168	7177	7185	7193	7202	7210	7218	7226	7235
53	7243	7251	7259	7267	7275	7284	7292	7300	7308	7316
54	7324	7332	7340	7348	7356	7364	7372	7380	7388	7396

TABLE III **COMMON LOGARITHMS** — Cont.

n	0	1	2	3	4	5	6	7	8	9
55	7404	7412	7419	7427	7435	7443	7451	7459	7466	7474
56	7482	7490	7497	7505	7513	7520	7528	7536	7543	7551
57	7559	7566	7574	7582	7589	7597	7604	7612	7619	7627
58	7634	7642	7649	7657	7664	7672	7679	7686	7694	7701
59	7709	7716	7723	7731	7738	7745	7752	7760	7767	7774
60	7782	7789	7796	7803	7810	7818	7825	7832	7839	7846
61	7853	7860	7868	7875	7882	7889	7896	7903	7910	7917
62	7924	7931	7938	7945	7952	7959	7966	7973	7980	7987
63	7993	8000	8007	8014	8021	8028	8035	8041	8048	8055
64	8062	8069	8075	8082	8089	8096	8102	8109	8116	8122
65	8129	8136	8142	8149	8156	8162	8169	8176	8182	8189
66	8195	8202	8209	8215	8222	8228	8235	8241	8248	8254
67	8261	8267	8274	8280	8287	8293	8299	8306	8312	8319
68	8325	8331	8338	8344	8351	8357	8363	8370	8376	8382
69	8388	8395	8401	8407	8414	8420	8426	8432	8439	8445
70	8451	8457	8463	8470	8476	8482	8488	8494	8500	8506
71	8513	8519	8525	8531	8537	8543	8549	8555	8561	8567
72	8673	8579	8585	8591	8597	8603	8609	8615	8621	8627
73	8633	8639	8645	8651	8657	8663	8669	8675	8681	8686
74	8692	8698	8704	8710	8716	8722	8727	8733	8739	8745
75	8751	8756	8762	8768	8774	8779	8785	8791	8797	8802
76	8808	8814	8820	8825	8831	8837	8842	8848	8854	8859
77	8865	8871	8876	8882	8887	8893	8899	8904	8910	8915
78	8921	8927	8932	8938	8943	8949	8954	8960	8965	8971
79	8976	8982	8987	8993	8998	9004	9009	9015	9020	9025
80	9031	9036	9042	9047	9053	9058	9063	9069	9074	9079
81	9085	9090	9096	9101	9106	9112	9117	9122	9128	9133
82	9138	9143	9149	9154	9159	9165	9170	9175	9180	9186
83	9191	9196	9201	9206	9212	9217	9222	9227	9232	9238
84	9243	9248	9253	9258	9263	9269	9274	9279	9284	9289
85	9294	9299	9304	9309	9315	9320	9325	9330	9335	9340
86	9345	9350	9355	9360	9365	9370	9375	9380	9385	9390
87	9395	9400	9405	9410	9415	9420	9425	9430	9435	9440
88	9445	9450	9455	9460	9465	9469	9474	9479	9484	9489
89	9494	9499	9504	9509	9513	9518	9523	9528	9533	9538
90	9542	9547	9552	9557	9562	9566	9571	9576	9581	9586
91	9590	9595	9600	9605	9609	9614	9619	9624	9628	9633
92	9638	9643	9647	9652	9657	9661	9666	9671	9675	9680
93	9685	9689	9694	9699	9703	9708	9713	9717	9722	9727
94	9731	9736	9741	9745	9750	9754	9759	9763	9768	9773
95	9777	9782	9786	9791	9795	9800	9805	9809	9814	9818
96	9823	9827	9832	9836	9841	9845	9850	9854	9859	9863
97	9868	9872	9877	9881	9886	9890	9894	9899	9903	9908
98	9912	9917	9921	9926	9930	9934	9939	9943	9948	9952
99	9956	9961	9965	9969	9974	9978	9983	9987	9991	9996

TABLE IV **LOGARITHMS OF TRIGONOMETRIC FUNCTIONS**

Degrees	Log Sine	Log Tangent	Log Cotangent	Log Cosine	
0° 00′					**90° 00′**
10′	.4637−3	.4637−3	2.5363	.0000	50′
20′	.7648−3	.7648−3	2.2352	.0000	40′
30′	.9408−3	.9409−3	2.0591	.0000	30′
40′	.0658−2	.0658−2	1.9342	.0000	20′
50′	.1627−2	.1627−2	1.8373	.0000	10′
1° 00′	.2419−2	.2419−2	1.7581	.9999−1	**89° 00′**
10′	.3088−2	.3089−2	1.6911	.9999−1	50′
20′	.3668−2	.3669−2	1.6331	.9999−1	40′
30′	.4179−2	.4181−2	1.5819	.9999−1	30′
40′	.4637−2	.4638−2	1.5362	.9998−1	20′
50′	.5050−2	.5053−2	1.4947	.9998−1	10′
2° 00′	.5428−2	.5431−2	1.4569	.9997−1	**88° 00′**
10′	.5776−2	.5779−2	1.4221	.9997−1	50′
20′	.6097−2	.6101−2	1.3899	.9996−1	40′
30′	.6397−2	.6401−2	1.3599	.9996−1	30′
40′	.6677−2	.6682−2	1.3318	.9995−1	20′
50′	.6940−2	.6945−2	1.3055	.9995−1	10′
3° 00′	.7188−2	.7194−2	1.2806	.9994−1	**87° 00′**
10′	.7423−2	.7429−2	1.2571	.9993−1	50′
20′	.7645−2	.7652−2	1.2348	.9993−1	40′
30′	.7857−2	.7865−2	1.2135	.9992−1	30′
40′	.8059−2	.8067−2	1.1933	.9991−1	20′
50′	.8251−2	.8261−2	1.1739	.9990−1	10′
4° 00′	.8436−2	.8446−2	1.1554	.9989−1	**86° 00′**
10′	.8613−2	.8624−2	1.1376	.9989−1	50′
20′	.8783−2	.8795−2	1.1205	.9988−1	40′
30′	.8946−2	.8960−2	1.1040	.9987−1	30′
40′	.9104−2	.9118−2	1.0882	.9986−1	20′
50′	.9256−2	.9272−2	1.0728	.9985−1	10′
5° 00′	.9403−2	.9420−2	1.0580	.9983−1	**85° 00′**
10′	.9545−2	.9563−2	1.0437	.9982−1	50′
20′	.9682−2	.9701−2	1.0299	.9981−1	40′
30′	.9816−2	.9836−2	1.0164	.9980−1	30′
40′	.9945−2	.9966−2	1.0034	.9979−1	20′
50′	.0070−1	.0093−1	.9907	.9977−1	10′
6° 00′	.0192−1	.0216−1	.9784	.9976−1	**84° 00′**
10′	.0311−1	.0336−1	.9664	.9975−1	50′
20′	.0426−1	.0453−1	.9547	.9973−1	40′
30′	.0539−1	.0567−1	.9433	.9972−1	30′
40′	.0648−1	.0678−1	.9322	.9971−1	20′
50′	.0755−1	.0786−1	.9214	.9969−1	10′
7° 00′	.0859−1	.0891−1	.9109	.9968−1	**83° 00′**
10′	.0961−1	.0995−1	.9005	.9966−1	50′
20′	.1060−1	.1096−1	.8904	.9964−1	40′
30′	.1157−1	.1194−1	.8806	.9963−1	30′
40′	.1252−1	.1291−1	.8709	.9961−1	20′
50′	.1345−1	.1385−1	.8615	.9959−1	10′
8° 00′	.1436−1	.1478−1	.8522	.9958−1	**82° 00′**
10′	.1525−1	.1569−1	.8431	.9956−1	50′
20′	.1612−1	.1658−1	.8342	.9954−1	40′
30′	.1697−1	.1745−1	.8255	.9952−1	30′
40′	.1781−1	.1831−1	.8169	.9950−1	20′
50′	.1863−1	.1915−1	.8085	.9948−1	10′
9° 00′	.1943−1	.1997−1	.8003	.9946−1	**81° 00′**
	Log Cosine	Log Cotangent	Log Tangent	Log Sine	Degrees

TABLE IV **LOGARITHMS OF TRIGONOMETRIC FUNCTIONS** — Cont.

Degrees	Log Sine	Log Tangent	Log Cotangent	Log Cosine	
9° 00′	.1943−1	.1997−1	.8003	.9946−1	**81° 00′**
10′	.2022−1	.2078−1	.7922	.9944−1	50′
20′	.2100−1	.2158−1	.7842	.9942−1	40′
30′	.2176−1	.2236−1	.7764	.9940−1	30′
40′	.2251−1	.2313−1	.7687	.9938−1	20′
50′	.2324−1	.2389−1	.7611	.9936−1	10′
10° 00′	.2397−1	.2463−1	.7537	.9934−1	**80° 00′**
10′	.2468−1	.2536−1	.7464	.9931−1	50′
20′	.2538−1	.2609−1	.7391	.9929−1	40′
30′	.2606−1	.2680−1	.7320	.9927−1	30′
40′	.2674−1	.2750−1	.7250	.9924−1	20′
50′	.2740−1	.2819−1	.7181	.9922−1	10′
11° 00′	.2806−1	.2887−1	.7113	.9919−1	**79° 00′**
10′	.2870−1	.2953−1	.7047	.9917−1	50′
20′	.2934−1	.3020−1	.6980	.9914−1	40′
30′	.2997−1	.3085−1	.6915	.9912−1	30′
40′	.3058−1	.3149−1	.6851	.9909−1	20′
50′	.3119−1	.3212−1	.6788	.9907−1	10′
12° 00′	.3179−1	.3275−1	.6725	.9904−1	**78° 00′**
10′	.3238−1	.3336−1	.6664	.9901−1	50′
20′	.3296−1	.3397−1	.6603	.9899−1	40′
30′	.3353−1	.3458−1	.6542	.9896−1	30′
40′	.3410−1	.3517−1	.6483	.9893−1	20′
50′	.3466−1	.3576−1	.6424	.9890−1	10′
13° 00′	.3521−1	.3634−1	.6366	.9887−1	**77° 00′**
10′	.3575−1	.3691−1	.6309	.9884−1	50′
20′	.3629−1	.3748−1	.6252	.9881−1	40′
30′	.3682−1	.3804−1	.6196	.9878−1	30′
40′	.3734−1	.3859−1	.6141	.9875−1	20′
50′	.3786−1	.3914−1	.6086	.9872−1	10′
14° 00′	.3837−1	.3968−1	.6032	.9869−1	**76° 00′**
10′	.3887−1	.4021−1	.5979	.9866−1	50′
20′	.3937−1	.4074−1	.5926	.9863−1	40′
30′	.3986−1	.4127−1	.5873	.9859−1	30′
40′	.4035−1	.4178−1	.5822	.9856−1	20′
50′	.4083−1	.4230−1	.5770	.9853−1	10′
15° 00′	.4130−1	.4281−1	.5719	.9849−1	**75° 00′**
10′	.4177−1	.4331−1	.5669	.9846−1	50′
20′	.4223−1	.4381−1	.5619	.9843−1	40′
30′	.4269−1	.4430−1	.5570	.9839−1	30′
40′	.4314−1	.4479−1	.5521	.9836−1	20′
50′	.4359−1	.4527−1	.5473	.9832−1	10′
16° 00′	.4403−1	.4575−1	.5425	.9828−1	**74° 00′**
10′	.4447−1	.4622−1	.5378	.9825−1	50′
20′	.4491−1	.4669−1	.5331	.9821−1	40′
30′	.4533−1	.4716−1	.5284	.9817−1	30′
40′	.4576−1	.4762−1	.5238	.9814−1	20′
50′	.4618−1	.4808−1	.5192	.9810−1	10′
17° 00′	.4659−1	.4853−1	.5147	.9806−1	**73° 00′**
10′	.4700−1	.4898−1	.5102	.9802−1	50′
20′	.4741−1	.4943−1	.5057	.9798−1	40′
30′	.4781−1	.4987−1	.5013	.9794−1	30′
40′	.4821−1	.5031−1	.4969	.9790−1	20′
50′	.4861−1	.5075−1	.4925	.9786−1	10′
18° 00′	.4900−1	.5118−1	.4882	.9782−1	**72° 00′**
	Log Cosine	Log Cotangent	Log Tangent	Log Sine	Degrees

TABLE IV **LOGARITHMS OF TRIGONOMETRIC FUNCTIONS** — Cont.

Degrees	Log Sine	Log Tangent	Log Cotangent	Log Cosine	
18° 00′	.4900−1	.5118−1	.4882	.9782−1	**72° 00′**
10′	.4939−1	.5161−1	.4839	.9778−1	50′
20′	.4977−1	.5203−1	.4797	.9774−1	40′
30′	.5015−1	.5245−1	.4755	.9770−1	30′
40′	.5052−1	.5287−1	.4713	.9765−1	20′
50′	.5090−1	.5329−1	.4671	.9761−1	10′
19° 00′	.5126−1	.5370−1	.4630	.9757−1	**71° 00′**
10′	.5163−1	.5411−1	.4589	.9752−1	50′
20′	.5199−1	.5451−1	.4549	.9748−1	40′
30′	.5235−1	.5491−1	.4509	.9743−1	30′
40′	.5270−1	.5531−1	.4469	.9739−1	20′
50′	.5306−1	.5571−1	.4429	.9734−1	10′
20° 00′	.5341−1	.5611−1	.4389	.9730−1	**70° 00′**
10′	.5375−1	.5650−1	.4350	.9725−1	50′
20′	.5409−1	.5689−1	.4311	.9721−1	40′
30′	.5443−1	.5727−1	.4273	.9716−1	30′
40′	.5477−1	.5766−1	.4234	.9711−1	20′
50′	.5510−1	.5804−1	.4196	.9706−1	10′
21° 00′	.5543−1	.5842−1	.4158	.9702−1	**69° 00′**
10′	.5576−1	.5879−1	.4121	.9697−1	50′
20′	.5609−1	.5917−1	.4083	.9692−1	40′
30′	.5641−1	.5954−1	.4046	.9687−1	30′
40′	.5673−1	.5991−1	.4009	.9682−1	20′
50′	.5704−1	.6028−1	.3972	.9677−1	10′
22° 00′	.5736−1	.6064−1	.3936	.9672−1	**68° 00′**
10′	.5767−1	.6100−1	.3900	.9667−1	50′
20′	.5798−1	.6136−1	.3864	.9661−1	40′
30′	.5828−1	.6172−1	.3828	.9656−1	30′
40′	.5859−1	.6208−1	.3792	.9651−1	20′
50′	.5889−1	.6243−1	.3757	.9646−1	10′
23° 00′	.5919−1	.6279−1	.3721	.9640−1	**67° 00′**
10′	.5948−1	.6314−1	.3686	.9635−1	50′
20′	.5978−1	.6348−1	.3652	.9629−1	40′
30′	.6007−1	.6383−1	.3617	.9624−1	30′
40′	.6036−1	.6417−1	.3583	.9618−1	20′
50′	.6065−1	.6452−1	.3548	.9613−1	10′
24° 00′	.6093−1	.6486−1	.3514	.9607−1	**66° 00′**
10′	.6121−1	.6520−1	.3480	.9602−1	50′
20′	.6149−1	.6553−1	.3447	.9596−1	40′
30′	.6177−1	.6587−1	.3413	.9590−1	30′
40′	.6205−1	.6620−1	.3380	.9584−1	20′
50′	.6232−1	.6654−1	.3346	.9579−1	10′
25° 00′	.6259−1	.6687−1	.3313	.9573−1	**65° 00′**
10′	.6286−1	.6720−1	.3280	.9567−1	50′
20′	.6313−1	.6752−1	.3248	.9561−1	40′
30′	.6340−1	.6785−1	.3215	.9555−1	30′
40′	.6366−1	.6817−1	.3183	.9549−1	20′
50′	.6392−1	.6850−1	.3150	.9543−1	10′
26° 00′	.6418−1	.6882−1	.3118	.9537−1	**64° 00′**
10′	.6444−1	.6914−1	.3086	.9530−1	50′
20′	.6470−1	.6946−1	.3054	.9524−1	40′
30′	.6495−1	.6977−1	.3023	.9518−1	30′
40′	.6521−1	.7009−1	.2991	.9512−1	20′
50′	.6546−1	.7040−1	.2960	.9505−1	10′
27° 00′	.6570−1	.7072−1	.2928	.9499−1	**63° 00′**
	Log Cosine	Log Cotangent	Log Tangent	Log Sine	Degrees

TABLE IV **LOGARITHMS OF TRIGONOMETRIC FUNCTIONS** – Cont.

Degrees	Log Sine	Log Tangent	Log Cotangent	Log Cosine	
27° 00′	.6570−1	.7072−1	.2928	.9499−1	**63° 00′**
10′	.6595−1	.7103−1	.2897	.9492−1	50′
20′	.6620−1	.7134−1	.2866	.9486−1	40′
30′	.6644−1	.7165−1	.2835	.9479−1	30′
40′	.6668−1	.7196−1	.2804	.9473−1	20′
50′	.6692−1	.7226−1	.2774	.9466−1	10′
28° 00′	.6716−1	.7257−1	.2743	.9459−1	**62° 00′**
10′	.6740−1	.7287−1	.2713	.9453−1	50′
20′	.6763−1	.7317−1	.2683	.9446−1	40′
30′	.6787−1	.7348−1	.2652	.9439−1	30′
40′	.6810−1	.7378−1	.2622	.9432−1	20′
50′	.6833−1	.7408−1	.2592	.9425−1	10′
29° 00′	.6856−1	.7438−1	.2562	.9418−1	**61° 00′**
10′	.6878−1	.7467−1	.2533	.9411−1	50′
20′	.6901−1	.7497−1	.2503	.9404−1	40′
30′	.6923−1	.7526−1	.2474	.9397−1	30′
40′	.6946−1	.7556−1	.2444	.9390−1	20′
50′	.6968−1	.7585−1	.2415	.9383−1	10′
30° 00′	.6990−1	.7614−1	.2386	.9375−1	**60° 00′**
10′	.7012−1	.7644−1	.2356	.9368−1	50′
20′	.7033−1	.7673−1	.2327	.9361−1	40′
30′	.7055−1	.7701−1	.2299	.9353−1	30′
40′	.7076−1	.7730−1	.2270	.9346−1	20′
50′	.7097−1	.7759−1	.2241	.9338−1	10′
31° 00′	.7118−1	.7788−1	.2212	.9331−1	**59° 00′**
10′	.7139−1	.7816−1	.2184	.9323−1	50′
20′	.7160−1	.7845−1	.2155	.9315−1	40′
30′	.7181−1	.7873−1	.2127	.9308−1	30′
40′	.7201−1	.7902−1	.2098	.9300−1	20′
50′	.7222−1	.7930−1	.2070	.9292−1	10′
32° 00′	.7242−1	.7958−1	.2042	.9284−1	**58° 00′**
10′	.7262−1	.7986−1	.2014	.9276−1	50′
20′	.7282−1	.8014−1	.1986	.9268−1	40′
30′	.7302−1	.8042−1	.1958	.9260−1	30′
40′	.7322−1	.8070−1	.1930	.9252−1	20′
50′	.7342−1	.8097−1	.1903	.9244−1	10′
33° 00′	.7361−1	.8125−1	.1875	.9236−1	**57° 00′**
10′	.7380−1	.8153−1	.1847	.9228−1	50′
20′	.7400−1	.8180−1	.1820	.9219−1	40′
30′	.7419−1	.8208−1	.1792	.9211−1	30′
40′	.7438−1	.8235−1	.1765	.9203−1	20′
50′	.7457−1	.8263−1	.1737	.9194−1	10′
34° 00′	.7476−1	.8290−1	.1710	.9186−1	**56° 00′**
10′	.7494−1	.8317−1	.1683	.9177−1	50′
20′	.7513−1	.8344−1	.1656	.9169−1	40′
30′	.7531−1	.8371−1	.1629	.9160−1	30′
40′	.7550−1	.8398−1	.1602	.9151−1	20′
50′	.7568−1	.8425−1	.1575	.9142−1	10′
35° 00′	.7586−1	.8452−1	.1548	.9134−1	**55° 00′**
10′	.7604−1	.8479−1	.1521	.9125−1	50′
20′	.7622−1	.8506−1	.1494	.9116−1	40′
30′	.7640−1	.8533−1	.1467	.9107−1	30′
40′	.7657−1	.8559−1	.1441	.9098−1	20′
50′	.7675−1	.8586−1	.1414	.9089−1	10′
36° 00′	.7692−1	.8613−1	.1387	.9080−1	**54° 00′**
	Log Cosine	Log Cotangent	Log Tangent	Log Sine	Degrees

TABLE IV **LOGARITHMS OF TRIGONOMETRIC FUNCTIONS** — Cont.

Degrees	Log Sine	Log Tangent	Log Cotangent	Log Cosine	
36° 00′	.7692−1	.8613−1	.1387	.9080−1	**54° 00′**
10′	.7710−1	.8639−1	.1361	.9070−1	50′
20′	.7727−1	.8666−1	.1334	.9061−1	40′
30′	.7744−1	.8692−1	.1308	.9052−1	30′
40′	.7761−1	.8718−1	.1282	.9042−1	20′
50′	.7778−1	.8745−1	.1255	.9033−1	10′
37° 00′	.7795−1	.8771−1	.1229	.9023−1	**53° 00′**
10′	.7811−1	.8797−1	.1203	.9014−1	50′
20′	.7828−1	.8824−1	.1176	.9004−1	40′
30′	.7844−1	.8850−1	.1150	.8995−1	30′
40′	.7861−1	.8876−1	.1124	.8985−1	20′
50′	.7877−1	.8902−1	.1098	.8975−1	10′
38° 00′	.7893−1	.8928−1	.1072	.8965−1	**52° 00′**
10′	.7910−1	.8954−1	.1046	.8955−1	50′
20′	.7926−1	.8980−1	.1020	.8945−1	40′
30′	.7941−1	.9006−1	.0994	.8935−1	30′
40′	.7957−1	.9032−1	.0968	.8925−1	20′
50′	.7973−1	.9058−1	.0942	.8915−1	10′
39° 00′	.7989−1	.9084−1	.0916	.8905−1	**51° 00′**
10′	.8004−1	.9110−1	.0890	.8895−1	50′
20′	.8020−1	.9135−1	.0865	.8884−1	40′
30′	.8035−1	.9161−1	.0839	.8874−1	30′
40′	.8050−1	.9187−1	.0813	.8864−1	20′
50′	.8066−1	.9212−1	.0788	.8853−1	10′
40° 00′	.8081−1	.9238−1	.0762	.8843−1	**50° 00′**
10′	.8096−1	.9264−1	.0736	.8832−1	50′
20′	.8111−1	.9289−1	.0711	.8821−1	40′
30′	.8125−1	.9315−1	.0685	.8810−1	30′
40′	.8140−1	.9341−1	.0659	.8800−1	20′
50′	.8155−1	.9366−1	.0634	.8789−1	10′
41° 00′	.8169−1	.9392−1	.0608	.8778−1	**49° 00′**
10′	.8184−1	.9417−1	.0583	.8767−1	50′
20′	.8198−1	.9443−1	.0557	.8756−1	40′
30′	.8213−1	.9468−1	.0532	.8745−1	30′
40′	.8227−1	.9494−1	.0506	.8733−1	20′
50′	.8241−1	.9519−1	.0481	.8722−1	10′
42° 00′	.8255−1	.9544−1	.0456	.8711−1	**48° 00′**
10′	.8269−1	.9570−1	.0430	.8699−1	50′
20′	.8283−1	.9595−1	.0405	.8688−1	40′
30′	.8297−1	.9621−1	.0379	.8676−1	30′
40′	.8311−1	.9646−1	.0354	.8665−1	20′
50′	.8324−1	.9671−1	.0329	.8653−1	10′
43° 00′	.8338−1	.9697−1	.0303	.8641−1	**47° 00′**
10′	.8351−1	.9724−1	.0278	.8629−1	50′
20′	.8365−1	.9747−1	.0253	.8618−1	40′
30′	.8378−1	.9772−1	.0228	.8606−1	30′
40′	.8391−1	.9798−1	.0202	.8594−1	20′
50′	.8405−1	.9823−1	.0177	.8582−1	10′
44° 00′	.8418−1	.9848−1	.0152	.8569−1	**46° 00′**
10′	.8431−1	.9874−1	.0126	.8557−1	50′
20′	.8444−1	.9899−1	.0101	.8545−1	40′
30′	.8457−1	.9924−1	.0076	.8532−1	30′
40′	.8469−1	.9949−1	.0051	.8520−1	20′
50′	.8482−1	.9975−1	.0025	.8507−1	10′
45° 00′	.8495−1	.0000	.0000	.8495−1	**45° 00′**
	Log Cosine	Log Cotangent	Log Tangent	Log Sine	Degrees

TABLE V **POWERS AND ROOTS**

Number	Square	Square Root	Cube	Cube Root
1	1	1.000	1	1.000
2	4	1.414	8	1.260
3	9	1.732	27	1.442
4	16	2.000	64	1.587
5	25	2.236	125	1.710
6	36	2.449	216	1.817
7	49	2.646	343	1.913
8	64	2.828	512	2.000
9	81	3.000	729	2.080
10	100	3.162	1,000	2.154
11	121	3.317	1,331	2.224
12	144	3.464	1,728	2.289
13	169	3.606	2,197	2.351
14	196	3.742	2,744	2.410
15	225	3.873	3,375	2.466
16	256	4.000	4,096	2.520
17	289	4.123	4,913	2.571
18	324	4.243	5,832	2.621
19	361	4.359	6,859	2.668
20	400	4.472	8,000	2.714
21	441	4.583	9,261	2.759
22	484	4.690	10,648	2.802
23	529	4.796	12,167	2.844
24	576	4.899	13,824	2.884
25	625	5.000	15,625	2.924
26	676	5.099	17,576	2.962
27	729	5.196	19,683	3.000
28	784	5.292	21,952	3.037
29	841	5.385	24,389	3.072
30	900	5.477	27,000	3.107
31	961	5.568	29,791	3.141
32	1,024	5.657	32,768	3.175
33	1,089	5.745	35,937	3.208
34	1,156	5.831	39,304	3.240
35	1,225	5.916	42,875	3.271
36	1,296	6.000	46,656	3.302
37	1,369	6.083	50,653	3.332
38	1,444	6.164	54,872	3.362
39	1,521	6.245	59,319	3.391
40	1,600	6.325	64,000	3.420
41	1,681	6.403	68,921	3.448
42	1,764	6.481	74,088	3.476
43	1,849	6.557	79,507	3.503
44	1,936	6.633	85,184	3.530
45	2,025	6.708	91,125	3.557
46	2,116	6.782	97,336	3.583
47	2,209	6.856	103,823	3.609
48	2,304	6.928	110,592	3.634
49	2,401	7.000	117,649	3.659
50	2,500	7.071	125,000	3.684
51	2,601	7.141	132,651	3.708
52	2,704	7.211	140,608	3.733
53	2,809	7.280	148,877	3.756
54	2,916	7.348	157,464	3.780
55	3,025	7.416	166,375	3.803
56	3,136	7.483	175,616	3.826
57	3,249	7.550	185,193	3.849
58	3,364	7.616	195,112	3.871
59	3,481	7.681	205,379	3.893
60	3,600	7.746	216,000	3.915
61	3,721	7.810	226,981	3.936
62	3,844	7.874	238,328	3.958
63	3,969	7.937	250,047	3.979
64	4,096	8.000	262,144	4.000
65	4,225	8.062	274,625	4.021
66	4,356	8.124	287,496	4.041
67	4,489	8.185	300,763	4.062
68	4,624	8.246	314,432	4.082
69	4,761	8.307	328,509	4.102
70	4,900	8.367	343,000	4.121
71	5,041	8.426	357,911	4.141
72	5,184	8.485	373,248	4.160
73	5,329	8.544	389,017	4.179
74	5,476	8.602	405,224	4.198
75	5,625	8.660	421,875	4.217
76	5,776	8.718	438,976	4.236
77	5,929	8.775	456,533	4.254
78	6,084	8.832	474,552	4.273
79	6,241	8.888	493,039	4.291
80	6,400	8.944	512,000	4.309
81	6,561	9.000	531,441	4.327
82	6,724	9.055	551,368	4.344
83	6,889	9.110	571,787	4.362
84	7,056	9.165	592,704	4.380
85	7,225	9.220	614,125	4.397
86	7,396	9.274	636,056	4.414
87	7,569	9.327	658,503	4.431
88	7,744	9.381	681,472	4.448
89	7,921	9.434	704,969	4.465
90	8,100	9.487	729,000	4.481
91	8,281	9.539	753,571	4.498
92	8,464	9.592	778,688	4.514
93	8,649	9.644	804,357	4.531
94	8,836	9.695	830,584	4.547
95	9,025	9.747	857,375	4.563
96	9,216	9.798	884,736	4.579
97	9,409	9.849	912,673	4.595
98	9,604	9.899	941,192	4.610
99	9,801	9.950	970,299	4.626
100	10,000	10.000	1,000,000	4.642

Answers to Selected Problems

Chapter 1

PROBLEM SET 1, Page 7

1. a) $\{x \mid x = 2n, n \in N\}$ c) $\{x \mid 2 < x < 13, x \in N\}$

2. a) T c) T e) T g) F

3. a) Proper subset c) Proper subset

4. a) $\{2\}$; proper subset is $\emptyset$
c) $\{a, b, c\}$; proper subsets are $\{a\}$, $\{b\}$, $\{c\}$, $\{a, b\}$, $\{a, c\}$, $\{b, c\}$, $\emptyset$
e) $\{5, 6, 7, 8\}$; proper subsets are $\{5\}$, $\{6\}$, $\{7\}$, $\{8\}$, $\{5, 6\}$, $\{5, 7\}$, $\{5, 8\}$, $\{6, 7\}$, $\{6, 8\}$, $\{7, 8\}$, $\{5, 6, 7\}$, $\{5, 6, 8\}$, $\{5, 7, 8\}$, $\{6, 7, 8\}$, $\emptyset$

5. a) $\{2\}$ c) $\{3, 5\}$ e) $\emptyset$
g) $\emptyset$ i) $\{1, 2, 3, 4, 5, 6, 8\}$ k) $\{2\}$

7. a) $\emptyset$ c) N e) R

PROBLEM SET 2, Page 14

1. a) Rational c) Irrational e) Rational
g) Irrational i) Rational

2. a) Additive Property of Inequalities c) Transitive
e) Trichotomy

3. a) T c) T e) T

4. a) $(1, 4) \cup (7, 10)$ c) $(-\infty, -1) \cup (5, 8)$ e) $(-\infty, -1) \cup (1, \infty)$

5. a) $(-\infty, 1)$

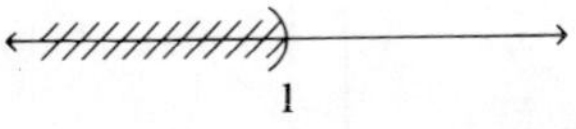

1

c) $[\frac{15}{2}, \infty)$

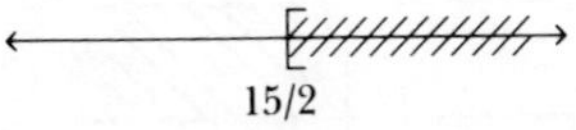

15/2

e) $(\frac{5}{3}, \infty)$

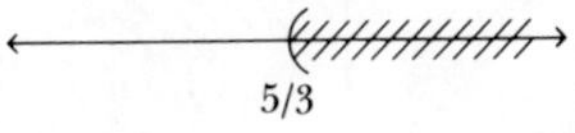

5/3

g) $(-\infty, -\frac{1}{3}]$

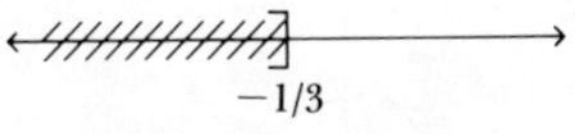

−1/3

i) $(-\infty, -1)$

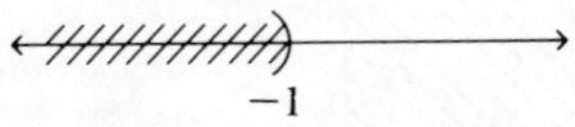

−1

PROBLEM SET 3, Page 21

1. a) 12 c) 12 e) 35
g) 49 i) 43

2. a) $\{-5, 5\}$ c) $\{-1, 5\}$ e) $\{-11, 1\}$
g) $\{1, 3\}$ i) $\{1, 3\}$

3. a) $(-3, 3)$

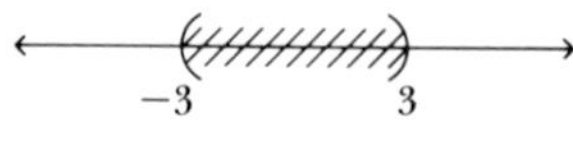

c) $[-2, 4]$

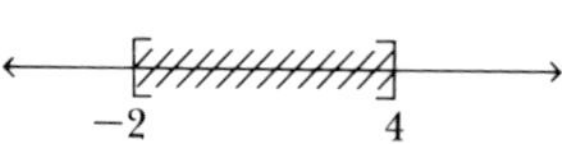

e) $(-\infty, 0) \cup (\frac{2}{3}, \infty)$

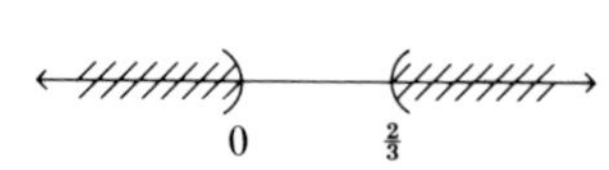

g) $(-\infty, -3] \cup [2, \infty)$

i) $(-\infty, -3)$

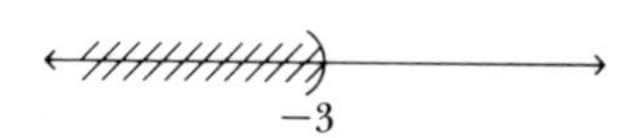

k) $(\frac{1}{6}, \frac{1}{4})$

4. a) 1 if $x > 0$, -1 if $x < 0$ c) $x \geqslant 0$ and $y \geqslant 0$; $x \leqslant 0$ and $y \leqslant 0$

5. a) $x \geqslant 0$ c) $x \geqslant 0$ e) $x \neq 0$

6. a) $x \geqslant 0$ c) $(\frac{1}{2}, \infty)$ e) R

7. a) $|x + \frac{3}{2}| < \frac{9}{2}$ c) $|x + 3.05| < 0.05$

PROBLEM SET 4, Page 28

1. a) $A = \{(-1, -2), (0, 1), (1, 4), (2, 7)\}$
c) $C = \{(1, 2), (2, 4), (3, 6), (4, 8), (5, 10)\}$

2. a) I c) I e) III
g) II i) IV

4. a)

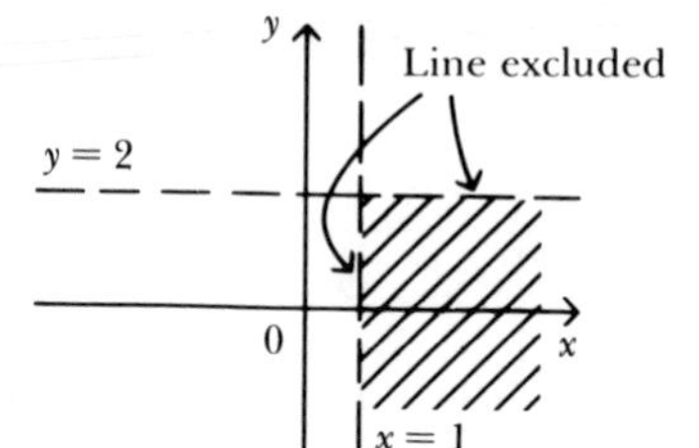

c)

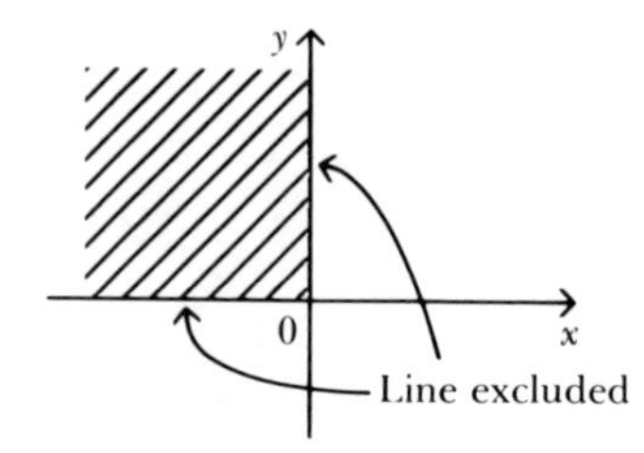

e)

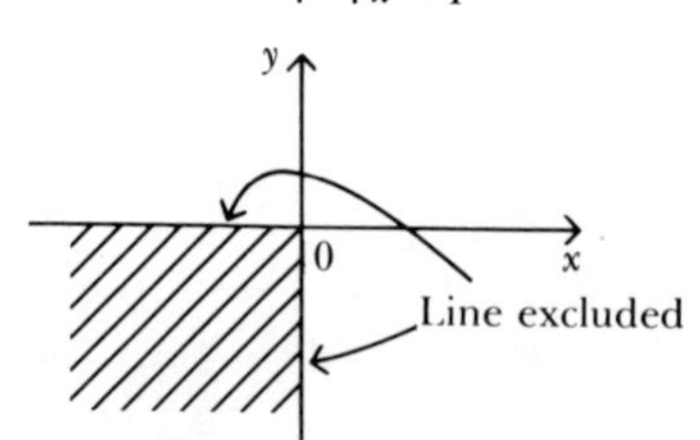

g)

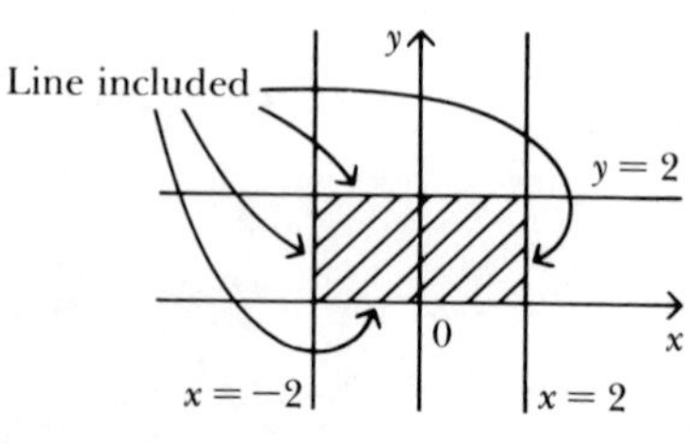

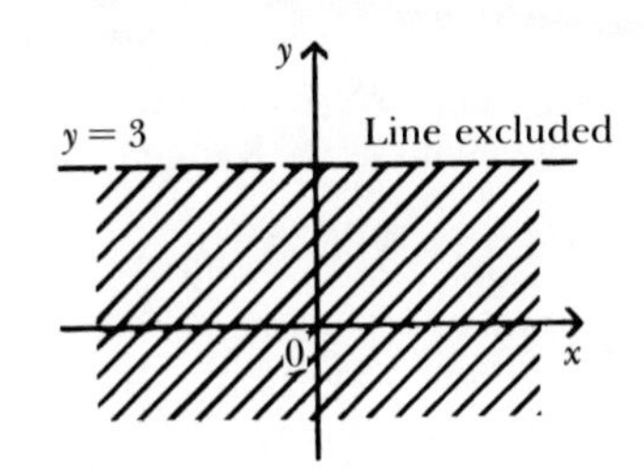

5. a) $\dfrac{\sqrt{41}}{2}$ c) $3\sqrt{17}$ e) $\sqrt{17}$

7. d) i) $(-1, 7)$ ii) $(-\frac{7}{2}, \frac{7}{2})$ iii) $(\frac{11}{2}, \frac{11}{2})$

PROBLEM SET 5, Page 39

1. a) Function
Domain = $\{2, 3, 7, 9\}$
Range = $\{4, 6, 2, -3\}$

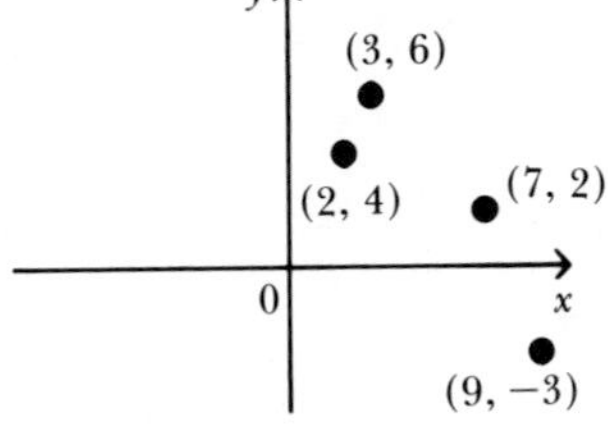

c) Function
Domain = R
Range = R

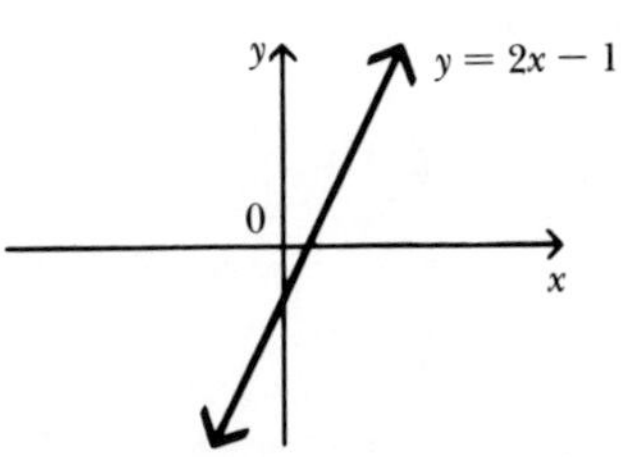

e) Function
Domain = R
Range = $\{y \mid y \geqslant 0\}$

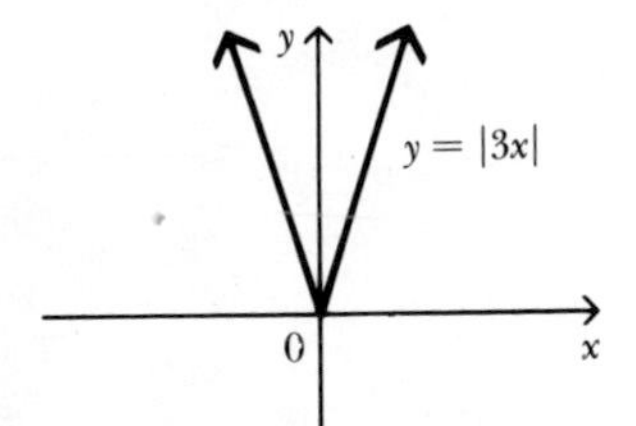

g) Function
Domain = $\{x \mid x > 0\}$
Range = R

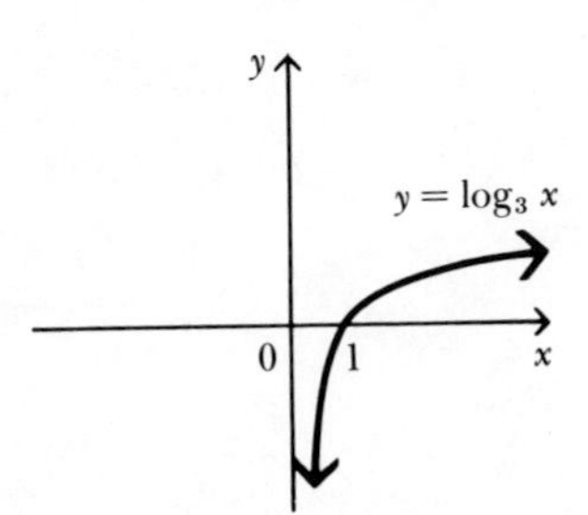

2. a) $x = 4$ c) $x = \pm \dfrac{3\sqrt{2}}{2}$

3. a) Function c) Not a function

4. a) $f(1) = 4, f(3) = 10, f(5) = 16$, Domain $= R$, Range $= R$, zero is $-\frac{1}{3}$

c) $f(1) = 0,\ f(3) = 2,\ f(5) = 4$, Domain $= R$, Range $= \{y \mid y \geqslant -1\}$, zeros are -1 and 1

e) $f(1) = 5, f(3) = 5^3 = 125, f(5) = 5^5 = 3125$, Domain $= R$, Range $= \{y \mid y > 0\}$, no zeros

g) $f(1) = 2,\ f(3) = \frac{2}{3},\ f(5) = \frac{2}{5}$, Domain $= \{x \mid x \neq 0\}$, Range $= \{y \mid y \neq 0\}$, no zeros

i) $f(1) = 0,\ f(3) = 0.4771,\ f(5) = 0.6990$, Domain $= \{x \mid x > 0\}$, Range $= R$, zero is 1

6. a) $(1, 0), (-2, 4), (-3, 6)$

c) $f(3) = 0, 2f(-3) = 12, f(2) = 0, f(-2) = 4, f(a+1) = |a+1| - a - 1$

7. $f[g(x)] = 3(-2x^2) - 1 = -6x^2 - 1$

$g[f(x)] = -2(3x-1)^2 = -18x^2 + 12x - 2$

PROBLEM SET 6, Page 46

1. a) $f(x) = -2x$ is symmetric with respect to the origin

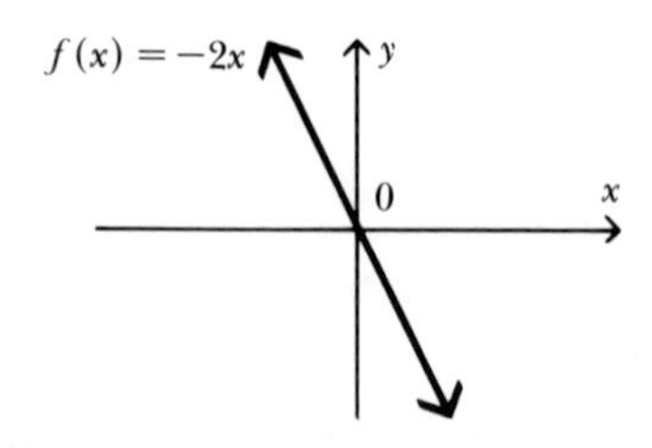

c) $f(x) = 5x^3$ is symmetric with respect to the origin

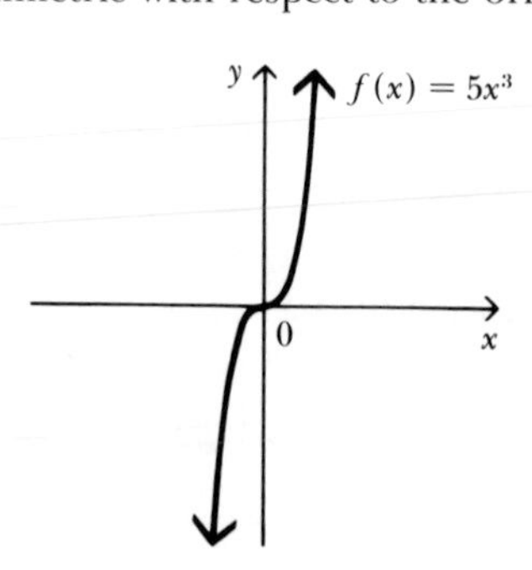

e) $f(x) = -2x^4$ is symmetric with respect to the y axis

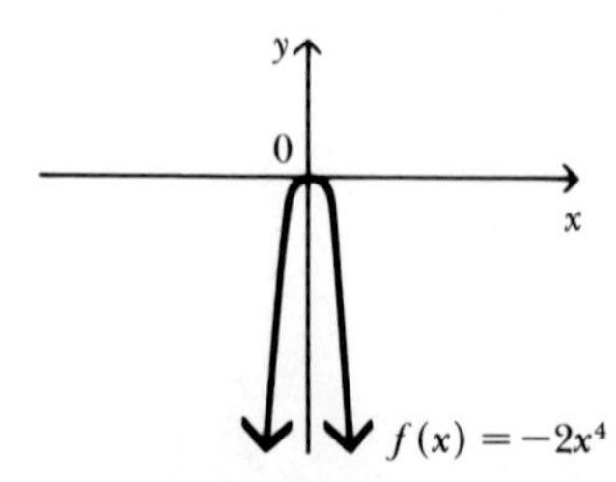

3. a) Odd c) Neither e) Even

5. Decreasing in [1, 4]

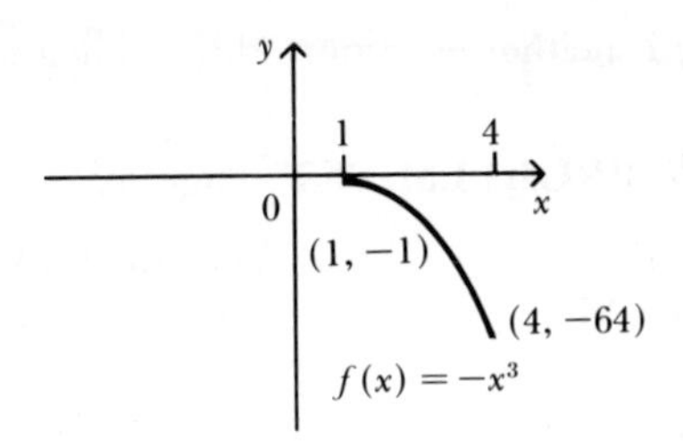

6. a) Decreasing on R

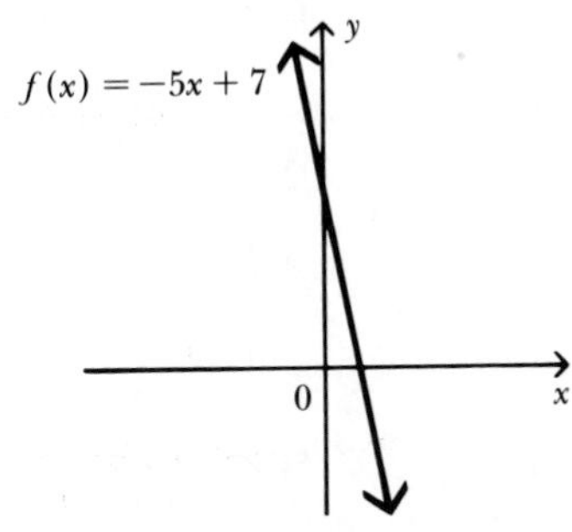

c) Increasing on R

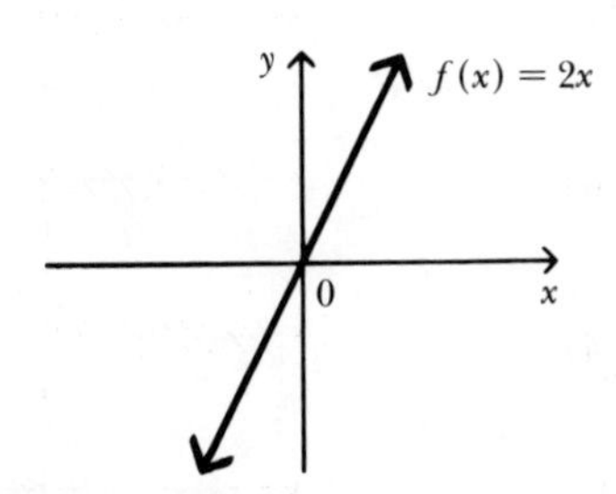

e) Increasing on R

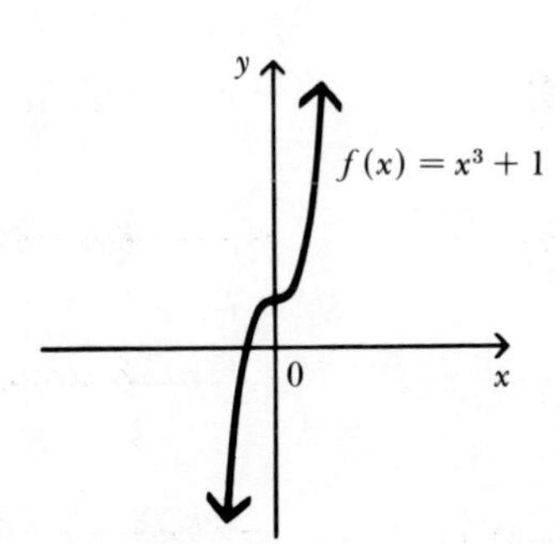

g) Increasing in $(0, \infty)$

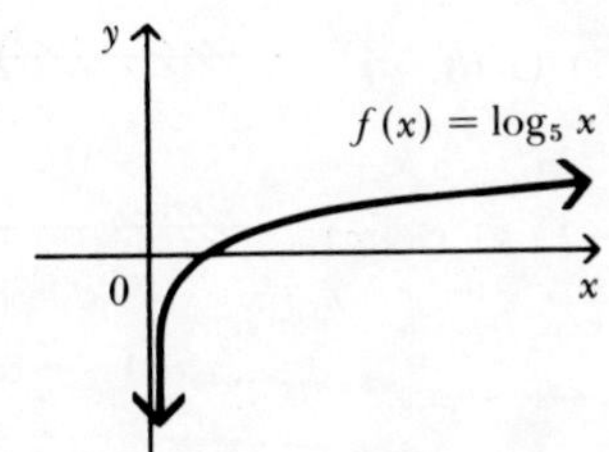

7. a) f is neither even nor odd; g is neither even nor odd

REVIEW PROBLEM SET, Page 47

1. a) $\{3\}$ c) $\{3, 4, 6\}$ e) $\{1, 3\}$
g) $\{1, 3\}$ i) $\{1, 2, 3, 4, 6\}$

2. a) $(-\infty, 1]$

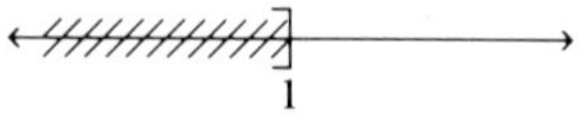

c) R

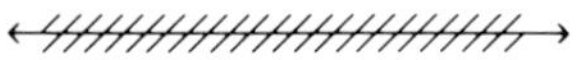

e) $[0, 2]$

3. a) F c) F e) T
g) T

4. a) F c) T e) T

5. a) $(-\infty, -4]$

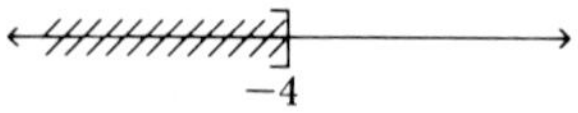

c) $(-\infty, \frac{7}{4})$

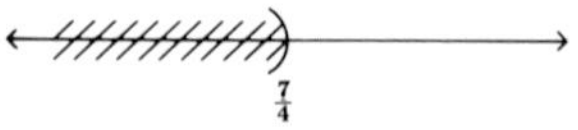

e) $(-\infty, 0]$

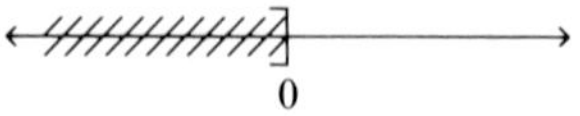

g) $(-\infty, 4)$

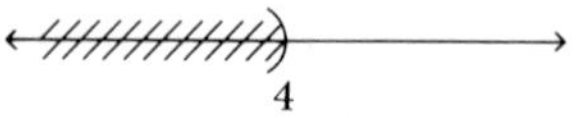

i) $[0, \infty)$

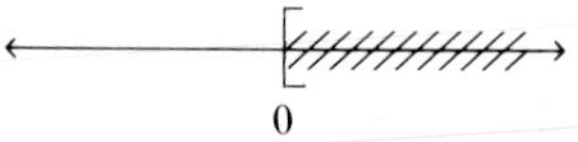

6. a) $\{-\frac{5}{2}, \frac{5}{2}\}$ c) $\{-\frac{16}{3}, \frac{8}{3}\}$ e) $\{-\frac{2}{7}, \frac{8}{7}\}$
g) $\{1, \frac{11}{5}\}$

7. a) $(-7, 7)$

c) $(\frac{7}{2}, \frac{13}{2})$

e) $(-\infty, -3) \cup (\frac{7}{3}, \infty)$

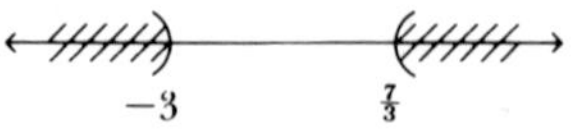

g) $(-\infty, 3) \cup (3, \infty)$

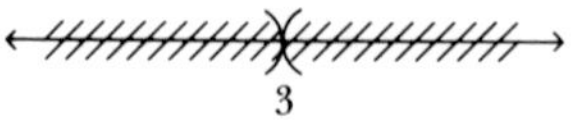

i) $(-\infty, -1) \cup (5, \infty)$

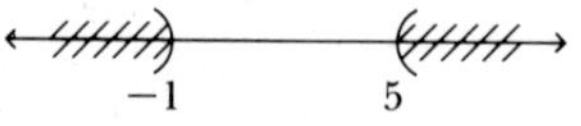

8. a) I c) II e) III
g) None

9. a) $\{(3, 3), (3, -3), (3, 4), (-3, 3), (-3, -3), (-3, 4), (4, 3), (4, -3), (4, 4)\}$
c) $\{(3, -1), (3, 2), (3, 4), (-3, -1), (-3, 2), (-3, 4), (4, -1), (4, 2), (4, 4)\}$

11. a) Function; Domain $= \{4, 3, 5, -1\}$, Range $= \{1, -1\}$
c) Not a function; Domain $= \{0, 1, 2\}$, Range $= \{1, -1, 3, 4\}$

12. a) Function c) Not a function e) Not a function

13. a) Symmetric with respect to y axis

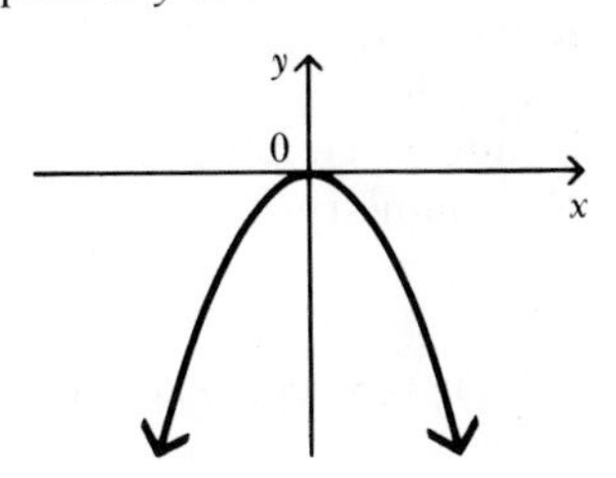

c) Symmetric with respect to the origin

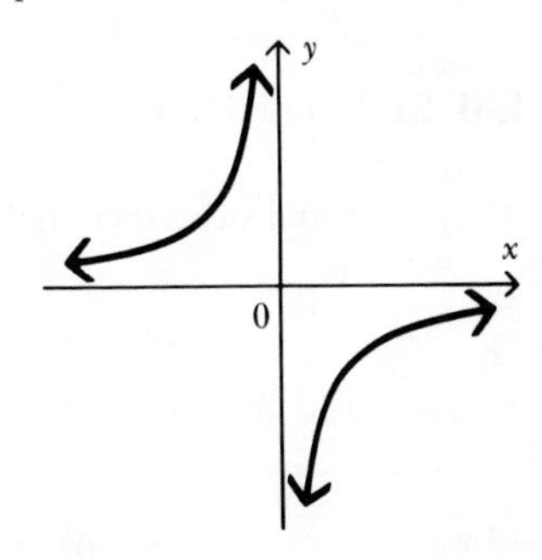

e) Symmetric with respect to the origin

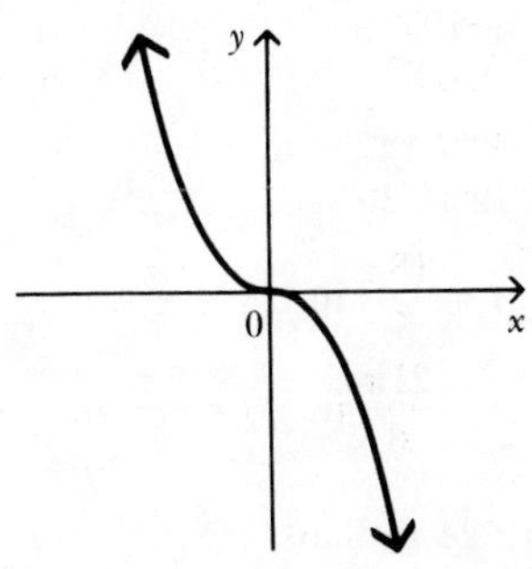

14. a) Domain $= R$
Range $= R$
Increasing
Neither even nor odd
No symmetry

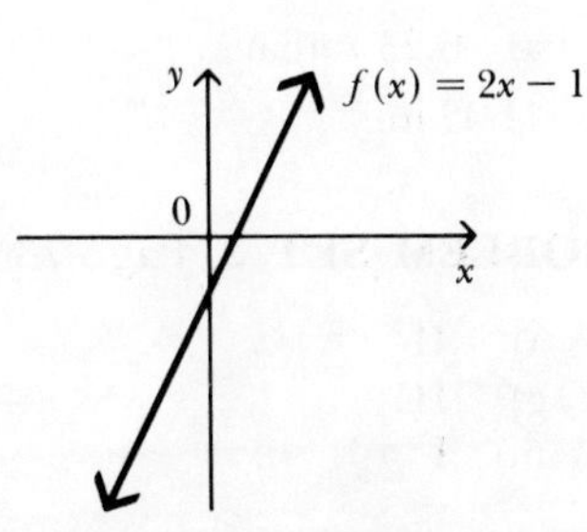

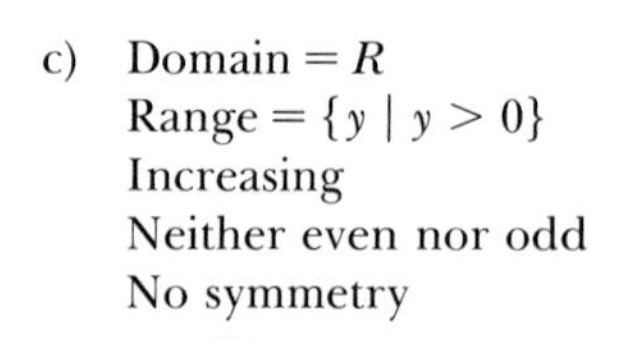

c) Domain $= R$
Range $= \{y \mid y > 0\}$
Increasing
Neither even nor odd
No symmetry

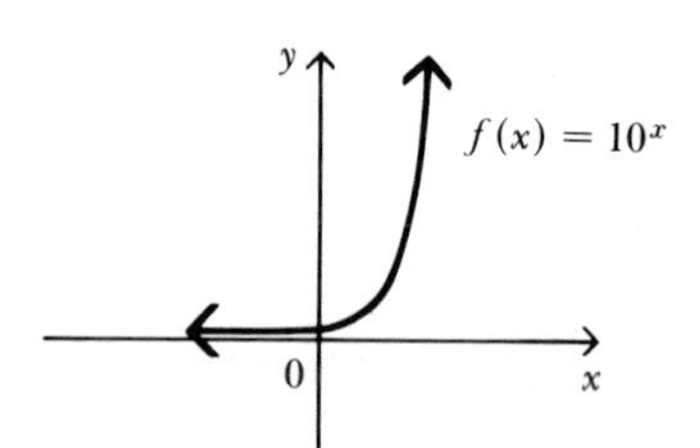

e) Domain $= \{x \mid x > 0\}$
Range $= R$
Increasing
Neither even nor odd
No symmetry

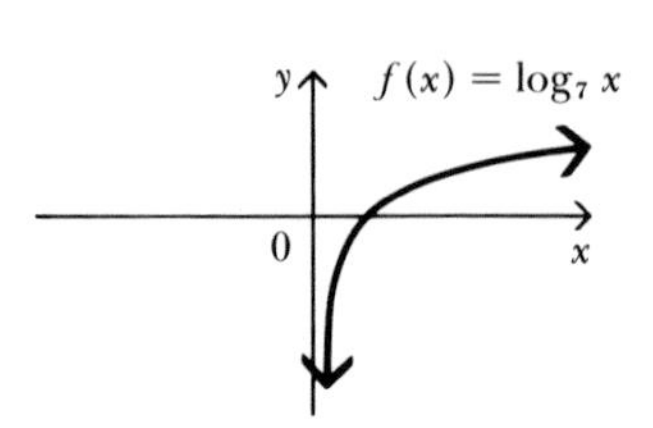

15. a) $f[g(x)] = x,\ g[f(x)] = x$

Chapter 2

PROBLEM SET 1, Page 62

1. a) Vertex; initial side; terminal side c) $\dfrac{180}{\pi}$ degrees

2. a) $\dfrac{\pi}{6}$ c) $-\dfrac{\pi}{4}$ e) $\dfrac{4\pi}{3}$
g) $-\dfrac{7\pi}{4}$ i) $\dfrac{35}{36}\pi$ k) $\dfrac{41\pi}{6}$
m) 3π o) $-\dfrac{11\pi}{9}$ q) $\dfrac{2\pi}{5}$

3. a) $-60°$ c) $-225°$ e) $45°$
g) $-105°$ i) $-157°30'$ k) $210°$
m) $1860°$ o) $72°$

4. a) $32°7'12''$ c) $110°15'$ e) $35°9'36''$
g) $72°18'36''$

5. a) $A = \dfrac{18\pi}{7}$ in.^{2}, $s = \dfrac{6\pi}{7}$ in. $A = \dfrac{40\pi}{3}$ in.^{2}, $s = \dfrac{10\pi}{3}$ in.
e) $A = \dfrac{21\pi}{8}$ in.^{2}, $s = \dfrac{7\pi}{4}$ in.

6. a) 2.24 radians c) 13.06 radians e) $\dfrac{13\pi}{5}$ radians

7. a) 6.23 radians

8. 11.42 in.

PROBLEM SET 2, Page 73

1. a) II c) None e) None
g) III i) IV k) I
m) I o) II q) I

3. a) I c) II e) IV

4. No

5.

	a)	c)	e)
$\sin\theta$	$\frac{\sqrt{3}}{2}$	$\frac{4}{5}$	$\frac{1}{2}$
$\cos\theta$	$-\frac{1}{2}$	$-\frac{3}{5}$	$\frac{\sqrt{3}}{2}$
$\tan\theta$	$-\sqrt{3}$	$-\frac{4}{3}$	$\frac{\sqrt{3}}{3}$
$\cot\theta$	$-\frac{\sqrt{3}}{3}$	$-\frac{3}{4}$	$\sqrt{3}$
$\sec\theta$	-2	$-\frac{5}{3}$	$\frac{2\sqrt{3}}{3}$
$\csc\theta$	$\frac{2\sqrt{3}}{3}$	$\frac{5}{4}$	2

6.

	a)	c)	e)
$\sin\theta$	0	$-\frac{10}{\sqrt{149}}$	$-\frac{\sqrt{3}}{2}$
$\cos\theta$	-1	$\frac{7}{\sqrt{149}}$	$\frac{1}{2}$
$\tan\theta$	0	$-\frac{10}{7}$	$-\sqrt{3}$
$\cot\theta$	Undefined	$-\frac{7}{10}$	$-\frac{\sqrt{3}}{3}$
$\sec\theta$	-1	$\frac{\sqrt{149}}{7}$	2
$\csc\theta$	Undefined	$-\frac{\sqrt{149}}{10}$	$-\frac{2\sqrt{3}}{3}$

7.

	a)	c)	e)
$\sin\theta$	—	$-\frac{1}{2}$	$\frac{12}{13}$
$\cos\theta$	$-\frac{8}{17}$	—	$-\frac{5}{13}$
$\tan\theta$	$-\frac{15}{8}$	$-\frac{\sqrt{3}}{3}$	$-\frac{12}{5}$
$\cot\theta$	$-\frac{8}{15}$	$-\sqrt{3}$	$-\frac{5}{12}$
$\sec\theta$	$-\frac{17}{8}$	$\frac{2\sqrt{3}}{3}$	—
$\csc\theta$	$\frac{17}{15}$	-2	$\frac{13}{12}$

PROBLEM SET 3, Page 86

1.

	a)	c)	e)	g)	i)
	$\theta = -180°$	$\theta = 360°$	$\theta = 810°$	$\theta = 1080°$	$\theta = \frac{11\pi}{2}$
$\sin\theta$	0	0	1	0	-1
$\cos\theta$	-1	1	0	1	0
$\tan\theta$	0	0	Undefined	0	Undefined
$\cot\theta$	Undefined	Undefined	0	Undefined	0
$\sec\theta$	-1	1	Undefined	1	Undefined
$\csc\theta$	Undefined	Undefined	1	Undefined	-1

2.

	a)	c)	e)	g)	i)	k)	m)
	$\theta = 135°$	$\theta = -300°$	$\theta = \frac{11\pi}{6}$	$\theta = -60°$	$\theta = -240°$	$\theta = \frac{13\pi}{6}$	$\theta = 585°$
$\sin\theta$	$\frac{\sqrt{2}}{2}$	$\frac{\sqrt{3}}{2}$	$-\frac{1}{2}$	$-\frac{\sqrt{3}}{2}$	$\frac{\sqrt{3}}{2}$	$\frac{1}{2}$	$-\frac{\sqrt{2}}{2}$
$\cos\theta$	$-\frac{\sqrt{2}}{2}$	$\frac{1}{2}$	$\frac{\sqrt{3}}{2}$	$\frac{1}{2}$	$-\frac{1}{2}$	$\frac{\sqrt{3}}{2}$	$-\frac{\sqrt{2}}{2}$
$\tan\theta$	-1	$\sqrt{3}$	$-\frac{\sqrt{3}}{3}$	$-\sqrt{3}$	$-\sqrt{3}$	$\frac{\sqrt{3}}{3}$	1
$\cot\theta$	-1	$\frac{\sqrt{3}}{3}$	$-\sqrt{3}$	$-\frac{\sqrt{3}}{3}$	$-\frac{\sqrt{3}}{3}$	$\sqrt{3}$	1
$\sec\theta$	$-\sqrt{2}$	2	$\frac{2\sqrt{3}}{3}$	2	-2	$\frac{2\sqrt{3}}{3}$	$-\sqrt{2}$
$\csc\theta$	$\sqrt{2}$	$\frac{2\sqrt{3}}{3}$	-2	$-\frac{2\sqrt{3}}{3}$	$\frac{2\sqrt{3}}{3}$	2	$-\sqrt{2}$

3.

	210°	225°	240°	300°	315°	330°
$\sin\theta$	$-\frac{1}{2}$	$-\frac{\sqrt{2}}{2}$	$-\frac{\sqrt{3}}{2}$	$-\frac{\sqrt{3}}{2}$	$-\frac{\sqrt{2}}{2}$	$-\frac{1}{2}$
$\cos\theta$	$-\frac{\sqrt{3}}{2}$	$-\frac{\sqrt{2}}{2}$	$-\frac{1}{2}$	$\frac{1}{2}$	$\frac{\sqrt{2}}{2}$	$\frac{\sqrt{3}}{2}$
$\tan\theta$	$\frac{\sqrt{3}}{3}$	1	$\sqrt{3}$	$-\sqrt{3}$	-1	$-\frac{\sqrt{3}}{3}$
$\cot\theta$	$\sqrt{3}$	1	$\frac{\sqrt{3}}{3}$	$-\frac{\sqrt{3}}{3}$	-1	$-\sqrt{3}$
$\sec\theta$	$-\frac{2\sqrt{3}}{3}$	$-\sqrt{2}$	-2	2	$\sqrt{2}$	$\frac{2\sqrt{3}}{3}$
$\csc\theta$	-2	$-\sqrt{2}$	$-\frac{2\sqrt{3}}{3}$	$-\frac{2\sqrt{3}}{3}$	$-\sqrt{2}$	-2

5. a) T c) F

6. a) $-\frac{1}{2}$ c) $\sqrt{3}$ e) $\frac{2\sqrt{3}}{3}$

g) -1 i) $-\frac{\sqrt{3}}{2}$ k) -2

PROBLEM SET 4, Page 99

1. a) 190° c) 20° e) 353°
g) 240°

2. a) 10° c) 20° e) 7°
g) 60°

3. a) −sin 70° c) −tan 70° e) −cos 20°
g) −csc 45° i) −cot 10° k) −cot 35°
m) cos 78° o) −sec 45°

4. a) −0.9397 c) −2.747 e) −0.9397
g) −1.414 i) −5.671 k) −1.428
m) 0.2079 o) −1.414

5. a) 0.7030 c) 0.1013 e) −1.140
g) 2.545 i) −1.134 k) −0.9975

REVIEW PROBLEM SET, Page 99

1. a) $-\frac{\pi}{6}$ c) $\frac{4\pi}{9}$ e) $\frac{3\pi}{20}$
g) $\frac{13\pi}{36}$ i) $-\frac{31\pi}{18}$

2. a) −330° c) 2100° e) 2250°
g) −585° i) 648°

3. a) $A = 10\pi$ ft²; $s = 2\pi$ ft c) $A = \frac{200\pi}{3}$ in.²; $s = \frac{20\pi}{3}$ in.

4. a) 1.46 c) $\frac{\pi}{4}$

5. a) I c) III e) I
g) I i) I

6.

	a)	c)	e)
sin θ	$\frac{12}{13}$	$\frac{2\sqrt{6}}{5}$	$-\frac{3\sqrt{13}}{13}$
cos θ	$\frac{5}{13}$	$\frac{1}{5}$	$\frac{2\sqrt{13}}{13}$
tan θ	$\frac{12}{5}$	$2\sqrt{6}$	$-\frac{3}{2}$
cot θ	$\frac{5}{12}$	$\frac{\sqrt{6}}{12}$	$-\frac{2}{3}$
sec θ	$\frac{13}{5}$	5	$\frac{\sqrt{13}}{2}$
csc θ	$\frac{13}{12}$	$\frac{5\sqrt{6}}{12}$	$-\frac{\sqrt{13}}{3}$

7. a) II c) III e) III

8.

	a)	c)	e)
sin θ	—	$-\frac{3\sqrt{13}}{13}$	—
cos θ	$\frac{12}{13}$	$-\frac{2\sqrt{13}}{13}$	$-\frac{5}{13}$
tan θ	$-\frac{5}{12}$	—	$\frac{12}{5}$
cot θ	$-\frac{12}{5}$	$\frac{2}{3}$	$\frac{5}{12}$
sec θ	$\frac{13}{12}$	$-\frac{\sqrt{13}}{2}$	$-\frac{13}{5}$
csc θ	$-\frac{13}{5}$	$-\frac{\sqrt{13}}{3}$	$-\frac{13}{12}$

9. a) $-\sin 50°$ c) $-\sec 70°$ e) $-\csc 40°$

10. a) −0.7660 c) −2.924 e) −1.556

11. a) 0.8958 c) −1.099 e) 0.7554
g) −0.4661 i) 0.3404 k) 3.470

Chapter 3

PROBLEM SET 1, Page 116

1. a) y axis c) $y = x$

2. a) II c) I e) III
g) I i) III

3. a) (−1, 0) c) (1, 0) e) (0, 1)

4.

	a)	c)	e)
sin t	0	−1	0
cos t	−1	0	1
tan t	0	Undefined	0
cot t	Undefined	0	Undefined
sec t	−1	Undefined	1
csc t	Undefined	−1	Undefined

5.

	a)	c)	e)
sin t	$\frac{2}{\sqrt{5}}$	$\frac{\sqrt{3}}{2}$	$-\frac{5}{13}$
cos t	$\frac{1}{\sqrt{5}}$	$-\frac{1}{2}$	$-\frac{12}{13}$
tan t	2	$-\sqrt{3}$	$\frac{5}{12}$
cot t	$\frac{1}{2}$	$-\frac{1}{\sqrt{3}}$	$\frac{12}{5}$
sec t	$\sqrt{5}$	−2	$-\frac{13}{12}$
csc t	$\frac{\sqrt{5}}{2}$	$\frac{2}{\sqrt{3}}$	$-\frac{13}{5}$

6. a)

	i)	iii)	v)	vii)
sin t	$\frac{\sqrt{2}}{2}$	$-\frac{\sqrt{2}}{2}$	$-\frac{\sqrt{3}}{2}$	$-\frac{\sqrt{2}}{2}$
cos t	$-\frac{\sqrt{2}}{2}$	$\frac{\sqrt{2}}{2}$	$-\frac{1}{2}$	$\frac{\sqrt{2}}{2}$
tan t	-1	-1	$\sqrt{3}$	-1
cot t	-1	-1	$\frac{\sqrt{3}}{3}$	-1
sec t	$-\sqrt{2}$	$\sqrt{2}$	-2	$\sqrt{2}$
csc t	$\sqrt{2}$	$-\sqrt{2}$	$-\frac{2\sqrt{3}}{3}$	$-\sqrt{2}$

7. Since $\cot t = \frac{\cos t}{\sin t}$ and $\csc t = \frac{1}{\sin t}$, and $\sin t \neq 0$ for $t \neq$ even multiple of $\frac{\pi}{2}$

9.

	Domain
sin t	R
cos t	R
tan t	$\left\{t \mid t \neq \frac{\pi}{2} + n\pi,\ n \in I\right\}$
cot t	$\{t \mid t \neq n\pi,\ n \in I\}$
sec t	$\left\{t \mid t \neq \frac{\pi}{2} + n\pi,\ n \in I\right\}$
csc t	$\{t \mid t \neq n\pi,\ n \in I\}$

11.

	a)	b)	c)	d)	e)	f)	g)	h)	i)	j)
Quadrant	II	I	IV	III	III	IV	I	IV	I	III
sin θ	+	+	−	−	−	−	+	−	+	−
cos θ	−	+	+	−	−	+	+	+	+	−
tan θ	−	+	−	+	+	−	+	−	+	+
cot θ	−	+	−	+	+	−	+	−	+	+
sec θ	−	+	+	−	−	+	+	+	+	−
csc θ	+	+	−	−	−	−	+	−	+	−

PROBLEM SET 2, Page 125

1. a) $\frac{\sqrt{2}}{2}, \frac{\sqrt{2}}{2}$ c) $\sin t$, periodic, 2π e) $-\csc t$, $\sec t$

2. a) $\cos\theta = -\frac{12}{13}$, $\tan\theta = -\frac{5}{12}$, $\cot\theta = -\frac{12}{5}$, $\sec\theta = -\frac{13}{12}$, $\csc\theta = \frac{13}{5}$
c) $\sin\theta = -\frac{7}{25}$, $\csc\theta = -\frac{25}{7}$, $\tan\theta = \frac{7}{24}$

3. a) $\sin t = \frac{1}{2}$, $\csc t = 2$ c) $\sin t = -\frac{12}{13}$, $\csc t = -\frac{13}{12}$

4. a) $\cos t = -\frac{5}{13}$, $\sec t = -\frac{13}{5}$ c) $\cos t = -\frac{\sqrt{551}}{24}$, $\sec t = -\frac{24\sqrt{551}}{551}$

5. a) $\cos(1.41) = 0.1601$

6.

	a)	c)	e)
tan t	$\frac{4}{3}$	$-\frac{5}{12}$	$-\frac{7}{24}$
cot t	$\frac{3}{4}$	$-\frac{12}{5}$	$-\frac{24}{7}$
sec t	$\frac{5}{3}$	$\frac{13}{12}$	$-\frac{25}{24}$
csc t	$\frac{5}{4}$	$-\frac{13}{5}$	$\frac{25}{7}$

7.

	a)	c)	e)	g)
cos t	$\frac{\sqrt{3}}{2}$	-1	0	$-\frac{\sqrt{3}}{2}$
sin t	$\frac{1}{2}$	0	1	$\frac{1}{2}$

8. a) $-\sqrt{3}$ c) $-\frac{2\sqrt{3}}{3}$ e) $\frac{\sqrt{3}}{3}$

g) -2

9. a) $\frac{2\pi}{5}$ c) 3π e) $\frac{8\pi}{3}$

10. a) F c) T

11. cos t and sec t are even, the remaining are odd

PROBLEM SET 3, Page 137

1. a) sin (3.82) c) sin (6.195) e) cot (1.04)
g) sec (1.565) i) cot (3.635) k) sin (5.3)
m) cot (3.745) o) sec (3.485)

2. a) −sin (0.68) c) −sin (0.085) e) cot (1.04)
g) sec (1.565) i) cot (0.495) k) −sin (0.98)
m) cot (0.605) o) −sec (0.345)

3. a) −0.6288 c) −0.0849 e) 0.5870
g) 674.213 i) 1.852 k) −0.8305
m) 1.447 o) −1.063

4. a) $\cos 6 > \cos 7$ c) $\cos 9 < \cos 10$

6. a) $\frac{\pi}{2}$ to π, sin, cos, cot are decreasing; tan, sec, and csc are increasing

c) $\frac{3\pi}{2}$ to 2π, sin, cos, tan are increasing; cot, sec and csc are decreasing

REVIEW PROBLEM SET, Page 138

1. a) $\frac{\pi}{2}$ c) $\frac{3\pi}{2}$

2.

	a)	c)
sin t	$-\frac{\sqrt{3}}{2}$	$\frac{12}{13}$
cos t	$\frac{1}{2}$	$\frac{5}{13}$
tan t	$-\sqrt{3}$	$\frac{12}{5}$
cot t	$-\frac{1}{\sqrt{3}}$	$\frac{5}{12}$
sec t	2	$\frac{13}{5}$
csc t	$-\frac{2}{\sqrt{3}}$	$\frac{13}{12}$

3. $y = \pm \dfrac{2\sqrt{2}}{3}$

4. a) 2 c) $\dfrac{\sqrt{3}}{2}$ e) $\sqrt{3}$
g) $\frac{1}{2}$ i) $-\sqrt{2}$ k) $-\dfrac{\sqrt{3}}{3}$

5. a) $\cos t = -\dfrac{2\sqrt{2}}{3}$, $\tan t = -\dfrac{\sqrt{2}}{4}$, $\cot t = -2\sqrt{2}$, $\sec t = -\dfrac{3\sqrt{2}}{4}$, $\csc t = 3$
c) $\sin t = -\dfrac{\sqrt{15}}{4}$, $\tan t = -\sqrt{15}$, $\cot t = -\dfrac{\sqrt{15}}{15}$, $\sec t = 4$, $\csc t = -\dfrac{4\sqrt{15}}{15}$

6.

	a)	c)
$\cos t$	$-\dfrac{\sqrt{5}}{3}$	$-\dfrac{2\sqrt{2}}{3}$
$\tan t$	$-\dfrac{2\sqrt{5}}{5}$	$\dfrac{\sqrt{2}}{4}$
$\cot t$	$-\dfrac{\sqrt{5}}{2}$	$2\sqrt{2}$
$\sec t$	$-\dfrac{3\sqrt{5}}{5}$	$-\dfrac{3\sqrt{2}}{4}$
$\csc t$	$\frac{3}{2}$	-3

7. a) $\dfrac{2\pi}{7}$ c) $\dfrac{2\pi}{3}$

8. a) sin (0.84) c) tan (6.17) e) sin (5.047)
g) sec (1.065) i) tan (4.895)

9. a) sin (0.84) c) −tan (0.11) e) −sin (1.233)
g) sec (1.065) i) −tan (1.385)

10. a) 0.8058 c) 0.9294 e) 4.256
g) −3.142 i) 3.341 k) 1.105

11. a) III c) III

Chapter 4

PROBLEM SET 1, Page 156

1. a)

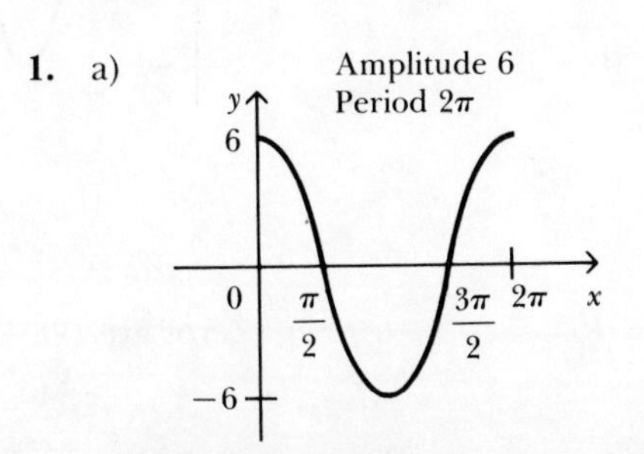

c)

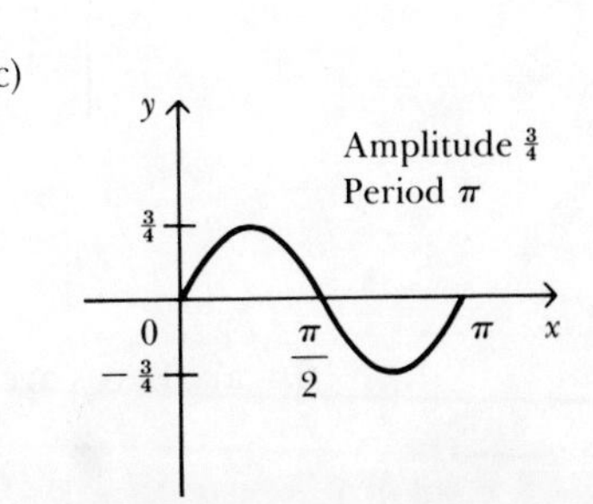

e)

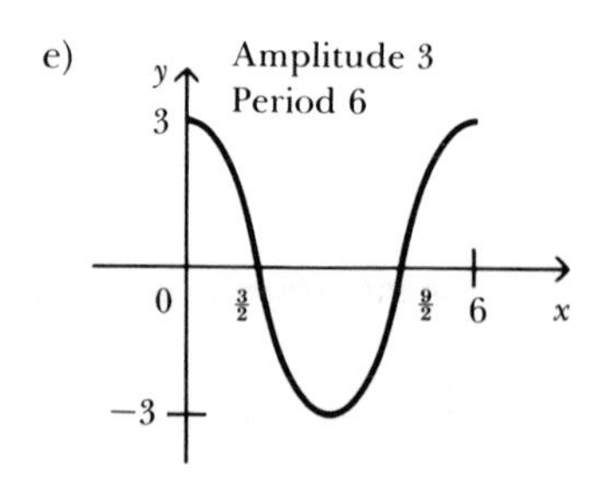

g)

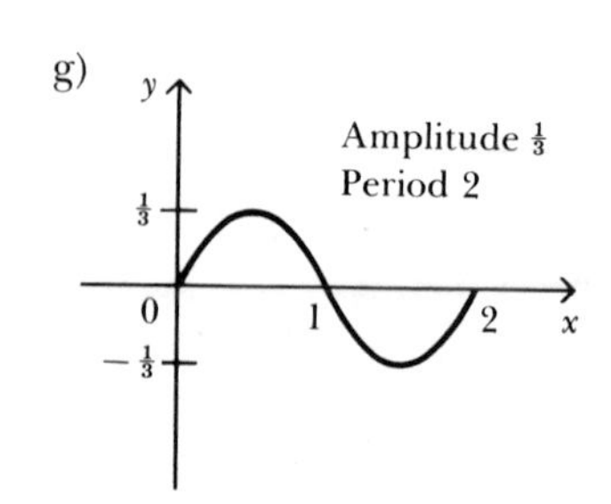

2. a)

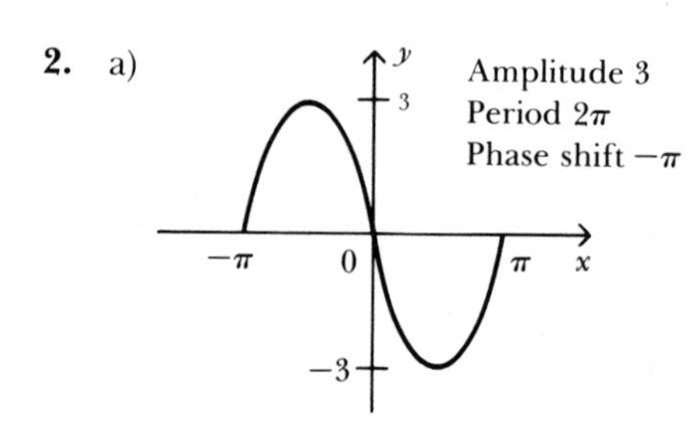

c)

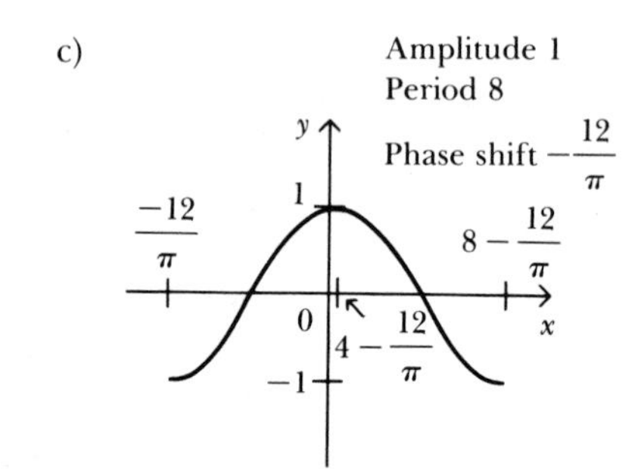

e)

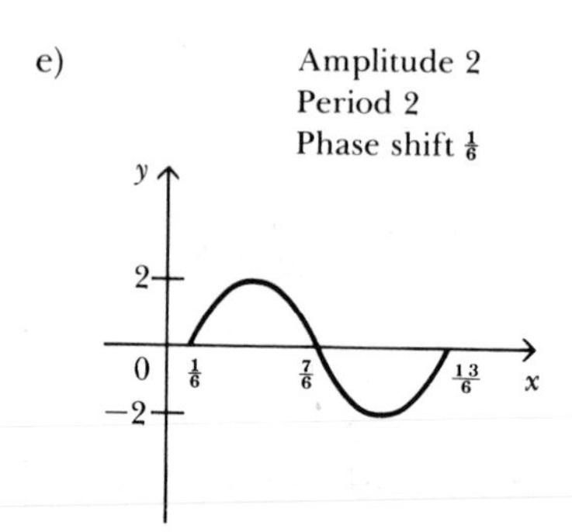

g)

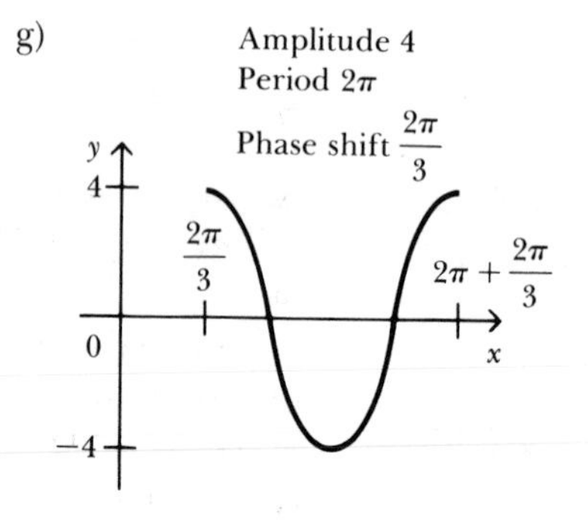

5. a)

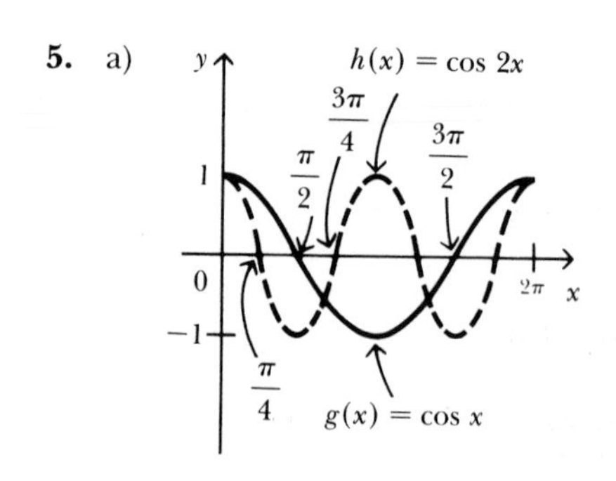

b)

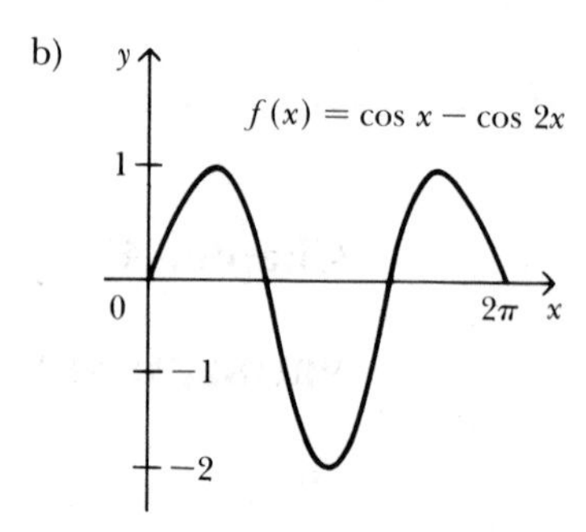

6. a) Zeros are $\frac{n\pi}{2}$, $n \in I$

c) Zeros are $(2n + 1)\frac{\pi}{8}$, $n \in I$

PROBLEM SET 2, Page 162

1.

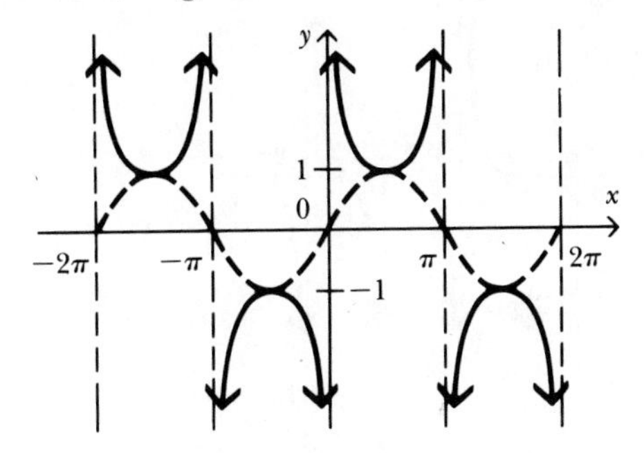

3. a) i) Domain of $f = \left\{x \mid x \neq \frac{\pi}{2} + n\pi,\ n \in I\right\}$

ii) Range of $f = R$

iii) Odd

iv) Symmetric with respect to the origin

v) π

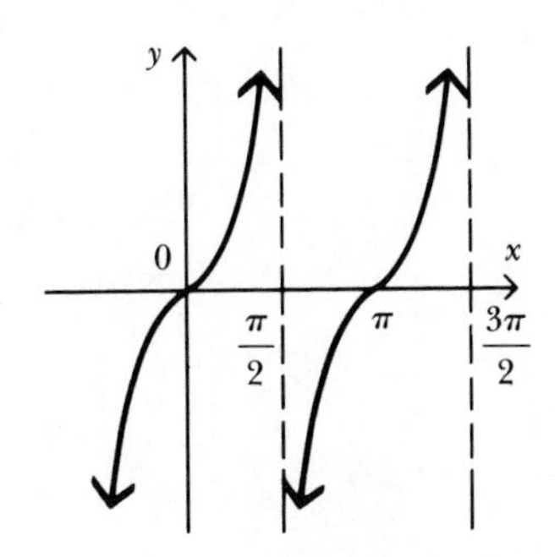

c) i) Domain of $f = \left\{x \mid x \neq \frac{\pi}{2} + n\pi,\ n \in I\right\}$

ii) Range of $f = (-\infty, -1] \cup [1, \infty)$

iii) Even

iv) Symmetric with respect to the y axis.

v) 2π

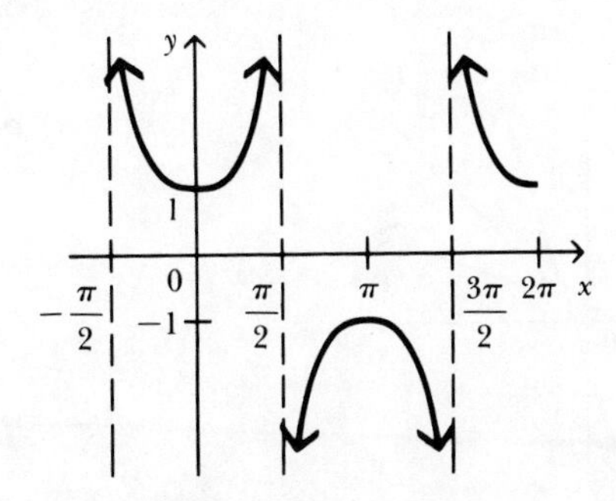

4. a)

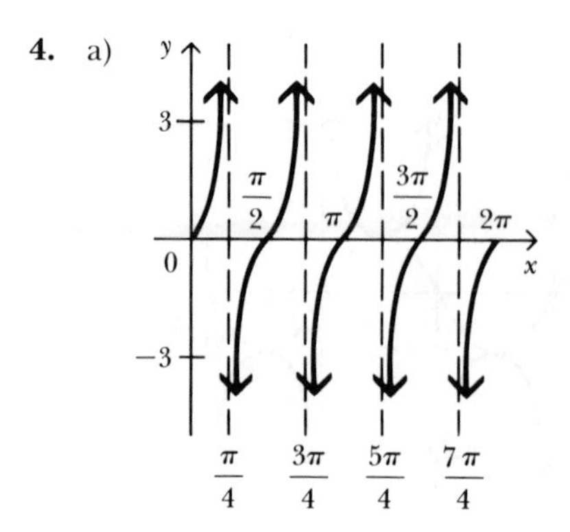

c)

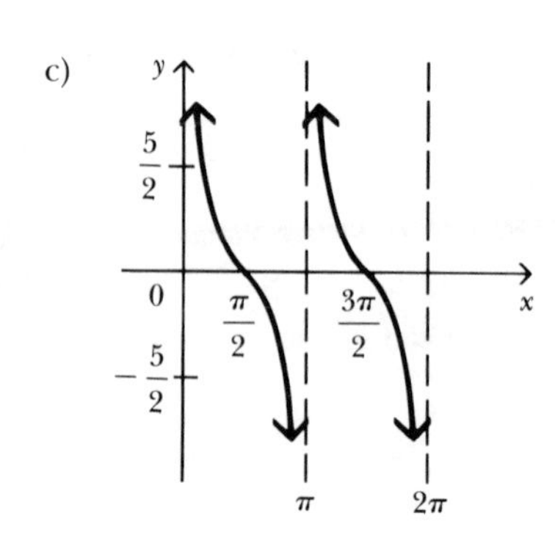

e)

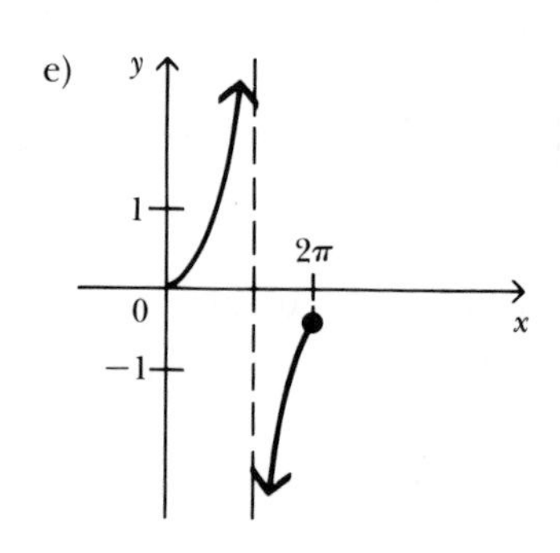

g)

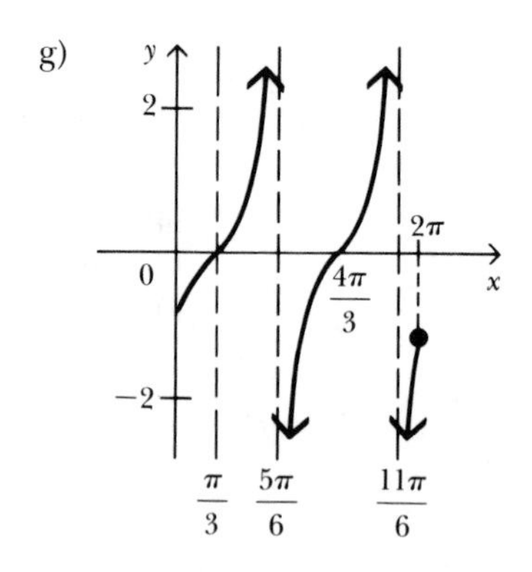

5. a) F c) T e) F

PROBLEM SET 3, Page 170

3. a) $f^{-1}(x) = \frac{1}{3}(1 - x)$ c) $f^{-1}(x) = \dfrac{3}{x}$

4. a) Yes

c) $f^{-1}(4) = 0, f^{-1}(0) = \sqrt[3]{-4}, f^{-1}(-4) = -2$

5. a) Yes c) $f^{-1}(0) = \frac{7}{3}, f^{-1}(1) = 2$

6. a) Yes b) Yes

8. a) Yes, $f^{-1}(x) = \frac{1}{3}(x + 1)$ c) No

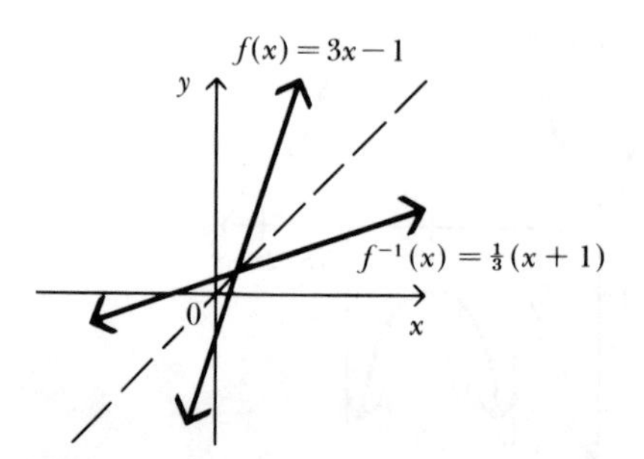

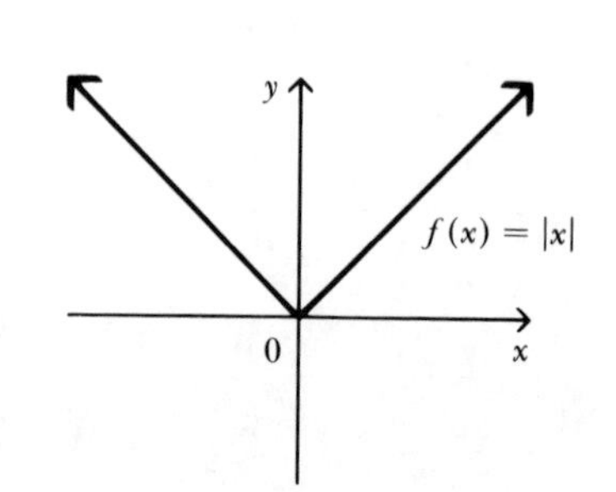

e) Yes, $f^{-1}(x) = \sqrt{x}$

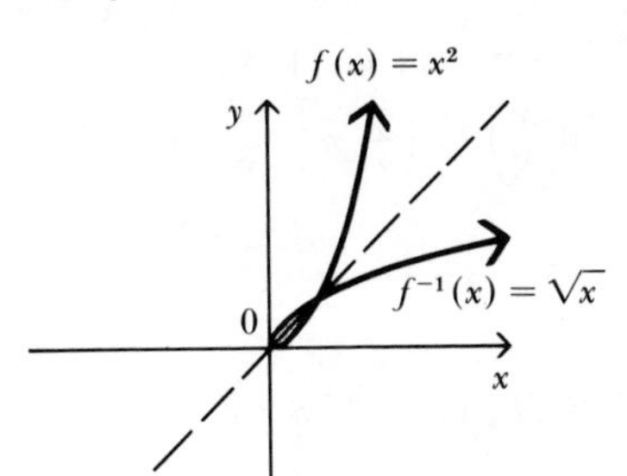

g) No

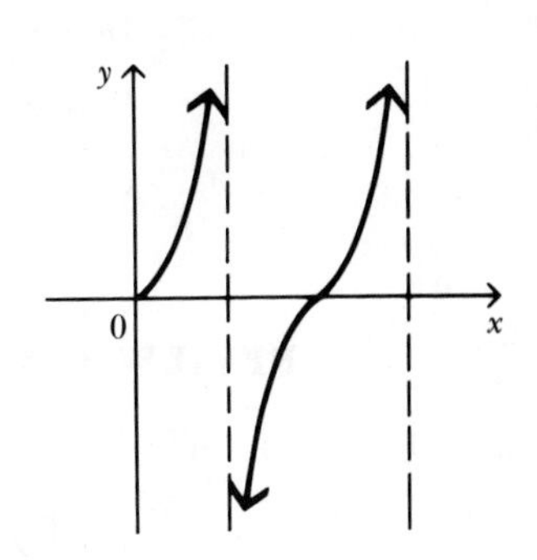

PROBLEM SET 4, Page 177

1. a) $-\frac{\pi}{2}, \frac{\pi}{2}$ e) x, x

2. a) $\frac{\pi}{4}$ c) $-\frac{\pi}{4}$ e) 0.52

g) $\frac{\pi}{3}$ i) $\frac{3\pi}{4}$ k) 1.38

3. a) $\frac{\sqrt{3}}{3}$ c) $\frac{\pi}{6}$ e) $\frac{7}{8}$

g) $\frac{2\sqrt{2}}{3}$ i) $\frac{\pi}{3}$

4.

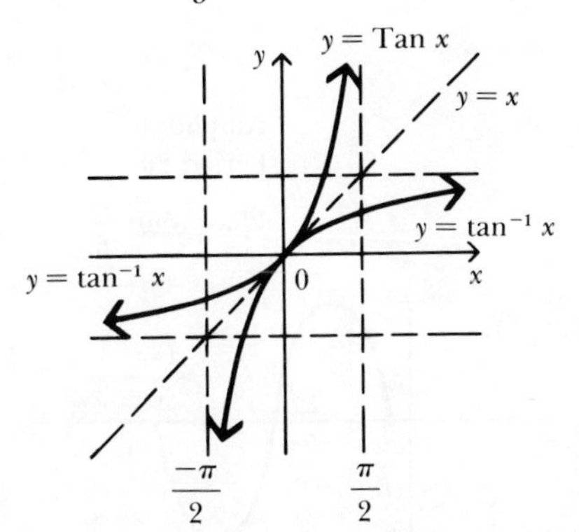

6. a)

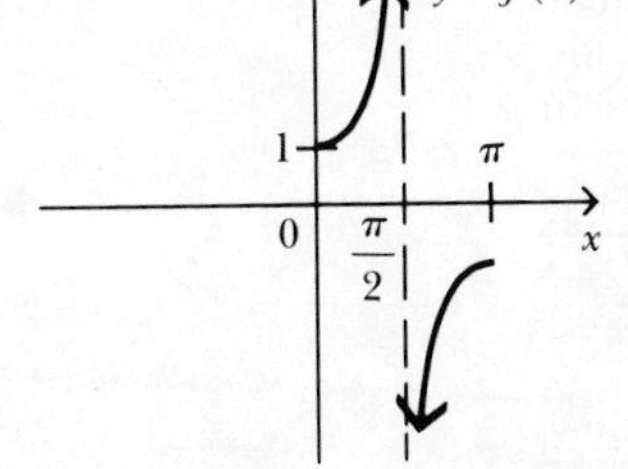

c)

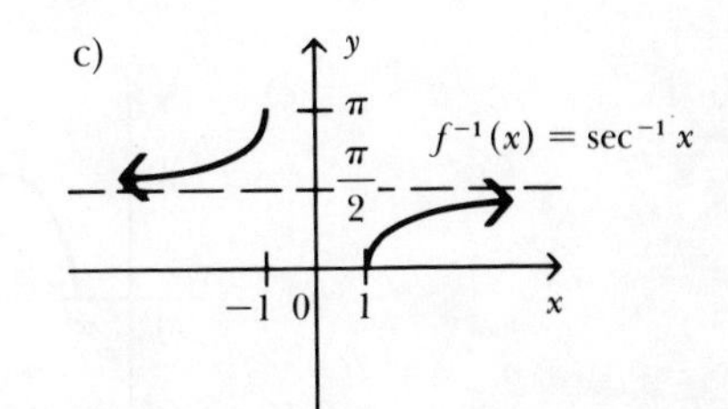

b) $f^{-1}(x) = \sec^{-1}x$
Domain of $f^{-1} = (-\infty, -1] \cup [1, \infty)$
Range of $f^{-1} = \left[0, \frac{\pi}{2}\right) \cup \left(\frac{\pi}{2}, \pi\right]$

d) $\sec^{-1} x = \cos^{-1}\left(\frac{1}{x}\right)$ for $|x| \geqslant 1$

9. a) $x = \frac{1}{2}\cos y$

c) $x = \frac{1}{4}\tan(5 - y)$

REVIEW PROBLEM SET, Page 178

1. a)

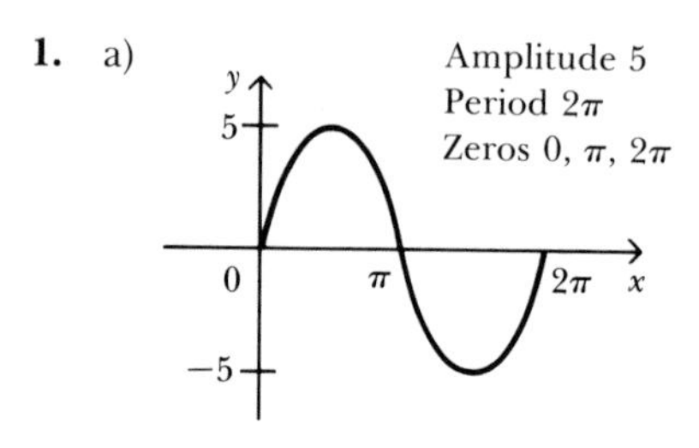

c)

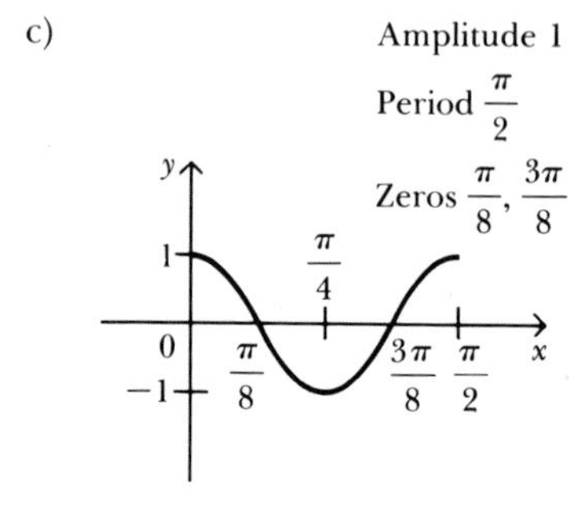

e)

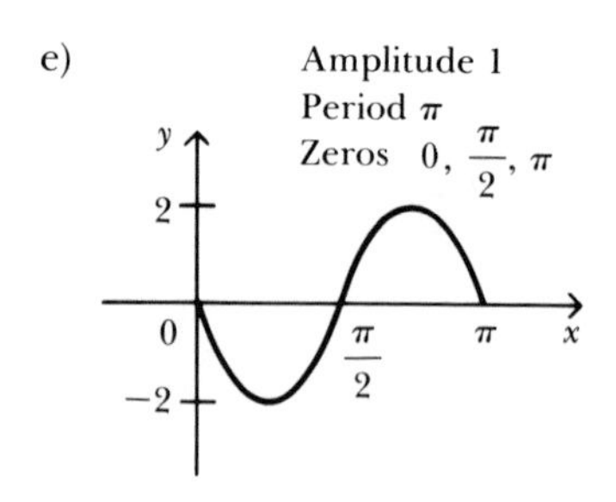

2. a)

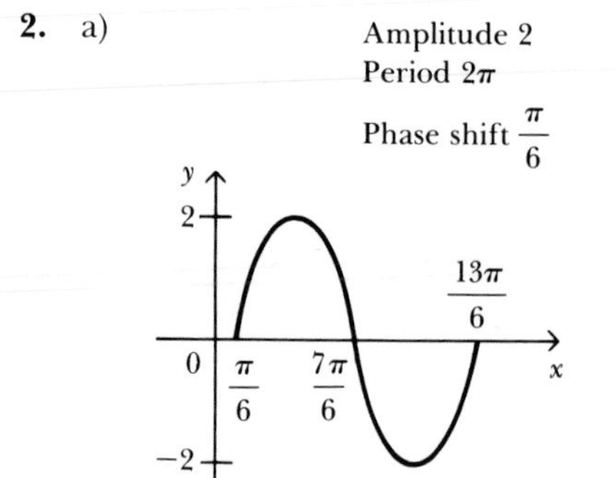

c)

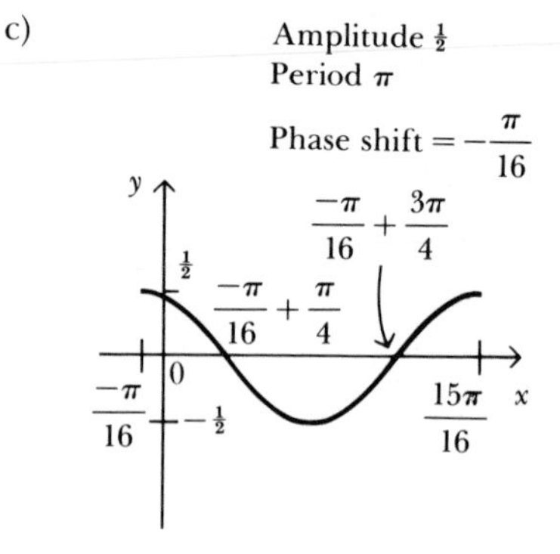

e)

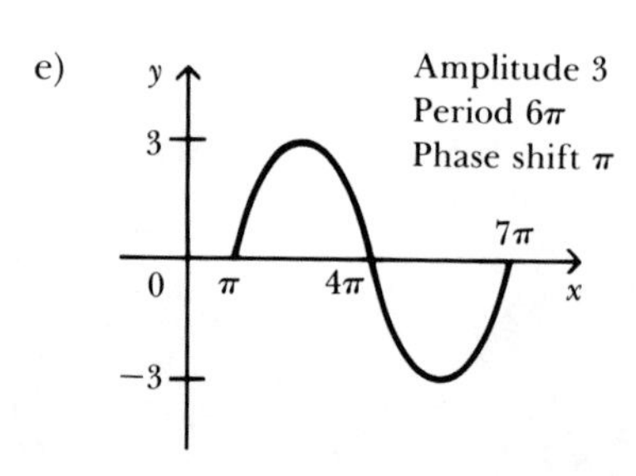

4. a)

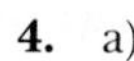

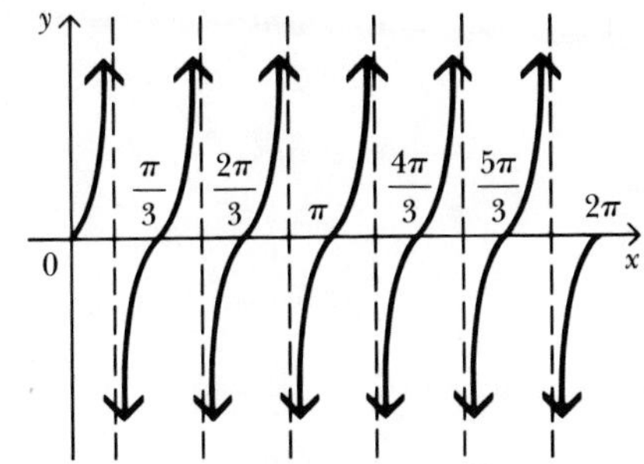

Fundamental period $\frac{\pi}{3}$

c)

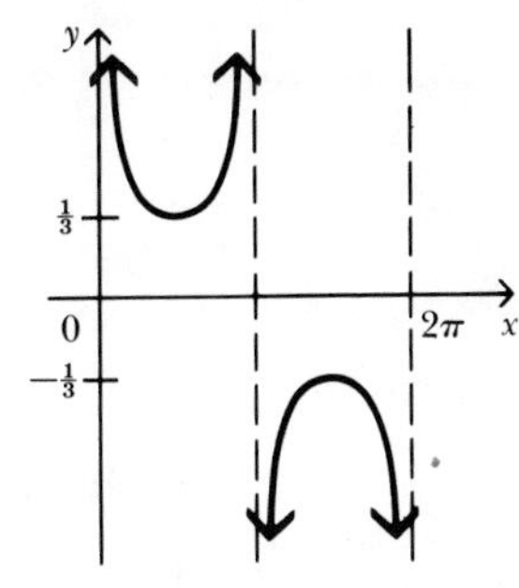

Fundamental period 2π

5. a)

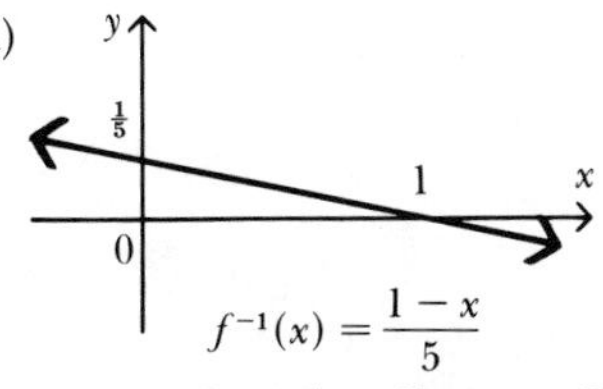

$f^{-1}(x) = \frac{1-x}{5}$

Domain = Range = R

c)

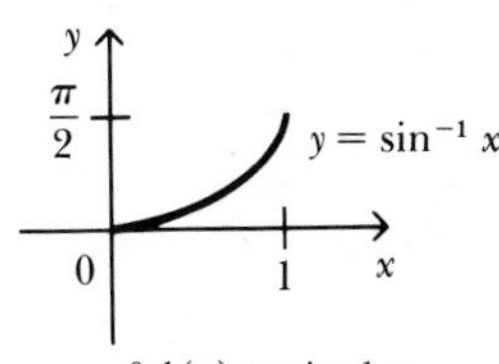

$f^{-1}(x) = \sin^{-1} x$

Domain = $[0, 1]$

Range = $\left[0, \frac{\pi}{2}\right]$

e)

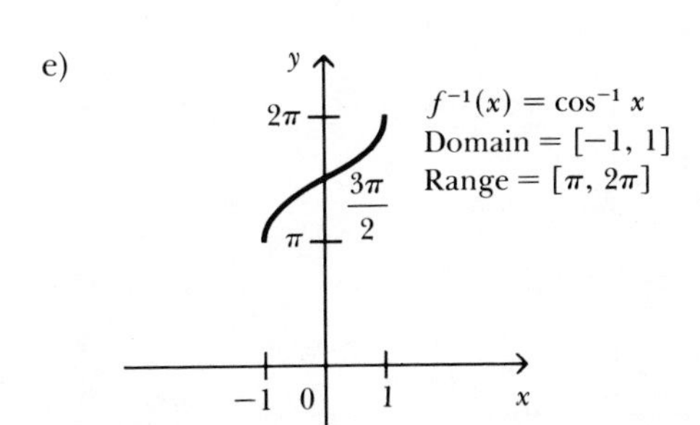

$f^{-1}(x) = \cos^{-1} x$

Domain = $[-1, 1]$

Range = $[\pi, 2\pi]$

6. a) $\frac{\pi}{4}$ c) $\frac{\pi}{4}$ e) $\frac{\pi}{6}$

7. a) 0.52 c) 2.25 e) −1.48

8. a) $\frac{24}{25}$ c) $\frac{7}{25}$ e) $-\frac{3\sqrt{7}}{7}$

9. a) $x = \frac{1}{3} \sin y$ c) $x = \frac{1}{4} \sec (y - 3)$

Chapter 5

PROBLEM SET 1, Page 188

1. a) $\sin \theta = \frac{\sqrt{x^2+1}}{x^2+1}$, $\cos \theta = \frac{x\sqrt{x^2+1}}{x^2+1}$, $\tan \theta = \frac{1}{x}$, $\sec \theta = \frac{\sqrt{x^2+1}}{x}$,

$\csc \theta = \frac{\sqrt{x^2+1}}{1}$

2. a) $\cos \theta = \sqrt{1 - \sin^2 \theta}$, $\tan \theta = \frac{\sin \theta}{\sqrt{1-\sin^2 \theta}}$, $\cot \theta = \frac{\sqrt{1 - \sin^2 \theta}}{\sin \theta}$,

$\sec \theta = \frac{1}{\sqrt{1-\sin^2 \theta}}$, $\csc \theta = \frac{1}{\sin \theta}$

3. a) $2\cos^2\theta - 1$ c) $\sin^2\theta$ e) $2\cos\theta$ g) $2\csc^2\theta$

4. a) $-\cos t$ c) $2\sin^2 t - 1$ e) $\dfrac{1+\cos\theta}{1-\cos\theta}$ g) $\dfrac{\cos t}{\sin t}$

PROBLEM SET 2, Page 199

1. a) $\sin 60°$ c) $\cos 80°$ e) $\cos t$

2. a) $\cos\theta$ c) $-\tan\theta$ e) $\sin\theta$ g) $\sec\theta$ i) $-\cot\theta$ k) $-\sin\theta$ m) $\cos\theta$

3. a) $\dfrac{\sqrt{2}(\sqrt{3}+1)}{4}$ c) $2+\sqrt{3}$

4. a) $-\dfrac{\sqrt{2}(\sqrt{3}+1)}{4}$ c) $2+\sqrt{3}$

5. a) $-\frac{63}{65}$ c) $-\frac{33}{65}$ e) $\frac{63}{16}$

6. a) $-\frac{16}{65}$ c) $-\frac{13}{85}$

8. $\dfrac{\cot s \cot t + 1}{\cot s - \cot t}$

PROBLEM SET 3, Page 205

1. a) $2\sin 35° \cos 35°$ c) $2\sin 55° \cos 55°$

2.

	$\sin\frac{t}{2}$	$\cos\frac{t}{2}$	$\tan\frac{t}{2}$
a)	$\frac{\sqrt{3}}{2}$	$-\frac{1}{2}$	$-\sqrt{3}$
c)	$-\frac{\sqrt{2}}{2}$	$-\frac{\sqrt{2}}{2}$	1

3. a) $\dfrac{\sqrt{2-\sqrt{2}}}{2}$ c) $-\left(\dfrac{\sqrt{2+\sqrt{3}}}{2}\right)$ e) $\dfrac{\sqrt{2+\sqrt{3}}}{2}$ g) $-\sqrt{7-4\sqrt{3}}$

4. a) $\sin 6t$ c) $\cos 14t$ e) $\cos t$

5. a) $\frac{336}{625}$ c) $-\frac{336}{527}$ e) $-\frac{3}{5}$

PROBLEM SET 4, Page 212

1. a) $\left\{\frac{7\pi}{6}, \frac{11\pi}{6}\right\}$ c) $\left\{\frac{\pi}{4}, \frac{5\pi}{4}\right\}$ e) $\left\{\frac{\pi}{4}, \frac{7\pi}{4}\right\}$ g) $\{1.14, 5.14\}$ i) $\{0.59, 3.73\}$

2. a) $\left\{\frac{\pi}{6}, \frac{5\pi}{6}\right\}$ c) $\left\{0, \frac{\pi}{3}, \pi, \frac{5\pi}{3}\right\}$ e) $\left\{\frac{\pi}{6}, \frac{5\pi}{6}, \frac{7\pi}{6}, \frac{11\pi}{6}\right\}$ g) $\left\{0, \frac{\pi}{6}, \frac{5\pi}{6}, \pi\right\}$

3. a) $\{240°, 300°\}$ c) $\{135°, 225°\}$
e) $\{0°, 30°, 90°, 150°, 180°, 210°, 270°, 330°\}$

4. $s = \frac{3\pi}{8}, t = \frac{\pi}{8}$

5. a) $\left\{0.64, \frac{3\pi}{2}\right\}$ c) $\left\{0, \frac{\pi}{6}, \frac{5\pi}{6}, \pi\right\}$ e) $\left\{\frac{\pi}{2}, \frac{7\pi}{6}, \frac{11\pi}{6}\right\}$
g) $\left\{\frac{2\pi}{3}, \frac{4\pi}{3}\right\}$ i) $\left\{\frac{\pi}{6}, \frac{5\pi}{6}\right\}$

6. a) $\{150°, 210°\}$ c) $\{60°, 90°, 270°, 300°\}$
e) $\{45°, 225°\}$ g) $\{90°, 120°, 240°, 270°\}$
i) $\{60°, 120°, 240°, 300°\}$

7. a) $\left\{t \mid t = \frac{\pi}{4} + n\pi, n \in I\right\}$
c) $\left\{t \mid t = n\pi \text{ or } t = \frac{\pi}{3} + 2n\pi \text{ or } t = \frac{2\pi}{3} + 2\pi n, n \in I\right\}$
e) $\{t \mid t = 0.34 + 2\pi n \text{ or } t = 2.8 + 2\pi n$
$\text{or } t = 0.73 + 2\pi n \text{ or } t = 2.41 + 2\pi n, n \in I\}$
g) $\left\{t \mid t = \frac{\pi}{6} + 2\pi n \text{ or } t = \frac{5\pi}{6} + 2n\pi, n \in I\right\}$

REVIEW PROBLEM SET, Page 213

1. a) 1 c) $\sin t$ e) $\sec t$

2. a) $\frac{1 + 2 \sin t \cos t}{\sin t \cos^3 t}$ c) $\frac{1 + \sin t}{1 - \sin t}$

4. a) $\sin 45°$ c) $\cos 1$

5. a) $-\cos \theta$ c) $\frac{\tan \theta - 1}{\tan \theta + 1}$ e) $\sin \theta$

6. a) $\frac{\sqrt{2}(\sqrt{3} - 1)}{4}$ c) $2 - \sqrt{3}$

7. a) 0 c) $-\frac{7}{25}$ e) 0

9. a) $-\frac{\sqrt{2 + \sqrt{3}}}{2}$ c) $2 - \sqrt{3}$ e) $\frac{2}{\sqrt{2 - \sqrt{2}}}$

11. a) $\left\{\frac{5\pi}{4}, \frac{7\pi}{4}\right\}$ c) $\left\{\frac{3\pi}{4}, \frac{7\pi}{4}\right\}$ e) $\left\{\frac{\pi}{3}, \frac{4\pi}{3}\right\}$
g) $\left\{\frac{3\pi}{2}, \frac{\pi}{6}, \frac{5\pi}{6}\right\}$

Chapter 6

PROBLEM SET 1, Page 223

1. a) $a = 3, b = 3\sqrt{3}$ c) $a = 111.8, b = 165.8$
e) $c = \frac{25}{4}, b = \frac{15}{4}$ g) $a = \frac{40}{7}, b = \frac{10\sqrt{33}}{7}, \cos \alpha = \frac{\sqrt{33}}{7}$
i) $a = 3.21, b = 3.83$

2. a) $36°50'; 53°10'$ c) $24°40'; 65°20'$ e) $45°; 45°$

3. a) $22.5°$ c) $16.36°$ e) $33.3°$

5. 225 feet **7.** 1866.2 feet **9.** 5.92 feet

11. 10.46 miles **13.** 942.35 feet

PROBLEM SET 2, Page 231

1. a) 8.89 c) 270.21 e) $104°30'$

2. a) $c = 5.53$, $\alpha = 105°40'$, $\beta = 42°20'$ c) $b = 41.7$, $\alpha = 112°30'$, $\gamma = 27°30'$
e) $\alpha = 10°30'$, $\beta = 114°40'$, $\gamma = 54°50'$

5. $100°50'$, $79°10'$

7. $125°10'$

9. a) $10\sqrt{3}$ c) 38.15

PROBLEM SET 3, Page 243

1. a) $c = 6.64$, $\gamma = 109°50'$ c) $c = 52.02$, $\gamma = 97°$ e) $\alpha = 21°10'$
g) $b = 39.5$

2. a) No triangle
c) One triangle, $\alpha = 51°$, $\beta = 18°$, $b = 19.87$
e) Two triangles; $c = 14.3$, $\beta = 57°$, $\gamma = 89°$ or $c = 5.6$, $\beta = 123°$, $\gamma = 23°$
g) One triangle, $a = 24.9$, $\alpha = 95°40'$, $\beta = 31°20'$

5. 710.4 feet **7.** $78°10'$

9. a) $\gamma = 35°$, $b = 12.34$, $c = 7.34$ c) $\beta = 28°$, $b = 10$, $c = 21.1$

11. 444.12 feet **13.** 92.54 feet **15.** 4.9 in. and 12.9 in.

REVIEW PROBLEM SET, Page 244

1. a) $c = 20$, $b = 10\sqrt{3}$, $\alpha = 30°$ c) $a = 10.7$, $c = 16.1$, $\alpha = 41°37'$

2. 180.4 feet, $33°40'$, $56°20'$

3. 126.4 feet

4. 185.1 feet

5. a) $c = 3$ c) $a = 23.7$

6. 241.5 feet or 408.2 feet

7. a) $b = 31$ c) $a = 121.5$

8. a) $\beta = 46°30'$, $\alpha = 76°10'$, $a = 66.9$ c) $\gamma = 7°10'$, $\beta = 127°50'$, $b = 54.7$

9. $c = 6\sqrt{2}$, $b = 16.4$, $\gamma = 30°$

10. 116

Chapter 7

PROBLEM SET 1, Page 254

1. a) $6 + 8i$ c) $-2 - 2i$ e) $8 - 6i$
g) $-19 - 4i$ i) $1 + i$ k) $-1 - 3i$
m) $-7 - 24i$ o) $-\dfrac{5i}{3}$ q) $\dfrac{3 - 6i}{5}$

s) $-\dfrac{3-11i}{10}$ u) $\dfrac{7-3i}{5}$ w) $-\dfrac{9+12i}{25}$

y) $\dfrac{(4+2\sqrt{3})+(4\sqrt{3}-2)i}{5}$

2. a) $x=4,\ y=\frac{1}{2}$ c) $x=-2,\ y=-\frac{3}{2}$

3.

	$\bar{z}$	Re (z)	Im (z)	$\dfrac{1}{z}$
a)	$-5-7i$	-5	7	$-\dfrac{5+7i}{74}$
c)	$5-12i$	5	12	$\dfrac{5-12i}{169}$

5. a) $-\dfrac{i}{2}$ c) $\dfrac{2-i}{5}$ e) $\dfrac{4-3i}{25}$

9. a) $13-8i$ c) $5-6i$

PROBLEM SET 2, Page 272

1. a) (2, 120°) c) (5, 36°52′) e) $(5\sqrt{2}, 45°)$
g) (6, 120°) i) $(5\sqrt{2}, 225°)$

2. a) $(3\sqrt{3}, 3)$ c) $\left(-\dfrac{7}{2}, \dfrac{7\sqrt{3}}{2}\right)$

3. a) $\sqrt{13}$ c) 1 e) 1
g) 1

4.

	Mod z	Arg z	Polar form
a)	$2\sqrt{2}$	$\dfrac{5\pi}{4}$	$2\sqrt{2}\left(\cos\dfrac{5\pi}{4}+i\sin\dfrac{5\pi}{4}\right)$
c)	2	π	$2(\cos\pi+i\sin\pi)$
e)	2	$\dfrac{\pi}{3}$	$2\left(\cos\dfrac{\pi}{3}+i\sin\dfrac{\pi}{3}\right)$
g)	4	$\dfrac{4\pi}{3}$	$4\left(\cos\dfrac{4\pi}{3}+i\sin\dfrac{4\pi}{3}\right)$
i)	2	$\dfrac{5\pi}{3}$	$2\left(\cos\dfrac{5\pi}{3}+i\sin\dfrac{5\pi}{3}\right)$

5. a) $\dfrac{3\sqrt{3}}{2}+\dfrac{3}{2}i$ c) $-2\sqrt{2}+2\sqrt{2}\,i$ e) $-10i$
g) $1+\sqrt{3}\,i$ i) $-5-5\sqrt{3}\,i$

6.

	zw (polar)	zw (rectangular)	$\dfrac{z}{w}$ (polar)	$\dfrac{z}{w}$ (rectangular)
a)	$30(\cos 270°+i\sin 270°)$	$-30i$	$\dfrac{5}{6}[\cos(-210°)+i\sin(-210°)]$	$-\dfrac{5\sqrt{3}}{12}+\dfrac{5i}{12}$
c)	$32\left(\cos\dfrac{\pi}{3}+i\sin\dfrac{\pi}{3}\right)$	$16+16\sqrt{3}\,i$	$\dfrac{1}{2}\left(\cos\dfrac{\pi}{3}+i\sin\dfrac{\pi}{3}\right)$	$\dfrac{1}{4}+\dfrac{\sqrt{3}}{4}i$
e)	$10(\cos 360°+i\sin 360°)$	10	$5[\cos(-270°)+i\sin(-270°)]$	$5i$

7. a) $zw = 60\left(\cos \frac{7\pi}{6} + i \sin \frac{7\pi}{6}\right)$; $zw = -30\sqrt{3} - 30i$

c) $\frac{w}{z} = \frac{5}{3}\left(\cos \frac{\pi}{2} + i \sin \frac{\pi}{2}\right)$; $\frac{w}{z} = \frac{5}{3}\, i$

PROBLEM SET 3, Page 282

1. a) $512 - 512\sqrt{3}\, i$ c) $\frac{729}{2} + \frac{729\sqrt{3}}{2}\, i$ e) $-\frac{243\sqrt{3}}{2} - \frac{243}{2}\, i$

g) -1024

2. a) $-512i$ c) 729 e) $-1024i$

g) -1 i) $-\frac{1}{2} - \frac{\sqrt{3}}{2}\, i$ k) $-16{,}384$

m) $(2\sqrt{2})^{16}$

3. $\cos 3\theta = \cos^3 \theta - 3 \cos \theta \sin^2 \theta$, $\sin 3\theta = 3 \cos^2 \theta \sin \theta - \sin^3 \theta$

4. a) $\frac{\sqrt{2}}{2} + \frac{\sqrt{2}}{2}\, i$

$-\frac{\sqrt{2}}{2} - \frac{\sqrt{2}}{2}\, i$

c) $\sqrt[4]{18}\left(\cos \frac{7\pi}{8} + i \sin \frac{7\pi}{8}\right)$

$\sqrt[4]{18}\left(\cos \frac{11\pi}{8} + i \sin \frac{11\pi}{8}\right)$

5. a) $z_k = (81)^{1/4}\left[\cos\left(\frac{240°}{4} + \frac{360°k}{4}\right) + i \sin\left(\frac{240°}{4} + \frac{360°k}{4}\right)\right]$, $k = 0, 1, 2, 3$

$z_0 = 3(\cos 60° + i \sin 60°)$
$z_1 = 3(\cos 150° + i \sin 150°)$
$z_2 = 3(\cos 240° + i \sin 240°)$
$z_3 = 3(\cos 330° + i \sin 330°)$

c) $z_k = (32)^{1/5}\left[\cos\left(\frac{200°}{5} + \frac{360°k}{5}\right) + i \sin\left(\frac{200°}{5} + \frac{360°k}{5}\right)\right]$, $k = 0, 1, 2, 3, 4$

$z_0 = 2(\cos 40° + i \sin 40°)$
$z_1 = 2(\cos 112° + i \sin 112°)$
$z_2 = 2(\cos 184° + i \sin 184°)$
$z_3 = 2(\cos 256° + i \sin 256°)$
$z_4 = 2(\cos 328° + i \sin 328°)$

e) $z_k = (32)^{1/5}\left[\cos\left(\frac{300°}{5} + \frac{360°k}{5}\right) + i \sin\left(\frac{300°}{5} + \frac{360°k}{5}\right)\right]$, $k = 0, 1, 2, 3, 4$

$z_0 = 2(\cos 60° + i \sin 60°)$
$z_1 = 2(\cos 132° + i \sin 132°)$
$z_2 = 2(\cos 204° + i \sin 204°)$
$z_3 = 2(\cos 276° + i \sin 276°)$
$z_4 = 2(\cos 348° + i \sin 348°)$

6. a) $z_k = (36)^{1/4}\left[\cos\left(\frac{\pi}{6} + \frac{2\pi k}{2}\right) + i \sin\left(\frac{\pi}{6} + \frac{2\pi k}{2}\right)\right]$, $k = 0, 1$

$z_0 = 6^{1/2}\left(\cos \frac{\pi}{6} + i \sin \frac{\pi}{6}\right)$; $z_1 = 6^{1/2}\left(\cos \frac{7\pi}{6} + i \sin \frac{7\pi}{6}\right)$

c) $z_k = (1)^{1/3}\left[\cos\left(\frac{3\pi}{6} + \frac{2\pi k}{3}\right) + i \sin\left(\frac{3\pi}{6} + \frac{2\pi k}{3}\right)\right]$, $k = 0, 1, 2$

$z_0 = \cos \frac{\pi}{2} + i \sin \frac{\pi}{2}$; $z_1 = \cos \frac{7\pi}{6} + i \sin \frac{7\pi}{6}$, $z_3 = \cos \frac{11\pi}{6} + i \sin \frac{11\pi}{6}$

e) $z_k = (1)^{1/4}\left[\cos\left(\frac{\pi}{4} + \frac{2\pi k}{4}\right) + i \sin\left(\frac{\pi}{4} + \frac{2\pi k}{4}\right)\right]$, $k = 0, 1, 2, 3$

$z_0 = \cos\frac{\pi}{4} + i\sin\frac{\pi}{4}$; $z_1 = \cos\frac{3\pi}{4} + i\sin\frac{3\pi}{4}$; $z_2 = \cos\frac{5\pi}{4} + i\sin\frac{5\pi}{4}$;

$z_3 = \cos\frac{7\pi}{4} + i\sin\frac{7\pi}{4}$

g) $z_k = (1)^{1/5}\left[\cos\left(\frac{\pi}{10} + \frac{2\pi k}{5}\right) + i\sin\left(\frac{\pi}{10} + \frac{2\pi k}{5}\right)\right]$, $k = 0, 1, 2, 3, 4$

$z_0 = \cos\frac{\pi}{10} + i\sin\frac{\pi}{10}$

$z_1 = \cos\frac{\pi}{2} + i\sin\frac{\pi}{2}$

$z_2 = \cos\frac{9\pi}{10} + i\sin\frac{9\pi}{10}$

$z_3 = \cos\frac{13\pi}{10} + i\sin\frac{13\pi}{10}$

$z_4 = \cos\frac{17\pi}{10} + i\sin\frac{17\pi}{10}$

REVIEW PROBLEM SET, Page 283

1. a) 23 c) $16 + i$ e) $-8 - 8i$
g) $1 + 4i$ i) $-11 - 17i$ k) $18 - 13i$
m) $8 + 6i$ o) $-1 + 13i$ q) $\frac{3 - 6i}{5}$
s) $\frac{48 + 29i}{85}$

2. a) $x = 3, y = -7$ c) $x = 5, y = 5$ e) $x = 0, y = \frac{1}{3}$

4. a) $\left(\frac{5\sqrt{2}}{2}, \frac{5\sqrt{2}}{2}\right)$ c) $(-1, -1)$ e) $(0, -3)$

5.

	Mod z	Arg z	Polar form
a)	5	π	$z = 5(\cos\pi + i\sin\pi)$
c)	8	$\frac{\pi}{6}$	$z = 8\left(\cos\frac{\pi}{6} + i\sin\frac{\pi}{6}\right)$
e)	7	$\frac{\pi}{2}$	$z = 7\left(\cos\frac{\pi}{2} + i\sin\frac{\pi}{2}\right)$
g)	12	$\frac{3\pi}{2}$	$z = 12\left(\cos\frac{3\pi}{2} + i\sin\frac{3\pi}{2}\right)$
i)	12	$\frac{\pi}{4}$	$z = 12\left(\cos\frac{\pi}{4} + i\sin\frac{\pi}{4}\right)$

6.

	zw (polar)	zw (rectangular)	$\frac{z}{w}$ (polar)	$\frac{z}{w}$ (rectangular)
a)	$15(\cos 60° + i\sin 60°)$	$\frac{15}{2} + \frac{15\sqrt{3}}{2}i$	$\frac{5}{3}(\cos 30° + i\sin 30°)$	$\frac{5\sqrt{3}}{6} + \frac{5i}{6}$
c)	$54(\cos 60° + i\sin 60°)$	$27 + 27\sqrt{3}\,i$	$\frac{1}{6}(\cos 26° + i\sin 26°)$	$0.1498 + 0.0731i$
e)	$40(\cos 240° + i\sin 240°)$	$-20 - 20\sqrt{3}\,i$	$\frac{5}{2}(\cos(-90°) + i\sin(-90°))$	$-\frac{5i}{2}$

7. a) $\cos 300° + i \sin 300°; \frac{1}{2} - i\frac{\sqrt{3}}{2}$

c) $\cos 990° + i \sin 990°; -i$

e) $\cos 70\pi + i \sin 70\pi;\ 1$

g) $\cos\left(-\frac{55\pi}{6}\right) + i \sin\left(-\frac{55\pi}{6}\right); -\frac{\sqrt{3}}{2} + i\frac{1}{2}$

9. $\cos 4\theta = \sin^4\theta - 6\cos^2\theta\sin^2\theta + \cos^4\theta$
$\sin 4\theta = 4\cos^3\theta\sin\theta - 4\cos\theta\sin^3\theta$

11. a) $\cos\frac{3\pi}{4} + i\sin\frac{3\pi}{4};\ \cos\frac{7\pi}{4} + i\sin\frac{7\pi}{4}$

c) $3\left(\cos\frac{5\pi}{9} + \sin\frac{5\pi}{9}\right);\ 3\left(\cos\frac{11\pi}{9} + i\sin\frac{11\pi}{9}\right);\ 3\left(\cos\frac{17\pi}{9} + i\sin\frac{17\pi}{9}\right)$

e) $\sqrt[8]{2^7}\left(\cos\frac{\pi}{16} + i\sin\frac{\pi}{16}\right)$

$\sqrt[8]{2^7}\left(\cos\frac{9\pi}{16} + i\sin\frac{9\pi}{16}\right)$

$\sqrt[8]{2^7}\left(\cos\frac{17\pi}{16} + i\sin\frac{17\pi}{16}\right)$

$\sqrt[8]{2^7}\left(\cos\frac{25\pi}{16} + i\sin\frac{25\pi}{16}\right)$

Chapter 8

PROBLEM SET 1, Page 292

1. a) **AC** c) -2**AB**

3. No

4. a) **AC** c) **AB**

5. $\mathbf{AD} = \frac{5}{8}\mathbf{u} + \frac{3}{8}\mathbf{v}$

6. $\frac{1}{2}\mathbf{u} + \mathbf{v}$

7. $\mathbf{AE} = \frac{\mathbf{v}}{2}$, $\mathbf{CD} = \frac{\mathbf{u}}{2} - \mathbf{v}$, $\mathbf{DE} = \frac{\mathbf{v}}{2} - \frac{\mathbf{u}}{2}$

PROBLEM SET 2, Page 300

1. a) $\langle -2, 3\rangle, \langle 8, 1\rangle$ c) $\langle -2, 15\rangle, \langle -8, 5\rangle$ e) $\langle 8, 8\rangle, \langle 6, -4\rangle$
g) $\langle 9, -4\rangle, \langle -3, -4\rangle$

2. a) $\langle -1, -3\rangle, \langle -1, -3\rangle$ c) $\langle -3, -9\rangle, \langle -3, -9\rangle$

3. a) $\langle -1, -1\rangle$ c) 10 e) $\langle 1, -6\rangle$
g) $\langle 2, -13\rangle$ i) 5

4. a) $\mathbf{u} = \sqrt{2}\left(\cos\frac{7\pi}{4}\mathbf{i} - \sin\frac{7\pi}{4}\mathbf{j}\right)$ c) $\mathbf{u} = 4(\cos \pi\mathbf{i} + \sin \pi\mathbf{j})$

5. $\frac{4\mathbf{i} + \mathbf{j}}{\sqrt{17}}$

6. $a = -\frac{19}{4}, b = -\frac{9}{4}$

7. a) $\mathbf{u} = -7\mathbf{i} - \mathbf{j}; -\dfrac{7\sqrt{2}}{10}\mathbf{i} - \dfrac{\sqrt{2}}{10}\mathbf{j}$ c) $\mathbf{u} = 0\mathbf{i} + 3\mathbf{j}; \mathbf{j}$

e) $\mathbf{u} = 3\mathbf{i} + 4\mathbf{j}; \frac{3}{5}\mathbf{i} + \frac{4}{5}\mathbf{j}$ g) $\mathbf{u} = -2\mathbf{i} + 2\mathbf{j}, -\dfrac{\sqrt{2}}{2}\mathbf{i} + \dfrac{\sqrt{2}}{2}\mathbf{j}$

8. a) $(\frac{5}{3}, 1)$

PROBLEM SET 3, Page 307

1. a) 10 c) $-10\sqrt{3}$

2. a) 2 c) $-2\sqrt{3}$

3. a) $0, \theta = 90°$ c) $0, \theta = 90°$ e) $2, \theta = \cos^{-1} \dfrac{2\sqrt{85}}{85}$

4. a) 0 c) 0 e) $\dfrac{2\sqrt{5}}{5}$

5. a) $x = -\frac{16}{3}$ c) $x = -3$

6. $\angle B = 90°$

7. a) $\sqrt{|\mathbf{u}|^2 + |\mathbf{v}|^2}$ c) $\sqrt{9|\mathbf{u}|^2 + 16|\mathbf{v}|^2}$ e) 1

PROBLEM SET 4, Page 314

1. a) $f(\mathbf{i}) = \dfrac{1}{2}\mathbf{i} + \dfrac{\sqrt{3}}{2}\mathbf{j}$

c) $f(3\mathbf{i}) = \dfrac{3}{2}\mathbf{i} + \dfrac{3\sqrt{3}}{2}\mathbf{j}; f(-5\mathbf{j}) = \dfrac{5\sqrt{3}}{2}\mathbf{i} - \dfrac{5}{2}\mathbf{j}$

$f(3\mathbf{i} - 5\mathbf{j}) = \left(\dfrac{3 + 5\sqrt{3}}{2}\right)\mathbf{i} + \left(\dfrac{3\sqrt{3} - 5}{2}\right)\mathbf{j}, f(6\mathbf{i} + 8\mathbf{j}) = (3 - 4\sqrt{3})\mathbf{i} + (3\sqrt{3} + 4)\mathbf{j}$

2. a) $f(\mathbf{i}) = \dfrac{\sqrt{3}}{2}\mathbf{i} + \dfrac{1}{2}\mathbf{j}$

c) $f(3\mathbf{i}) = \dfrac{3\sqrt{3}}{2}\mathbf{i} + \dfrac{3}{2}\mathbf{j}, f(-5\mathbf{j}) = \dfrac{5}{2}\mathbf{i} - \dfrac{5\sqrt{3}}{2}\mathbf{j}$

$f(3\mathbf{i} - 5\mathbf{j}) = \dfrac{3\sqrt{3} + 5}{2}\mathbf{i} + \dfrac{3 - 5\sqrt{3}}{2}\mathbf{j}, f(6\mathbf{i} + 8\mathbf{j}) = (3\sqrt{3} - 4)\mathbf{i} + (3 + 4\sqrt{3})\mathbf{j}$

3. a) $60°$ c) $f[g(\mathbf{i})] = \mathbf{j}$ or $90°$ rotation

REVIEW PROBLEM SET, Page 315

1. a) **AC** c) **BA**

2. a) $\mathbf{u} = -5\sqrt{3}\,\mathbf{i} + 5\mathbf{j}$ c) $\mathbf{u} = -4\sqrt{2}\,\mathbf{i} + 4\sqrt{2}\,\mathbf{j}$

3. a) $\langle 1, 0 \rangle$ c) $\langle 4, -4 \rangle$ e) $\langle -15, -32 \rangle$

4. a) $\mathbf{u} = 10\mathbf{i} - 5\mathbf{j}, |\mathbf{u}| = 5\sqrt{5}$ c) $\mathbf{u} = -\mathbf{i} + 4\mathbf{j}, |\mathbf{u}| = \sqrt{17}$
e) $\mathbf{u} = 4\mathbf{i} + 2\mathbf{j}, |\mathbf{u}| = 2\sqrt{5}$

5. a) $\sqrt{29}$ c) $\sqrt{101}$ e) $\sqrt{74}$
g) 5 i) 17

6. a) $\langle 2, -2 \rangle$ c) $\langle -3, \frac{1}{2} \rangle$

7. a) 1 c) 78 e) -16
g) 18

9. a) $90°$ c) $\cos^{-1}\frac{2}{5} = 66°25'$

10. a) -4 c) 1

11. a) $\left\langle\frac{4}{5}, \frac{3}{5}\right\rangle$ or $\left\langle-\frac{4}{5}, -\frac{3}{5}\right\rangle$ c) $\left\langle\frac{7}{\sqrt{53}}, \frac{2}{\sqrt{53}}\right\rangle$ or $\left\langle-\frac{7}{\sqrt{53}}, -\frac{2}{\sqrt{53}}\right\rangle$

12. a) $f(\mathbf{i}) = -\frac{\sqrt{3}}{2}\mathbf{i} + \frac{1}{2}\mathbf{j}$ c) $f(\mathbf{i}) = -\frac{\sqrt{2}}{2}\mathbf{i} - \frac{\sqrt{2}}{2}\mathbf{j}$

13. a) $\mathbf{u} = \frac{\sqrt{2}}{2}\mathbf{i} + \frac{\sqrt{2}}{2}\mathbf{j}$ c) $f(\mathbf{j}) = -\frac{\sqrt{2}}{2}\mathbf{i} + \frac{\sqrt{2}}{2}\mathbf{j}$

APPENDIX A PROBLEM SET, Page 330

1. a) $9^2 = 81$ c) $(\frac{1}{3})^{-2} = 9$ e) $(\sqrt{16})^{1/2} = 2$
g) $x^4 = 2$ i) $x^0 = 1$

2. a) $x = 11$ c) $x = 7$ e) $x = 3$

4. a) 0.6020 c) 1.2552 e) 0.1761
g) 0.1505 i) 3.4771

5. a) $x = 9$ c) $x = 63\frac{1}{2}$ e) $x = \frac{8}{3}$
g) $x = -25$ or $x = 4$

6. a) $x = 7$ c) $x = -27$ or $x = 3$

7. a) $1.50 \times 10^{-2}, -1.8239, 0.1761, -2$
c) $5.31 \times 10^{-1}, -0.2749, 0.7251, -1$
e) $3.333 \times 10^{1}, 1.5228, 0.5228, 1$

8. a) $x = 1397$ c) $x = 1.413$ e) $x = 3.14 \times 10^{-10}$

9. a) 16.28 c) 4.029 e) 4.626
g) 9.49 i) 2.029 k) 17.26

Index